Selections from
MODERN ABSTRACT ALGEBRA

Selections from

MODERN ABSTRACT ALGEBRA

SECOND EDITION

RICHARD V. ANDREE

The University of Oklahoma

HOLT, RINEHART AND WINSTON, INC.

New York Chicago San Francisco Atlanta
Dallas Montreal Toronto London Sydney

T_0

$$\begin{vmatrix} J_0 & -ph \\ \dfrac{\pi i}{e^2 n_e} & s\, ln^{-1}\ 1 \end{vmatrix}$$

who has the acumen to help when help is needed

and has the sagacity to preserve silence when help is of no avail.

Preface

A modern text on abstract algebra tends to become a ten-volume series. This brief volume is not designed to replace such a series, but rather whet the student's appetite for the series, to help him decide which portions of the series are most suitable for him to take, and to help bridge the possible gap between his preparation and the abstract thinking required in higher mathematics. Students enjoy the work—enthusiasm runs high. The more advanced courses now contain a larger percentage of engineers, physicists, and chemists than ever before. Applications from these fields, as well as from psychology and social science, are indicated in this volume, and the student is provided with an opportunity to explore those regions nearest his own interests.

It is currently fashionable to require "a certain amount of mathematical maturity" as a prerequisite for advanced mathematics courses. This assumption is *not* made in this text. Indeed, *one important purpose of this text is to develop the "mathematical maturity" which many authors require.*

In accord with the author's conviction that students should be encouraged to use the mathematical library, there are suggestions for further reading from other texts and from the *American Mathematical Monthly* and *Mathematics Magazine*. A sincere effort has been made to suggest articles which are both palatable and authoritative.

Abstract algebra now occupies about the same relative position to mathematics in general as mathematics does to engineering and the physical sciences. In addition to being a fascinating discipline in its own right, abstract algebra provides the vocabulary and many of the general techniques used in the larger body of knowledge. It therefore seems quite appropriate to introduce abstract algebra early. A mathematics major at the University of Oklahoma usually takes this course in his freshman or sophomore year, concurrently with calculus. Engineering and science majors often fit it into their junior year. The text is suitable for two, three, or four semester-hours of work, depending upon student preparation and the selection of material to be presented. Chapter 3 (Boolean Algebra) can be studied independently of the rest of the text. However, it has been found more effective if preceded by Sections 1.1 to 1.6. The approach to Boolean algebra via switching circuits has wide appeal to students and illustrates how closely mathematical theory can parallel physical reality.

In a short course, optional ($\star$) sections may be omitted or used as project material. Likewise Chapters 6 and 7 may be covered rapidly or omitted entirely if the student is familiar with their contents.

Although the author personally likes the vector space approach to matrices, it has been avoided here for two reasons. First, it seems unfair to spoil the elegance of the vector space approach for the student who later takes a course in linear algebra, and, second, experience shows that the first introduction to matrices is easier if a matrix is considered as an entity— as an element of a matric algebra.

Chapter 9, which contains more advanced work on matrices, can be taken directly after Chapter 5 if the student is already familiar with determinant theory (Chapter 7), but the author's experience suggests that a better rounded course is obtained if Chapter 8 (Fields, Rings, and Ideals) is studied before Chapter 9. In a short course, it may be well to omit Chapter 9 entirely. If this is done, you may still wish to discuss Section 9.7, "What Mathematics to Take Next," with your students.

One marked difference between this book and certain other recent texts is that it selects interesting and important ideas from *various* parts of modern abstract algebra rather than being mostly devoted to the theory of matrices. Matric theory is vital, but it is only one facet of modern abstract algebra. For many students this book may well provide their maiden voyage into the abstract thinking which is the heart of mathematics. Hence, special care has been taken in the development of basic concepts such as equivalence relations and their corresponding equivalence classes. Structure and abstract patterns are brought to the student's attention throughout the text.

Every student deserves the thrill of making mathematical discoveries of his own, and then of proving or disproving his conjectures. If these discoveries happen already to be known to others, this in no way need detract from his accomplishment—it may merely mean that the others were born sooner.

This text contains many indications of where and how abstract algebra is applied in the world of today, but this is not the reason students study it. They study *Selections from Modern Abstract Algebra* because it is interesting and fun. Every effort has been made to produce a text that will be both useful to and understandable by the student. Blank pages have been left at the beginning of chapters and at the end of the book for additional teacher or student remarks.

The author will welcome an opportunity to correspond with you concerning the use of this text. It is his sincere hope that you and your students will enjoy *Selections from Modern Abstract Algebra*.

R. V. A.

Norman, Oklahoma
May 1971

Acknowledgments

The lecture notes which sparked this book were first used in 1954 by the author for a two-hour sophomore-level course in abstract algebra at the University of Oklahoma. The engineering students quickly discovered the advantages of an undergraduate course containing Boolean algebra, and introductions to the theories of groups and matrices. The demand for the course increased and, as it was offered more frequently, its revision became an almost continuous process. Colleagues at various institutions used portions of the notes in their classes and offered helpful suggestions. Among those who used the notes are: J. C. Brixey (University of Oklahoma), Emil Grosswald (University of Pennsylvania), V. O. McBrien (Holy Cross), C. O. Oakley (Haverford College), G. E. Schweigert (University of Pennsylvania), O. T. Shannon (Arkansas A, M and N College), and R. J. Swords (Holy Cross). In addition to this, portions of these notes were used at graduate summer institutes for high-school mathematics teachers at the University of Oklahoma (Norman), and at Montana State College (Bozeman). Professor E. Grosswald used them for a similar group at the University of Pennsylvania. Chapter 3 on Boolean algebra was duplicated separately by Professor F. E. McFarlin for use by the Department of Electrical Engineering at Oklahoma State University (Stillwater), and the University of Pennsylvania (Philadelphia). Mr. E. L. Walters (Wm. Penn High School, York, Pennsylvania) used them for enrichment material with a group of advanced high-school students.

Many friends read and made constructive suggestions on the notes; among them were: Bess E. Allen (Wayne University), J. H. Barrett (University of Utah), B. H. Bissinger (Lebanon Valley College), R. B. Crouch (New Mexico College of A. and M. Arts), J. C. Eaves (University of Kentucky), C. L. Farrar (University of Oklahoma), R. A. Good (University of Maryland), D. W. Hall (Harpur College), R. W. House (Pennsylvania State University), M. Gweneth Humphreys (Randolph-Macon Women's College), C. F. Koehler (Loyola College), Violet H. Larney (New York State College for Teachers), D. R. Lintvedt (Upsala College), C. C. MacDuffee (University of Wisconsin), J. E. Maxfield (Naval Ordnance Test Station), Margaret W. Maxfield (Naval Ordnance Test Station), B. E. Meserve (New Jersey State Teachers College), A. L. Mullikin (University of Oklahoma), D. A. Norton (University of California), R. L. San Soucie (Sylvania Electric), Augusta L. Schurrer (Iowa State Teacher's College), W. R. Utz (University of Missouri), R. J. Wisner (Haverford College), and J. L. Zemmer, Jr. (University of Missouri).

The most important contribution was certainly that of the author's wife, Josephine Peet Andree who combines a sound mathematical preparation with the rare qualities of patience, pedagogical judgment, and understanding.

Important contributions were also made, sometimes under duress, by the several hundred students who used this text in its various duplicated forms, and by the excellent editorship of Professor B. W. Jones (University of Colorado).

Galley proof was read by Professor Walter Stuermann (University of Tulsa), Professor D. J. Lewis (Notre Dame), Mrs. R. V. Andree, and Mrs. R. A. Andree in addition to the author. Each merits sincere thanks, both from the author and from the reader. The first edition has been carefully revised, taking into consideration the comments and suggestions of many colleagues who are using this text. Again particular thanks go to my wife without whose help this revision would have been impossible. Special thanks are also extended to Jeanne MacKay whose careful typing and common sense has proved a blessing beyond estimate.

Contents

Preface
Acknowledgments
Chapter 1   Number Theory and Proof

  1.1   Introduction   1
  1.2   The Modulo 7 System   2
  1.3   The Modulo 6 System   6
  1.4   Integral Domains   9
  1.5   The Nature of Definitions, Postulates, and Axioms   18
  1.6   The Nature of Proof   19
  1.7   Integral Domains   29
  1.8   Classification of the Integers   32
  1.9   Greatest Common Divisor   37
  1.10   Archimedes' Axiom and Euclid's Algorithm   38
 *1.11   Perfect Numbers   45
 *1.12   Number Systems   48
 *1.13   The Binary System   49
 *1.14   Nim   49

*1.15 Two wee Research Problems 52
*1.16 A Problem for Further Investigation 54
*1.17 Suggested Topics for Further Independent Study 56
Selected Reading List 59

Chapter 2 Equivalence and Congruence

2.1 Equivalence Relation 61
2.2 Equivalence Classes 66
2.3 Congruences 69
2.4 Linear Congruences 74
*2.5 A More Sophisticated Version 80
*2.6 Selected Topics for Independent Study 82
Selected Reading List 85

Chapter 3 Boolean Algebra

3.1 Introduction 87
3.2 Duality 87
3.3 Binary Boolean Arithmetic 89
3.4 Boolean Functions 90
3.5 Isomorphic Systems 91
3.6 Binary Boolean Algebra 94
3.7 The Negative or Complementary or Relay Relationship 101
3.8 Applications to Electrical Networks 107
3.9 Other Applications 109
3.10 Circuit Design 112
3.11 Point Set Interpretation 115
*3.12 An Algebraic Approach 118
Selected Reading List 121

Chapter 4 Groups

4.1 Introduction 123
4.2 Mathematical Systems 123
4.3 Group 124
4.4 Elementary Properties of Groups 137
4.5 Isomorphism 143
4.6 Cosets and LaGrange Theorem 149
4.7 Free Groups 157
*4.8 Quotient Groups. Jordan–Hölder Theorem 158
*4.9 Suggestions for Independent Investigation 161
Selected Reading List 163

Chapter 5 Matrices

5.1   Introduction   165
5.2   Matrix Product   167
5.3   Pauli Matrices   168
5.4   Square Matrices   171
5.5   A Proof of the Associativity of Matrices, Using $\sum$ Notation   175
5.6   Elementary Row Operations   178
5.7   Addition of Matrices   184
5.8   Domain Properties of Square Matrices   184
5.9   Proofs and Their Structure   189
5.10   More General Matrices and Vectors   194
5.11   Applications of Matrix Notation   202
5.12   Mappings and Transformations   211
Selected Reading List   221, 309

Chapter 6 Linear Systems

6.1   Systems of Linear Equations   223
Selected Reading List   241

Chapter 7 Determinants

7.1   Determinants   243
7.2   Minors and Cofactors   250
7.3   The Transpose of a Matrix   254
★7.4   The Adjoint Matrix   261
★7.5   Determinants and Linear Systems   265
Selected Reading List   269

Chapter 8 Fields, Rings, and Ideals

8.1   Field   271
8.2   Rings   275
8.3   Ideals   279
8.4   Residue Class Rings   281
8.5   Polynomials Modulo $(x^2 + 1)$—Complex Numbers   283
Selected Reading List   287

Chapter 9 More Matrix Theory

9.1   Characteristic Equations   289
9.2   Hamilton–Cayley Theorem   291

9.3   Characteristic Roots and Characteristic Vectors   292
9.4   Minimum Functions   298
9.5   Infinite Series with Matrix Elements   300
9.6   Derivatives and Integrals of Matrices   303
9.7   What Mathematics to Take Next   306
     Selected Reading List   221, 309

Answers and Hints   311

Index   353

Index of Symbols   361

NUMBER THEORY AND PROOF[1]

1

1.1 INTRODUCTION

The *integers* consist of the "counting numbers" or natural whole numbers 1, 2, 3, 4, $\cdots$ (positive integers), their negatives -1, -2, -3, -4, -5, $\cdots$ (negative integers), and zero 0. In later chapters, when rational numbers, real numbers, or complex numbers are used, it will be assumed that you know the meanings of these terms. Briefly: A *rational number* is a quotient of two integers a/b with $b \neq 0$.[2] A *real number* is a number which represents a distance or its negative. A *complex number* is an ordered pair of real numbers, (a, b) or equivalently, a number of the form $a + bi$, where a and b are real and $i^2 = -1$. More complete discussions of the real number system are presented in the book *What is Mathematics?* by R. Courant and H. Robbins (London: Oxford, 1941) or in *Numbers: Rational and Irrational* by I. Niven (New York: New Math Library, Random House, 1961).

The complex numbers contain all the real numbers, rational numbers,

---

[1] If you haven't read the preface, you should. It tells you what to expect and what the author hopes to accomplish.

[2] The symbol "$\neq$" is read "not equal to."

and integers as subsets. The real numbers contain all the rational numbers and the integers as subsets (but not all the complex numbers). The rational numbers contain all the integers as a subset (but not all the real nor complex numbers). The integers do not contain all of any of the other sets. This entire paragraph may be expressed in one line using the symbol "$\subset$" to mean "contained in" or "form a subset of":

$$\text{Integers} \subset \text{Rationals} \subset \text{Reals} \subset \text{Complex Numbers}.$$

It may be of interest to note that the properties of the rational numbers, real numbers, and even of the complex numbers can be derived from those of the integers by using logical reasoning. L. Kronecker (1823–1891, German) is reputed to have said, "God gave us the integers, all else is the work of man."

1.2 THE MODULO 7 SYSTEM

This section introduces a new arithmetic. To remind you that this is a new system, the congruence sign, $\equiv$, will be used in place of the usual equal sign, $=$. This system has only seven numbers in it: 0, 1, 2, 3, 4, 5, 6. It is called the modulo 7, or "mod 7," system.

The rules for addition in the mod 7 system are the same as those for ordinary addition *except that, if the sum is larger than 6, the sum is divided by 7, the quotient discarded, and the remainder is used in place of the ordinary sum.* Thus, $1 + 3 \equiv 4 \pmod 7$ and $2 + 3 \equiv 5 \pmod 7$; but $5 + 4 \equiv 2 \pmod 7$, since when 9 is divided by 7 the remainder is 2. Also, $6 + 5 \equiv 4 \pmod 7$, since the remainder 4 is obtained when 11 is divided by 7. In a similar fashion: $5 + 2 \equiv 0 \pmod 7$, $4 + 1 + 3 + 5 \equiv 6 \pmod 7$, and $4 + 0 + 2 + 3 + 6 \equiv 1 \pmod 7$.

The rules for multiplication in the mod 7 system are also like those of ordinary multiplication except that, if the product is larger than 6, the product is divided by 7 and the *remainder* is used in place of the ordinary product. Thus: $2 \times 2 \equiv 4 \pmod 7$, but $5 \times 2 \equiv 3 \pmod 7$, since if 10 is divided by 7, a remainder of 3 results. Also, $6 \times 3 \equiv 4 \pmod 7$, since, when 18 is divided by 7, the remainder is 4. In a similar fashion, $4 \times 3 \equiv 5 \pmod 7$, $5 \times 6 \equiv 2 \pmod 7$, and $2 \times 3 \times 4 \times 5 \equiv 1 \pmod 7$, the remainder when 120 is divided by 7. Practice until you can do sums and products easily in the mod 7 system. *Briefly:* $a \equiv b \pmod 7$ *means* $a = b + 7k$ *for some integer k.* (Why?)

There are no negative numbers in the mod 7 system. None are needed. The ordinary negative number -2 is the solution of the ordinary equation $x + 2 = 0$. In the mod 7 system, the number 5 is a solution of the equation

$x + 2 \equiv 0 \pmod 7$, since $(5 + 2)$ has the remainder 0 when divided by 7. In other words: 5, in the mod 7 system, plays a role similar to that of -2 in the ordinary system. In the mod 7 system, the number 6 plays a role similar to -1 in the ordinary numbers, since $6 + 1 \equiv 0 \pmod 7$ and $-1 + 1 = 0$.

There are no fractions in the mod 7 system and none are needed. The ordinary fraction 5/3 is the solution of the equation $3x = 5$. In the mod 7 system, the equation $3x \equiv 5 \pmod 7$ has the solution $x \equiv 4 \pmod 7$. (Try it and see.) The mod 7 equation $5x \equiv 2 \pmod 7$ has $x \equiv 6 \pmod 7$ as solution, while $4x \equiv 6 \pmod 7$ has the solution $x \equiv 5 \pmod 7$.

The equation $5x^3 + x^2 + 5x + 2 \equiv 0 \pmod 7$ may be shown to have $x \equiv 3 \pmod 7$ as a solution by direct substitutions. (Try it.) Can you find two other solutions?

To reiterate, there are only seven distinct "numbers" in the entire mod 7 system. There are no negative numbers and no fractions, yet equations can be solved. Best of all, since there are only seven numbers, *all* the solutions of a given equation can be found by merely substituting each of the seven numbers, in turn, for x to see which, if any, of them satisfy the equation.

The mod 7 system is a finite set of numbers, whereas the integers, rational numbers, and real numbers discussed above are each infinite sets.

A word of warning: There exist equations, such as $x^2 \equiv 6 \pmod 7$, which have no solution at all. This is not particularly surprising. The ordinary equation $x^2 = -1$ has no solution in the set of *real numbers*. In this book, the word "solve" will mean "find all possible solutions or prove that none exist."

It would also be possible to define the mod 7 system over the set of all integers rather than the set consisting only of 0, 1, 2, 3, 4, 5, 6. In this case $a \equiv b \bmod 7$ again means that there exists an integer k such that $a = b + 7 \cdot k$. Thus $5 + 4 = 9 \equiv 2 \bmod 7$

since
$$9 = 2 + 7 \cdot 1$$
$$a = b + 7 \cdot k.$$

In a similar fashion $4 + 0 + 2 + 3 + 6 = 15 \equiv 1 \bmod 7$

since
$$15 = 1 + 7 \cdot 2$$
$$a = b + 7 \cdot k.$$

In effect this partitions the integers into seven disjoint subsets (baskets).

Chapter 2 explores this important concept in greater detail and you will see how the two definitions are related. Modular systems are introduced here to provide laboratory material for your algebraic experiments in Chapter 1 and to lay the foundation needed for the successful study of equivalence classes in Chapter 2.

The reader may wish to show by direct substitution that for all x's in the mod 7 system

$$(3) \cdot (x) \equiv (x) + (x) + (x) \ (\text{mod } 7)$$

The distinction between 3 as an element of the mod 7 system which is multiplied by x in $3 \cdot x$ and 3 as a coefficient in $3x$ meaning $x + x + x$ becomes important in certain more advanced courses, but need not upset us here, since the results are identical. Nevertheless, the observation that a mathematical truth seems reasonable, or even trite, does not obviate the need for a proof, if one can be found. The more sophisticated reader may wish to create a proof that for each value of x in the mod 7 system and for each integer N, $N \cdot x \equiv \underbrace{x + x + x + \cdots + x}_{N \text{ terms}}$ by using mathematical induction. He may show by induction or otherwise that if $B \equiv N \bmod 7$ (that is, if there exists an integer k such that $B = N + 7 \cdot k$) then $N \cdot X \equiv B \cdot X \equiv \underbrace{X + X + X + \cdots + X}_{B \text{ terms}}$ for each X in the mod 7 system.

It should be noted that exponents cannot, in general, be handled in the same cavalier fashion. This is easy to demonstrate by noting that while $8 \equiv 1 \bmod 7$, $2^8 \not\equiv 2^1 \bmod 7$, since $256 \equiv 4 \not\equiv 2 \bmod 7$. The interested reader may study this further as an application of the Fermat–Euler Theorem in the *Selected Topics for Independent Study* at the end of Chapter 2, if he wishes to do some independent investigation.

Problem Set 1.2

1. Add: $4 + 3 + 6 + 5 + 2 + 4 \ (\text{mod } 7)$.

2. Add: $1 + 2 + 3 + 4 + 5 + 6 \ (\text{mod } 7)$.

3. Solve: $3x \equiv 5 \ (\text{mod } 7)$.

4. Solve: $6x - 5 \equiv 3 \ (\text{mod } 7)$.

5. Solve: $297x + 6 \equiv 0 \ (\text{mod } 7)$. Although 297 does *not* occur in the mod 7 system, $297x$ still has meaning, since $297x$ represents the

sum of $\underbrace{x + x + \cdots + x}_{297 \text{ terms}}$. This problem emphasizes the possible need for distinguishing between the set from which the unknowns of the equation are taken and the set from which the coefficients of the equation are taken. Fortunately this distinction need not be stressed in modular systems.

6. Solve: $x^2 \equiv 4 \pmod 7$.

7. Solve: $x^2 \equiv 2 \pmod 7$.

8. Solve: $x^2 \equiv 3 \pmod 7$.

9. Solve: $x^3 \equiv 6 \pmod 7$.

10. Solve: $x^3 \equiv 5 \pmod 7$.

11. (a) Make a table listing the seven numbers in the mod 7 system. Next to each number x, place its square, x^2; cube, x^3; fourth power, x^4; fifth power, x^5; sixth power, x^6; seventh power, x^7; and eighth power, x^8; all mod 7.
    (b) Compute, using the table of part (a), the values $(5)^{236} \pmod 7$, and $(3)^{179} \pmod 7$.
    (c) Will $x^4 \equiv 5 \pmod 7$ have a solution?
    (d) For what values of b will $x^5 \equiv b \pmod 7$ have solutions?

12. Solve: $4x^2 + 3x + 4 \equiv 0 \pmod 7$. Notice that in the mod 7 system the solutions are not complex.

13. Construct addition and multiplication tables for the mod 7 system.

14. If the symbol $\equiv$ is to be an equals (or equivalence) relation, it must be true that for every pair of values a and b in the set, either $a \equiv b$ or $a \not\equiv b$, and furthermore the following postulates must be satisfied:

    1. *Reflexive*: $a \equiv a \pmod 7$.
    2. *Symmetric*: If $a \equiv b \pmod 7$, then $b \equiv a \pmod 7$.
    3. *Transitive*: If $a \equiv b \pmod 7$ and $b \equiv c \pmod 7$, then $a \equiv c \pmod 7$.

    Use the definition "$a \equiv b \pmod 7$ means $a = b + 7k$, for some integer k" to show that the mod 7 system does satisfy these requirements. [HINT: Given $a \equiv b \pmod 7$, to prove that $b \equiv a \pmod 7$. This means that, if one assumes that there exists a k, such that $a = b + 7k$, then one may deduce that $b = a + 7(-k)$, and hence that $b \equiv a \pmod 7$. (Why?)]

15. Show from the definition of $a \equiv b \pmod 7$ that, if $b \equiv 2 \pmod 7$ and $c \equiv 5 \pmod 7$, then $b + c \equiv 2 + 5 \equiv 0 \pmod 7$, and that $b \times c \equiv 2 \times 5 \equiv 3 \pmod 7$. [HINT: Since $b = 2 + k \cdot 7$ and $c = 5 + j \cdot 7$, then $b + c = (2 + k7) + (5 + j7) = (2 + 5) + (k + j)7$. Also, $b \times c = (2 + k7) \times (5 + j7) = 2 \times 5 + (2j + 5k + 7jk)7$.]

16. The days of the week can be thought of as forming a mod 7 system in which the names of the days are replaced by integers mod 7. If Sunday $\leftrightarrow 0$, Monday $\leftrightarrow 1$, $\cdots$, Saturday $\leftrightarrow 6$, solve the following problem. If Christmas, the 359th day of the year, falls on Sunday, on what day does July 4, the 185th day, fall? On what day does September 1, the 244th day, fall?

17. How many different congruences (equations) of the form $Ax \equiv B \pmod 7$ with $A \not\equiv 0$ are there? Since $9 \equiv 2 \pmod 7$, we shall not consider $2x \equiv 5 \bmod 7$ and $9x \equiv 5 \bmod 7$ as different.

18. Does the relation $\sim$ (is similar to) satisfy the reflexive, symmetric, and transitive postulates given in Problem 14, if the elements $(a, b, c, \cdots)$ are triangles? [HINT: Replace "$\equiv$" by "$\sim$" and see.]

19. Does the relation $\neq$ satisfy the three postulates of Problem 14, if the elements a, b, c are integers?

20. Which of the postulates of Problem 14 are satisfied if "$\equiv$" is replaced by "$|$" (divides with remainder zero) where a and c are positive integers?

21. Find a relationship, other than those mentioned in the text, which satisfies the three postulates of Problem 14. [HINT: Try "Is a brother or half brother of," "Is a descendant of," "Has the same parents as," "Is the same color (or age) as," "Has long blond hair like," and other similar relationships. Does it make a difference whether the relation is defined over the set of all people or merely the set of all male humans?]

1.3 THE MODULO 6 SYSTEM

It is reasonable and prudent to inquire whether other positive integers, say 6, also yield a modular system similar to the mod 7 system. Certainly it is feasible to define addition and multiplication mod 6 just as we did mod 7.

$$a \equiv b \pmod 6 \text{ means } a = b + 6k \text{ for some integer } k.$$

The modulo 6 system may be thought of as containing only six distinct numbers: 0, 1, 2, 3, 4, 5. There are no negative numbers, for none are needed. (Why?) There are no fractions. (Why are fractions unnecessary?) If an equation has solutions, they can be found by direct substitution, since there are only six numbers in the mod 6 system. There are, however, important differences between the mod 6 system and either the real number system or the mod 7 system. In the real number system (and also in the mod 7 system, as you will prove in Chapter 2), a product is equal to zero *if, and only if*, at least one of its factors is zero, that is,

if either $A = 0$ or $B = 0$, then $A \cdot B = 0$ and conversely

if $A \cdot B = 0$, then either $A = 0$ or $B = 0$ (or both).

This important property is basic in the solution of equations.

In the mod 6 system, the "if" part—"*If either $A = 0$ or $B = 0$, then $A \cdot B = 0$*"—is still satisfied; but the "only if" part—"*If $A \cdot B = 0$, then either $A = 0$ or $B = 0$*"—does not hold. A single counterexample is sufficient to show this. (Why?) Take $A = 4$ and $B = 3$, neither of which is equivalent to 0 modulo 6. However, $4 \cdot 3 \equiv 12 \equiv 0 \pmod 6$.

An important difference between the mod 7 system and the mod 6 system is that, in the mod 7 system (as in the real numbers), the congruence (equation) $Ax \equiv B \pmod 7$, with $A \not\equiv 0$, always has a solution. (You can prove this now by examining the 42 possible cases. In Chapter 2, the problem is solved more easily.) In the mod 6 system, there are linear equations such as $4x \equiv 5 \pmod 6$ and $2x \equiv 3 \pmod 6$, which have no solution at all. (Try all possible values: 0, 1, 2, 3, 4, 5, and see.)

The proof that a polynomial equation has no more solutions than its degree uses the fact that a product of two factors is zero if, and only if, at least one of the factors is zero. Since the mod 6 system does not have this property, it is possible that an equation mod 6 may have more solutions than its degree. Indeed, this proves to be the case. Both $x \equiv 2 \pmod 6$ and $x \equiv 5 \pmod 6$ are solutions of $2x \equiv 4 \pmod 6$ while $4x^2 \equiv 4 \pmod 6$ has $x \equiv 1, 2, 4, 5$ as solutions. However, $5x^2 \equiv 4 \pmod 6$ has no solution. It is interesting to note that, while $4x^2 \equiv 4 \pmod 6$ has four solutions, $x^2 \equiv 1 \pmod 6$, obtained by dividing the previous equation by 4, has only two solutions.

Apparently these modular systems need closer examination before general conclusions can be drawn. Before doing so, let us investigate certain properties of ordinary integers and consider some remarks on the nature of proofs. Modular systems in general are considered in Chapter 2.

Problem Set 1.3

1. Add: $4 + 3 + 0 + 5 + 2 + 4 \pmod 6$.

2. Add: $1 + 2 + 3 + 4 + 5 + 0 \pmod 6$.

3. Solve: $3x \equiv 5 \pmod 6$.

4. Solve: $6x - 5 \equiv 3 \pmod 6$. [Show that there is no value of x in the mod 6 system such that $x + x + x + x + x + x - 5 \equiv 3 \pmod 6$.]

5. Solve: $297x + 6 \equiv 0 \pmod 6$. Although 297 does *not* occur in the mod 6 system, $297x$ still has meaning, since $297x$ represents the sum of $\underbrace{x + x + \cdots + x}_{297 \text{ terms}}$. This problem emphasizes the possible need for distinguishing between the set from which the *unknowns* of the equation are taken and the set from which the *coefficients* of the equation are taken. Fortunately this distinction need not be emphasized in modular systems. However the reader should show that even though $8 \equiv 2 \bmod 6$, there exists values of x for which $x^8 \equiv x^2 \bmod 6$.

6. Solve: $x^2 \equiv 4 \pmod 6$.

7. Solve: $x^2 \equiv 2 \pmod 6$.

8. Solve: $x^2 \equiv 3 \pmod 6$.

9. Solve: $x^3 \equiv 0 \pmod 6$.

10. Solve: $x^3 \equiv 5 \pmod 6$.

11. Solve: $4x \equiv 3 \pmod 6$.

12. Solve: $2x \equiv 0 \pmod 6$.

13. Solve: $4x \equiv 0 \pmod 6$.

14. (a) Make a table listing the six numbers in the mod 6 system. Next to each number x, place its square, x^2; cube, x^3; fourth power, x^4; fifth power, x^5; sixth power, x^6; seventh power, x^7; and eighth power, x^8; all mod 6.
    (b) Using the table of part (a), compute the values $(5)^{236}$ and $(3)^{179}$ will have in the mod 6 system.
    (c) Will $x^4 \equiv 5 \pmod 6$ have a solution?
    (d) For what values of b will $x^5 \equiv b \pmod 6$ have solutions?

15. Solve: $4x^2 + 3x + 4 \equiv 0 \pmod{6}$. Notice that in the mod 6 system the solutions are *not* complex or imaginary numbers.

16. Construct addition and multiplication tables for the mod 6 system.

17. If the symbol $\equiv$ is to be an equals (or equivalence) relation, it must satisfy the following postulates:

    1. *Reflexive*: $a \equiv a \pmod{6}$.
    2. *Symmetric*: If $a \equiv b \pmod{6}$, then $b \equiv a \pmod{6}$.
    3. *Transitive*: If $a \equiv b \pmod{6}$ and $b \equiv c \pmod{6}$, then $a \equiv c \pmod{6}$.

    and be true that for every pair of elements a, b in the set, either $a \equiv b$ or $a \not\equiv b$ mod 6, but not both. Use the definition "$a \equiv b \pmod{6}$ means $a = b + 6k$ for some integer k" to show that the mod 6 system does satisfy these requirements.

18. Show that, if $b \equiv 2 \pmod{6}$ and $c \equiv 5 \pmod{6}$, then

$$b + c \equiv 2 + 5 \equiv 1 \pmod{6}$$

    and

$$b \times c \equiv 2 \times 5 \equiv 4 \pmod{6}$$

    [HINT: Set $b = 2 + 6k$ and $c = 5 + 6j$.]

19. Discover three different congruences of the form $Ax \equiv B \pmod{6}$ with $A \not\equiv 0$ which have *no* solution. Do not use examples from the text.

20. Discover two different congruences of the form $Ax \equiv B \pmod{6}$ with $A \not\equiv 0$ which have *more than one* solution. Do not use examples from the text.

21. Make up a congruence of the form $Ax \equiv B \pmod{6}$, with $A \not\equiv 0$, which has exactly one solution. Prove that only one solution exists by actually substituting the six possible values.

1.4 INTEGRAL DOMAINS

An *integral domain* is defined to be a set of elements $a, b, c, \cdots$ having two operations, $+$ and $\cdot$, and an equals relation,[3] which satisfies the following postulates. The integers serve as one example of a set which

---

[3] In addition to the postulates given in Problem 17, Set 1.3, an equals relation must also be well defined with respect to the given operations; that is, $a = b$ must imply $a + c = b + c$ and $a \cdot c = b \cdot c$.

satisfies these postulates; there are other examples. In each postulate it is assumed that a, b, c are elements of the integral domain.

1. *Closure*: For each pair a, b of elements of the integral domain, $a + b$ and $a \cdot b$ are also elements of the integral domain and are unique.
2. *Commutative Laws*: For each pair a, b of elements of the domain, $a + b = b + a$ and $a \cdot b = b \cdot a$.
3. *Associative Laws*: For each set of three elements a, b, c,

$$a + (b + c) = (a + b) + c$$

and

$$a \cdot (b \cdot c) = (a \cdot b) \cdot c$$

4. *Additive Identity (Zero)*: There exists an element z such that, for every element $b, b + z = z + b = b$, and $b \cdot z = z \cdot b = z$. (In the case of integers, $z =$ zero, it is customary to use the symbol 0 in place of z.)
5. *Multiplicative Identity (Unity)*: There exists an element u such that, for every element b, $b \cdot u = u \cdot b = b$. (In the case of integers, $u = 1$, some texts use the symbol 1 in place of u.)
6. *Additive Inverse (Negative)*: For each element b there exists an element b^* such that $b + b^* = b^* + b = z$, where z is the zero of Postulate 4. (Clearly, $b^* = -b$, in the case of the integers.)
7. *Cancellation Law*: If a and b are elements, and if $c \neq z$ is an element such that $c \cdot a = c \cdot b$, then $a = b$.
8. *Distributive Law*: If a, b, and c are elements, then

$$a \cdot (b + c) = a \cdot b + a \cdot c$$

and

$$(a + b) \cdot c = a \cdot c + b \cdot c$$

These postulates may seem "obvious" at this point—they are, after all, rules of algebra which you have been using for years. Later in the course, mathematical systems—important mathematical systems—in which various of these postulates are *not* satisfied will appear. At that time, postulational systems will be discussed in more detail.

You will discover interesting and important mathematical systems other than the integers which do satisfy these postulates. The question arises, quite naturally, "Do all the laws of arithmetic which apply to integers also apply to integral domains?" The answer is "No." For example, given two unequal integers a and b, it is always true that either $a < b$ or $b < a$, but not both. Integral domains have been discovered in which there is no such order relationship, and furthermore it is impossible[4] to define an order relationship which obeys the usual laws: "If $k > 0$ and $a > b$, then $ak > bk$" and, "If $a > 0$ and $b > 0$, then $a + b > 0$." The complex numbers, $a + bi$, form one such domain.

The question next arises, "What properties of ordinary integral arithmetic *will* every integral domain have?" Parts of the answer are contained in the postulates already listed. For example, the system must be commutative under addition (that is, $a + b = b + a$, for all a, b).

The following theorem is valid for the integers. It will be derived here *from the listed postulates without using any other properties of integers*, thus showing that the theorem is valid in every system which satisfies this list of postulates, even if the system is not the integers. Also, you can see here one type of reasoning which is important in modern mathematics. Mathematicians often use the symbols $+$ and $\times$ rather than $+$ and $\cdot$ to indicate the two operations. We use the more familiar symbol 0 in place of the general symbol z used in Postulate 4.

THEOREM 1.1   If $a \cdot b = 0$, then *the postulates listed for the integers imply that either $a = 0$, or $b = 0$, or both.*

Proof: Given: Postulates 1 to 8 and $a \cdot b = 0$.
      To Prove: Either $a = 0$, or $b = 0$, or both.

There are two possibilities: either $a = 0$, or $a \neq 0$.

If $a = 0$, the theorem is satisfied.

If $a \neq 0$, then, since $0 = a \cdot 0$,

$a \cdot b = 0 = a \cdot 0$	Hypothesis and Postulate 4
$a \cdot b = a \cdot 0$	Quantities equal to the same thing are equal to each other. (The Transitive Law (3) of equivalence relationship. See Problem 14, Set 1.2.)
$b = 0$	Postulate 7, since $a \neq 0$.

[4] The reader may well ponder this statement. It does *not* state that no one has yet been able to discover an order relationship in these more general domains. It states rather, that someone has discovered a proof that *no one ever will be able to discover an order relationship of the desired type*—a rather remarkable statement. Can *you* prove that the complex numbers cannot be ordered in any manner which preserves these laws? Can you prove that the mod 7 system cannot be so ordered?

Hence, if $a \cdot b = 0$, either $a = 0$ or $b = 0$, or both. Note that *each step* is justified by a postulate or by a rule of logic. Explain why the statement "multiply both sides by $1/a$" would *not* be a valid justification for the last step.

The mod 6 system does *not* satisfy this theorem. Discover one of the above postulates which is not satisfied by the mod 6 system. Can you decide beforehand which postulates it would be good to check? (Which postulates were used in the proof of Theorem 1.1?)

Before continuing, let us prove a *lemma*. (A *lemma* is a "helping theorem" used in the proof of a major theorem.) The reader should recognize that if c and b are elements of an integral domain D, then $(c \cdot b)$ is also an element of the integral domain D. (Which postulate guarantees this?) Since $(c \cdot b)$ is a single element of the integral domain D, then by postulate 6 there is an additive inverse (negative) of the element $(c \cdot b)$. If we denote the additive inverse of $(c \cdot b)$ by the symbols $(c \cdot b)^*$ then $(c \cdot b) + (c \cdot b)^* = z$, the zero of Postulate 4. There also exists an additive inverse b^* of the element b. It would be nice if $(c \cdot b)^*$ were equal to $c \cdot (b^*)$, but we are *not* at liberty to make such an assumption without proof. [It is quite possible that you have never even *proved* the equivalent rule of ordinary algebra, namely $-(c \cdot b) = c \cdot (-b)$ even though you use it regularly—it *can be proved*, and should be.] We prove it as *Lemma* 1.1.

LEMMA 1.1 $(c \cdot b)^* = c \cdot (b^*)$

where b^* denotes the additive inverse of b and $(c \cdot b)^*$ the additive inverse of $(c \cdot b)$.

Proof: We begin by proving that two different expressions each equal zero and hence that the expressions are equal.

Using z to denote the zero (additive identity) of Postulate 4, we have

$$(c \cdot b) + (c \cdot b)^* = z \qquad \text{Postulate 6.}$$

It is also true, by Postulate 8, that

$$(c \cdot b) + [c \cdot (b^*)] = c \cdot (b + b^*)$$

$$= c \cdot (z) \qquad \text{by Postulate 6}$$

$$= z \qquad \text{by Postulate 4}$$

Since each expression is equal to z, by the transitive law ("Things equal to the same thing are equal to each other"—see Problem 17, Section 1.3) we have

$$[(c \cdot b) + (c \cdot b)^*] = \{(c \cdot b) + [c \cdot (b^*)]\}$$

Adding $(c \cdot b)^*$ to the left edge of each member we have

$$(c \cdot b)^* + [(c \cdot b) + (c \cdot b)^*] = (c \cdot b)^* + \{(c \cdot b) + [c \cdot (b^*)]\}$$

and by the associative law (Postulate 3) this becomes

$$[(c \cdot b)^* + (c \cdot b)] + (c \cdot b)^* = [(c \cdot b)^* + (c \cdot b)] + [c \cdot (b^*)]$$

$$(z) + (c \cdot b)^* = (z) + [c \cdot (b^*)] \qquad \text{by Postulate 6}$$

$$(c \cdot b)^* = c \cdot (b^*) \qquad \text{by Postulate 4}$$

This completes the proof.

We now turn our attention to a more subtle theorem.

THEOREM 1.2   *The cancellation law (postulate 7) is equivalent to the statement "$x \cdot y = 0$ implies either $x = 0$, or $y = 0$, or both" in the presence of the other integral domain postulates.*

Theorem 1.1 shows that the theorem "$a \cdot b = 0$ *implies either* $a = 0$, *or* $b = 0$, *or both*" may be derived from the given postulates 1 to 8. To complete Theorem 1.2 we must also show that *if* the cancellation law (Postulate 7) is replaced[5] by the new postulate "$x \cdot y = 0$ implies either $x = 0$, or $y = 0$, or both," then it is possible to prove the cancellation law (old Postulate 7) as a theorem in the new system. Thus the same system is defined by either set of postulates.

---

[5] Other changes in the system of postulates are also possible; for example, if we are given that $a + b = b + a$ for all elements, then it follows from $b + z = b$ that $z + b = z$ also and postulate 4 could have been shortened. It is also possible to prove from the remaining postulates that $a \cdot z = z \cdot a = z$. At present we are interested in the basic structure of an integral domain rather than in the most condensed set of postulates.

<table>
<tr><td align="center">Old Postulates</td><td align="center">New Postulates</td></tr>
<tr><td>1. Closure</td><td>1. Same</td></tr>
<tr><td>2. Commutative Laws</td><td>2. Same</td></tr>
<tr><td>3. Associative Laws</td><td>3. Same</td></tr>
<tr><td>4. Additive Identity (zero)</td><td>4. Same</td></tr>
<tr><td>5. Multiplicative Identity (Unity)</td><td>5. Same</td></tr>
<tr><td>6. Additive Inverse (Negative)</td><td>6. Same</td></tr>
<tr><td>7. Cancellation Law</td><td>7. $x \cdot y = 0$ implies either $x = 0$, or $y = 0$, or both</td></tr>
<tr><td>8. Distributive Laws</td><td>8. Same</td></tr>
</table>

We now turn our attention to the needed proof:

Given: Postulates 1 through 6, Postulate 8 and the postulate:

$$\text{``}x \cdot y = 0 \text{ implies either } x = 0, \text{ or } y = 0, \text{ or both.''}$$

To Prove: The cancellation law—"If $c \neq 0$ and $c \cdot a = c \cdot b$, then $a = b$" follows as a theorem.

Proof: By Postulate 6 there exists a single element $(c \cdot b)^*$ such that $c \cdot b + (c \cdot b)^* = z$. Add this element $(c \cdot b)^*$ to each member of the given statement $c \cdot a = c \cdot b$ obtaining $(c \cdot a) + (c \cdot b)^* = (c \cdot b) + (c \cdot b)^*$.

Since $(c \cdot b) + (c \cdot b)^* = z$ and since, by Lemma 1.1 $(c \cdot b)^* = c \cdot (b^*)$, this reduces to

$$c \cdot a + c \cdot (b^*) = z$$

$$c \cdot [a + (b^*)] = z \qquad \text{by Postulate 8}$$

Since $c \neq z$ by hypothesis and also by hypothesis a product is zero only if at least one of the factors is zero, it follows that

$$a + (b^*) = z$$

Upon adding b to each member and using Postulates 3 and 6 and 4 we obtain:

$$[a + (b^*)] + b = z + b$$

$$a + [b^* + b] = b \qquad \text{Postulate 3}$$

$$a + [z] = b \qquad \text{Postulate 6}$$

$$a = b \qquad \text{Postulate 4}$$

This completes the proof of the second part of Theorem 1.2.

To recapitulate: In the presence of Postulates 1 to 6, and 8, we have just shown that it is possible to deduce the cancellation law from the statement, "$a \cdot b = 0$ implies either $a = 0$ or $b = 0$, or both."

By Theorem 1.1, "$a \cdot b = 0 \Rightarrow a = 0$ or $b = 0$ or both" may be deduced from the integral domain postulates including the cancellation law.

Evidently, in the presence of Postulates 1–6, and 8, the cancellation law "$c \neq 0$, and $c \cdot a = c \cdot b$ imply $a = b$" and the law "$a \cdot b = 0$ implies either $a = 0$, or $b = 0$, or both" are logically equivalent, since either may be deduced from the other.

Problem Set 1.4

In each of Problems 1–12, determine which of Postulates 1–8 are satisfied by the given set of numbers. You may assume, in each case, that the associative laws hold.

1. All rational numbers, a/b, where a and b are integers, $b \neq 0$.

2. (a) All odd integers.
   (b) All even integers.

3. (a) The mod 7 system.
   (b) The mod 6 system.

4. All numbers of the form $a + b\sqrt{3}$, where a and b are integers.

5. The set consisting of zero alone. [Do not be misled by the notation. Postulates 5 and 7 *are* (vacuously) satisfied in this system. Why?]

6. The set consisting of the two numbers 0 and 1.

7. All numbers of the form $a/2 + (b/2)\sqrt{5}$, where a and b are integers. Note that $4 + 3\sqrt{5}$ *is* in this set, since it is permissible to have $a = 8$, $b = 6$.

8. (a) All positive integers.
   (b) All nonnegative integers.

9. All rational numbers having nonnegative integral powers of 2 (including the possibility $2^0 = 1$) as denominator.

10. A set consisting of the two numbers, 0 and 1, where we define $1 + 1 = 0$ instead of 2, but where other sums and products are defined as usual.

11. All integers, with addition defined as usual, but with multiplication defined so that the product of two integers is always zero.

12. All multiples of 13.

13. Prove from the postulates that, if W is an element of an integral domain, and if W has the property that for every element a, $a + W = a$, then $W = z$ of Postulate 4. Be careful that you do not prove the converse of this theorem, which is contained in Postulate 5. The theorem of this problem states, essentially, that the system of integers (or any other integral domain) contains *only one* element which can function as the identity element of addition. List the postulates used in each step of your proof. [HINT: Since Postulate 4 hypothesizes the existence of an element z, we may take $a = z$ in the equation $a + W = a$.]

14. The cancellation law states that, if $c \neq 0$, then $c \cdot a = c \cdot b$ implies $a = b$. A similar law for addition (that is, $\{c + a = c + b\} \Rightarrow \{a = b\}$) could also have been postulated. Show that it is unnecessary, since "$c + a = c + b$ implies $a = b$" may be proved as a theorem from the other postulates. Examine your completed proof to see why a similar proof could *not* be used to prove Postulate 7 from the other postulates.

15. Show from the postulates that for each integer there exists one, *and only one*, solution of the equation $B + x = 0$. [HINT: By Postulate 6, there exists a solution. To prove there exists only one solution, assume x and y are both solutions. Then $B + x = 0 = B + y$. Complete the proof, listing all postulates needed. Compare Problem 14.]

16. Show that any subset of an associative system is also associative. Can you prove a similar theorem with "closed" substituted for "associative"?

17. Invent or discover a system that does *not* satisfy the postulates for an integral domain.

18. Prove Lemma 1.1, using the symbol $\times$ to denote multiplication.

19. Prove Theorem 1.2, using the symbol $\times$ to denote multiplication.

20. Note that the integral domain postulates do *not* permit the use of the operation of subtraction. It is not necessary to hypothesize a third operation (subtraction) since one can use $a + x^*$ rather than $a - x$. What postulate guarantees the existence of x^*?

21. In order to help you think in more general terms, prove that in an integral domain (that is, Postulates 1–8 of this section) if R and S are elements of the integral domain, then $R \cdot S = z$ implies that either $R = z$ or $S = z$, where z is the additive identity of Postulate 4. It is easy to check your proof by comparing it with the proof of Theorem 1.1, but do *not* look at the latter until your own proof is completed, then rework the proof of Theorem 1.1.

22. Explain the difference between the contents of Theorem 1.1 and Theorem 1.2 to a friend who has not taken this course. Record his reaction to your explanation.

23. Give an example of a mathematical system which is not closed.

24. Give an example of a mathematical system which is not associative.

25. Determine several mathematical systems, each of which violates at least one of the eight postulates of an integral domain. When you are finished, you should have examples of systems that violate, in turn, each of the eight postulates. In each case you should specify the set of elements, the two operations involved, and what you mean by equality of elements. Thus you might specify the set of elements to be the ordinary integers and the two operations to be subtraction and multiplication with ordinary equality of integers being used for the equals relation. On the other hand, a different system would evolve if you used congruence modulo 6 (that is, $A \equiv B$ means there exists an integer k such that $A = B + k \cdot 6$) as your equals relation with the two operations of subtraction and multiplication of integers.

26. (a) The following algorithm or procedure produces a meaningful result which is a function of the two positive integers selected in Step 1. What is this result called?
    1. Select two positive integers M and N.
    2. Divide the current value of M by the current value of N obtaining a remainder R where $0 \leq R < N$.
    3. If $R = 0$, terminate the procedure. N is the result. If $R \neq 0$, go to Step 4.
    4. Replace the value of M by that of N and the value of N by that of R and go to Step 2.

    (b) Follow the algorithm to its termination (i.e., $R = 0$) with several choices of starting values M, N. If you have a computer or programmable desk calculator available you may wish to write a program for this algorithm.

1.5 THE NATURE OF DEFINITIONS, POSTULATES, AND AXIOMS

Did you have trouble with the last section? If so, Sections 1.5, and 1.6 were written especially for you. If you did not have trouble, you may read these sections rapidly, since you already understand much they contain.

1. *Undefined Terms*: Every definition must eventually depend upon words and ideas which have not been defined. It took many generations for men to realize this simple truth. If it comes as a shock to you that certain mathematical words and phrases are not, and *cannot*, be defined, then consider the plight of the person who seeks the meaning of the word *shadow* in a dictionary.

> *shadow*—shade or semidarkness
> *shade*—shadow or semidarkness
> *darkness*—lack of light
> *light*—a form of energy capable of casting a shadow
> *shadow*—......

An endless circular process is set up unless certain words are left undefined.

In plane geometry, for example, it is usual to take the words *point* and *line* as undefined terms. It is possible to "pseudodefine" a *point* as the intersection of two *lines*, and then turn around and say that, in a plane, a *line* is the shortest path between two *points*. However, the circle again closes about us. In the definition of integral domain given in the last section, the word *element* is undefined. Some people say that the operations $+$ and $\times$ are also undefined; others feel an operational definition of $+$ and $\times$ is contained in the postulates. Each viewpoint has advantages. For our purpose, both eventually reduce to the same thing.

In this course, the integers are also undefined, although for certain other purposes it is more desirable to define the integers in terms of Peano's "successor operation," which itself remains undefined. In short, there must be certain undefined terms which can be used as basic raw material from which to fashion other definitions.

2. *Postulates*: Postulates are statements which are assumed to be valid. One cannot generate theorems out of nothing, any more than he can definitions. Certain statements are assumed, and from them, and the undefined terms, an entire mathematical system is built. If the mathematical system is to be meaningful, it is desirable that the postulates be consistent

(that is, contain no contradictory statements). It is important that you realize that there is no one *correct* set of postulates. Different sets of postulates apply in different situations. In measuring land or navigating short distances, the postulates of plane geometry (trigonometry) may be suitable, but for long distances the postulates of spherical geometry (trigonometry) are better suited.

3. *Theorems*: A theorem is a statement which can be proved using the laws of logic, the undefined terms, and the postulates of a given system. Any theorem or definition which has already been so established may also be used in proving a subsequent theorem. It is meaningless to speak of the truth or falsity of a theorem without telling what system is being used as a frame of reference. "Does the equation $4x = 3$ have a solution?" is a meaningless question since the answer is "yes" in the system of rational numbers, "no" in the system of integers, "yes" in the mod 7 system, and "no" in the mod 6 system.

1.6 THE NATURE OF PROOF

A few remarks concerning the nature and construction of mathematical (and nonmathematical) proofs seem appropriate at this point. Several things should be clearly established before a statement can be proved or disproved:

(1) The exact statement to be proved or disproved must be stated in unambiguous language. (Surprisingly enough, this is often a point which causes difficulty. Hypothesis and conclusion should both be concise.)

(2) The "frame of reference" in which the statement is to be proved, that is, the permitted working tools (assumptions, postulates, definitions, and established theorems), must be understood. Many statements are true in one reference frame and false in another. "The sum of the three interior angles of a triangle is 180°" is true in plane trigonometry but false in spherical trigonometry. "Two lines which do not intersect are parallel" is valid in plane geometry but not in solid geometry. (Why not?)

(3) A plan of attack must be devised. Few theorems of interest today are proved by random juggling of symbols. Decide what is to be proved, and then see if you can find a step which, if you could establish it, would enable you to reach the desired conclusion.

Several different methods of proof will be discussed. The specific examples given are not of as great importance as the methods used.

The most familiar type of proof, of course, establishes, through a series of logical steps, that a conclusion or theorem follows from its hypothesis using only those assumptions permitted in the desired frame of reference.

Let A and B represent statements. In particular, let A represent the hypothesis of a theorem and B represent the conclusion of the theorem, where the theorem is of the form, "If A, then B," or "A implies B," or "$A \Rightarrow B$."

EXAMPLE 1

$A_1 \equiv$ triangle I and triangle II are similar.
$B_1 \equiv$ the corresponding sides of triangles I and II are proportional.
In this case, "If A_1, then B_1" becomes a familiar geometric theorem.

EXAMPLE 2

$A_2 \equiv$ the quadrilateral $ABCD$ is a square.
$B_2 \equiv$ the sides of quadrilateral $ABCD$ are equal.
In this case, "If A_2, then B_2" also becomes a familiar theorem.

The meanings of the words converse, opposite (or inverse), and contrapositive will become apparent upon examining the following diagrams.

THEOREM $A \Rightarrow B$ If A, then B.	CONVERSE $B \Rightarrow A$ If B, then A.
OPPOSITE (INVERSE) $\sim A \Rightarrow \sim B$ If *not A*, then *not B*.	CONTRAPOSITIVE $\sim B \Rightarrow \sim A$ If *not B*, then *not A*.

<table>
<tr><td colspan="2" align="center">THEOREM

$A \Rightarrow B$

If triangle I and triangle II are similar, then the corresponding sides of triangles I and II are proportional.
<br align="right">TRUE</td><td colspan="2" align="center">CONVERSE

$B \Rightarrow A$

If the corresponding sides of triangles I and II are proportional, then triangle I and triangle II are similar.
TRUE</td></tr>
<tr><td colspan="2" align="center">OPPOSITE   TRUE

$\sim A \Rightarrow \sim B$

If triangle I and triangle II are NOT similar, then the corresponding sides of triangles I and II are NOT proportional.</td><td colspan="2" align="center">TRUE   CONTRAPOSITIVE

$\sim B \Rightarrow \sim A$

If the corresponding sides of triangles I and II are NOT proportional, then triangle I and triangle II are NOT similar.</td></tr>
</table>

<table>
<tr><td colspan="2" align="center">THEOREM

$A \Rightarrow B$

If the quadrilateral ABCD is a square, then the sides of quadrilateral ABCD are equal.
<br align="right">TRUE</td><td colspan="2" align="center">CONVERSE

$B \Rightarrow A$

If the sides of quadrilateral ABCD are equal, then the quadrilateral ABCD is a square.
FALSE</td></tr>
<tr><td colspan="2" align="center">OPPOSITE   FALSE

$\sim A \Rightarrow \sim B$

If the quadrilateral ABCD is NOT a square, then the sides of quadrilateral ABCD are NOT equal.</td><td colspan="2" align="center">TRUE   CONTRAPOSITIVE

$\sim B \Rightarrow \sim A$

If the sides of quadrilateral ABCD are NOT equal, then the quadrilateral ABCD is NOT a square.</td></tr>
</table>

The latter diagram shows that the converse of a theorem is not necessarily valid just because the theorem is valid! However, a theorem and its contrapositive are equivalent. (Each implies the other.)

<table>
<tr><td align="center">Theorem</td><td align="center">Contrapositive</td></tr>
<tr><td align="center">$A \Rightarrow B$</td><td align="center">$\sim B \Rightarrow \sim A$</td></tr>
</table>

Assume $A \Rightarrow B$. If "not B" were true and "A" were also true, then the theorem $A \Rightarrow B$ yields "B" and both "not B" and "B" would hold. This is untenable in our logical system. Hence, $A \Rightarrow B$ implies that $\sim B \Rightarrow \sim A$. Conversely, assume the contrapositive $\sim B \Rightarrow \sim A$. If A and $\sim B$ are both true, then the contrapositive yields $\sim B \Rightarrow \sim A$ and both "not A" and "A" would hold. This is untenable in our logical system. Hence, $\sim B \Rightarrow \sim A$ implies $A \Rightarrow B$. Thus, a theorem and its contrapositive each imply the other (i.e., are *equivalent*). This is the basis of "indirect proof," familiar to geometry students everywhere.

The converse and the opposite of a theorem are also equivalent. (Can you prove this?) If you have trouble proving "If A, then B," it may be easier to prove "If not B, then not A." If either a theorem or its contrapositive is proved, then both are valid. (Why?) In a similar fashion, it may be easier to prove (or disprove) one, than the other, of the equivalent statements "If B, then A" or "If not A, then not B." If either the converse or the opposite is proved, then both are valid.

The words "sufficient" and "necessary" occasionally cause trouble. If the theorem "If A, then B" is valid, we say that A is a *sufficient* condition for B, since if A is satisfied it automatically follows that B is valid. If A is sufficient, then A is all that is needed to guarantee the validity of statement B. On the other hand, if a condition N is *necessary* for a conclusion C, then N must hold before C can be valid. We say that B is a *necessary* condition for A, because A cannot be valid unless B is satisfied. (Consult the contrapositive, whose validity follows from the validity of the theorem.) But we can *not* say that A is a necessary condition for B, unless the converse (if B, then A) is also valid.

THEOREM	CONVERSE
If A, then B. *A is sufficient for B.* *B is necessary for A.*	If B, then A. *A is necessary for B.* *B is sufficient for A.*
OPPOSITE (INVERSE)	CONTRAPOSITIVE
If *not A*, then *not B*. (equivalent to the converse)	If *not B*, then *not A*. (equivalent to the theorem)

To prove that A is a *necessary and sufficient* condition for B, we may prove any pair of nondiagonally opposite statements in the above box. (Why?)

If A is necessary and sufficient for B, then the conditions of A are sufficient to ensure the validity of B. Furthermore, each of the conditions of A is necessary before B can be valid. If one of the conditions of A is violated, then B is not valid (if not A, then not B).

It is important to note that, even though 10 million (or even infinitely many) examples which do satisfy a theorem may *not* prove that it is true, *a single counterexample is sufficient to show it is false.*

EXAMPLE 3

Conjecture to be proved or disproved: "Every integer greater than 7 is even."

It is easy to find as many even integers greater than 7 as desired, yet they do not prove the conjecture. However, the presence of any one counterexample, say 15, is sufficient to disprove the conjecture.

Before considering the next example, let us review the meaning of *prime* positive integer. An integer $N > 1$ is said to be *prime* if the only possible factors of N are ± 1 and $\pm N$. Thus, 2, 13, 41, 1009, 7432339208719 are examples of prime numbers, while 6, 24, 1003 $= (17)(59)$ are not prime.

EXAMPLE 4

Statement to be proved or disproved:

"The formula $P(n) = n(n + 1) + 41$ always yields prime numbers when a positive integer is substituted for n."

By actual substitution, you will find that $P(n)$ is prime for $n = 1, 2, 3, 4, 5, 6$. Before reading on, you, the reader, should try to prove or disprove the above conjecture. *Do not read on until you have tried!*

It is true that $P(n)$ is prime when n is an integer between 1 and 39, inclusive. For $n = 40$, $P(n)$ is *not* prime, since $P(40) > 1$, $P(40) \neq 41$, and 41 divides $P(40)$. (Note how this uses the definition of "prime.") This one counterexample is sufficient to disprove the statement. The statement is false! We need not bother to look for other counterexamples, although infinitely many other exist, since *one counterexample is sufficient to prove a statement is false.*

Did you, yourself, prove that the statement of Example 4 was false *before* reading the last paragraph? If so, congratulations. If not, ask yourself "Why not?" Was it because you were too lazy to work it out for yourself with the answer so handy? If so, you have a serious problem which may have far-reaching effects. Now is a good time to analyze your nonmathematical difficulties and try to correct those which are a result of your not being willing to put forth the effort needed to gain an end.

EXAMPLE 5

Although one counterexample will disprove a statement, a thousand examples illustrating the theorem do not prove it. A checkerboard has the two diagonally opposite corner (black) squares removed. A set of 31 rectangular cards, each two checkerboard squares long and one square wide (dominoes), are provided. Statement to be proved or disproved: "It is impossible to completely cover the remaining squares of the checkerboard with the 31 rectangles."

Possibly the first step should be to experiment and try to so cover a deleted checkerboard. If a method of covering the checkerboard is discovered, the statement is disproved, and your worries are over. If, after four or five hours of work, no such covering pattern is found, you may be ready to concede that it is impossible—*but that is not enough*! This does not *prove* that it is actually impossible—someone else may find a proof next week. It is possible to devise a proof that it is actually impossible to so cover the checkerboard (i.e., that no one will ever find a way to cover it according to the rules stated). This may be a new type of thinking, but you should make a serious effort to find a solution before going on.

If you found a solution, then you have already done independent mathematical research. (Someone else had also done it. Still, since you did not know of the earlier results, they do not detract from the glory of your achievement!) One possible argument to establish the impossibility of this covering depends upon the observation that every rectangle (domino) must cover one red and one black square. Since there are two fewer black squares than red squares, it is impossible to exactly cover the deleted checkerboard in the prescribed manner.

If this method of proof did not occur to you, you should still be able to follow and understand it, now that it has been suggested. Explain the problem and its solution to a friend who has not taken this course. This is always good practice, for it helps you to understand what you are doing.

EXAMPLE 6

Reductio ad absurdum or "indirect proof." It is often possible to prove a statement is true by assuming that the statement is untrue and showing, by a logical argument, that this assumption along with the other postulates leads to an explicit contradiction. If there is no error in reasoning, then the statement which was assumed false must actually be true. Hence, the original statement is valid. Your high-school geometry text will furnish examples of such indirect proofs. The proof often given in high school texts to show that $\sqrt{2}$ is not a rational number is another example of an indirect proof that may be familiar. It begins, "Assume that $\sqrt{2}$ is a rational number, then there exist relatively prime integers a and $b \neq 0$ such that $\sqrt{2} = a/b$ is in lowest terms." Can you complete the proof? *Don't* say, "I can't remember it." Mathematics is *not* a memory course. You have the ability to figure out proofs for yourself once you are on a suitable track. The next step leads to the statement "$2b^2 = a^2$, hence 2 divides a." Try your hand before you continue.

EXAMPLE 7

Mathematical induction. In later mathematical work, you will certainly need the technique of proof by mathematical induction. Indeed, you have probably already used induction, at least implicitly, to prove the compound interest formula or the binomial expansion, or to prove that $dx^n/dx = nx^{n-1}$. The idea is simple.

To prove that some statement $P(n)$ is valid for every positive integer n:

First, prove that $P(n)$ is valid for $n = 1$.

Next, show that, *if* you assume $P(n)$ is valid for $n = k$, it then follows that $P(n)$ is valid for $n = k + 1$.

The principle of mathematical induction then states that $P(n)$ is valid for all positive integers n. Does this seem like a reasonable postulate? The first statement puts you on the bottom rung of a ladder. The second statement shows you that from any rung it is possible to climb to the next higher rung. Hence, you may climb the ladder. One of the Peano postulates from which the integers may be derived is that mathematical induction is valid. Hence, this method of proof is basic in arithmetic.

Illustration: Let $P(n) \equiv dx^n/dx = nx^{n-1}$ for n, a positive integer.

(1) $P(n)$ for $n = 1$ becomes $dx^1/dx = 1$, a fact which is easily proved from the definition of derivative.

(2) Assume $dx^k/dx = kx^{k-1}$, and from this prove that

$$dx^{k+1}/dx = (k + 1)x^k$$

The latter may be accomplished by writing x^{k+1} as $x \cdot x^k$ and differentiating by the product rule. Consult any calculus text for details.

The reader should note that a theorem must deal with integers (in this case the exponent n) before mathematical induction can be used.

Although the theorem in question is valid for nonintegral exponents, a separate proof is required. The well-known formula

$$1 + 2 + 3 + \cdots + n = n(n + 1)/2$$

can also be proved by mathematical induction. Try your hand at it now.

Problem Set 1.6

1. How many solutions does the equation $2x = 5$ have in (a) the system of integers; (b) the system of rational numbers; (c) the system of complex numbers; (d) the mod 6 system; (e) the mod 7 system?

2. The statement, "If it doesn't rain this afternoon, I shall go downtown" is of the form, "If A, then B." Write its contrapositive, converse and opposite. Assume the statement is valid. Explain why the converse need *not* be valid. Is the contrapositive valid? Is the opposite valid?

3. Carry out the suggestions of Problem 2, using the statement, "If she wears low heels, she is shorter than he is." Note that, while in common conversation one might conclude that the opposite (inverse) was *intended*, it was not stated. In mathematics, it is essential that you state exactly what you mean, and mean exactly what you state.

4. Is the final sentence of Problem 3 redundant?

5. Find examples of indirect (*reductio ad absurdum*) proofs in some other text and bring one to class.

6. Make a false conjecture (not taken from this text) which can be disproved by a known counterexample.

7. Make a conjecture (not taken from this text) which has at least 50 examples which suggest it might be true, but which is actually false.

8. Find a theorem in which you are required to prove that something is impossible. Discuss your theorem with another class member.

9. Analyze the proof of Theorem 1.1. Be sure to discuss each of the three points mentioned at the beginning of Section 1.6.

10. Same as Problem 9 for Theorem 1.2.

11. The theorem, "If two lines are parallel, then they do not intersect," is valid in both plane geometry (two dimensions) and in solid geometry (three dimensions). State its opposite, converse, and contrapositive, and determine which of these is valid in (a) plane geometry; (b) solid geometry.

12. The statement, "The integer 17 is a prime" is *not* of the type, "If A, then B." However, it *is* possible to prove this statement by noting that each possible positive integral divisor is less than or equal to 17. Complete the proof. (Later, you will be asked to show that if an integer $N > 1$ is *not* prime, then N must have at least one divisor d such that $1 < d \leqq (\sqrt{|N|}.)$

13. Peter Bertdel's teacher asked him to write the converse of the statement, "If a dog barks, then he will not bite." Peter wrote, "A dog will not bite if he barks." Was the teacher justified in marking the answer incorrect?

14. A text on logic contains the statement, "The converse of every definition is true. If it were false, it would be an unsatisfactory definition and should be replaced by a better one." Discuss this.

15. Show that the contrapositive of the contrapositive of a theorem is the theorem.

16. Is the opposite of the converse of a theorem the same thing as the converse of the opposite of the theorem?

17. Many advertisements depend upon the human failing of supplying and assuming the validity of the *converse* of a statement. "If you don't buy Bloxo, you are wasting your money" really says very little. Almost everyone occasionally wastes money, whether or not he buys Bloxo. Find similar ads and bring them to class.

18. Skilled speakers often use the contrapositive of a statement when they wish to make a point. For example, a speaker might say, "If you don't vote for Wintergreen, then you are not interested in honest government," in the belief that the listener will prefer to supply the equivalent direct statement. Thus, the listener becomes the authority for the statement, "If you are interested in sound government, vote for Wintergreen," and is inclined to believe himself. See if you can find a contrapositive statement in a speech. Newspapers, radio, and television each provide possible sources of such material.

19. Two statements are said to be logically equivalent if each statement implies the other. Show that logical equivalence satisfies the three postulates for an equivalence (or equals) relation given in Problem 14, Set 1.2.

20. (With apologies to Gilbert and Sullivan) A barrister said, "I'll have you appointed a judge only if you marry my elderly ugly daughter." If you do marry his "elderly ugly daughter," has he promised to have you appointed a judge?

21. Bob told Jan, "If I win the office baseball pool, I'll take you to the 21 Club!" Has Bob violated his promise to Jan in any of the following situations?
    (a) Bob loses pool, and refuses to take Jan to the 21 Club.
    (b) Bob loses pool, and takes Jan to the 21 Club.
    (c) Bob wins pool, and refuses to take Jan to the 21 Club.
    (d) Bob wins pool, and takes Jan to the 21 Club.
    (e) Bob wins pool, and takes Jan to the door of the 21 Club but refuses to go in. (Does the answer depend upon the definition of certain words or phrases?)

22. Answer the questions of Problem 21 if Bob's original statement is, "If I don't win the office baseball pool, we won't go to the 21 Club!"

23. Henry said, "It will be a cold day in July when I date Alice."
    (a) July 25 was a very cold day. Can Alice *logically* count on having a date with Henry, assuming that Henry is a man of his word?
    (b) Can Henry date Alice for the Christmas dance without violating his statement?

24. A school pageant has the following announcement printed on its tickets, "This pageant will not be held only if the weather is not fair." State, in other words, what, and only what, the printed statement implies.
    (a) If there is no performance, what can you say about the weather?
    (b) If there is a performance, what can you say about the weather?
    (c) If the weather is fair, will a performance be given?
    (d) If the weather is not fair, will a performance be given?

25. Discuss the validity of the following reasoning:
    Given: (a) If country B is communist dominated, the United States is in danger.
    (b) If the United States does not feed the people of country B, it will become communist dominated.

Conclusion: If the United States feeds the people of country B, then the United States will not be in danger from country B. (It may be interesting to discuss this problem with friends who have strong feelings on the subject of foreign aid. Note whether they use logic or emotion to come to their conclusion. Even the more astute are apt to change the subject from the logical reasoning involved to the subject of foreign aid.)

26. Prove, by mathematical induction, that the formula

$$1 + 2 + 3 + \cdots + n = \tfrac{1}{2}n(n + 1)$$

holds for all positive integers n.

27. Read one of the following articles in the *American Mathematical Monthly* and prepare a short written or oral report thereon.

    C. G. Hempel, "The Nature of Mathematical Proof," **52,** p. 543.
    S. MacLane, "Symbolic Logic," **46,** p. 289.
    H. Weyl, "Mathematics and Logic," **53,** p. 2.
    R. L. Wilder, "The Nature of Mathematical Proof," **51,** p. 309.

28. Prepare a brief report on the indicated portion of one of the following books available in your library. If you can also use the report in another class (English or Speech, for example) you may use the carbon copy as part of this assignment.

    J. Anderson and H. Johnstone, *Natural Deduction*, (Belmont, Calif.: Wadsworth, 1962, Chap. 1.
    F. B. Fitch, *Symbolic Logic*, (New York: Ronald Press, 1952), pp. 12ff.
    P. Suppes and S. Hill, *First Course in Mathematical Logic*, (Waltham, Mass.: Blaisdell, 1964), pp. 29–40.
    R. L. Wilder, *Introduction to the Foundations of Mathematics*, (New York: Wiley, 1965), Chap. 1.

1.7 INTEGRAL DOMAINS

For convenience, the definition of *integral domain* given in Section 1.4 is repeated here.

An *integral domain D* is a set of elements $a, b, c, \cdots$ having a well-defined equals relation, and two operations, $+$ and $\cdot$, which satisfy the following postulates:

1. *Closure*: For each pair a, b of elements of the domain D, $a + b$ and $a \cdot b$ are unique elements of D.

2. *Commutative Laws*: For each pair a, b, of elements of D, $a + b = b + a$ and $a \cdot b = b \cdot a$.
3. *Associative Laws*: For each set a, b, c of three elements of D, $a + (b + c) = (a + b) + c$ and $a \cdot (b \cdot c) = (a \cdot b) \cdot c$.
4. *Additive Identity (Zero)*: There exists z, an element of D such that, for every b in D, $b + z = z + b = b$ and $b \cdot z = z \cdot b = z$.
5. *Multiplicative Identity (Unity)*: The domain D contains an element u such that, for every b in D, $b \cdot u = u \cdot b = b$.
6. *Additive Inverse (Negative)*: For each element b of D, there exists an element b^* in D such that $b + b^* = b^* + b = z$, where z is the zero of Postulate 4.
7. *Cancellation Law*: If a and b are elements of D, and if $c \neq z$ is an element of D such that $c \cdot a = c \cdot b$, then $a = b$.
8. *Distributive Law*: If a, b, and c are elements of D, then
$$a \cdot (b + c) = (a \cdot b) + (a \cdot c), \quad \text{and} \quad (a + b) \cdot c = (a \cdot c) + (b \cdot c).$$

It should be pointed out that these postulates are *not* independent. For example, $z + b = b$ follows from $b + z = b$ in the presence of Postulate 2. The second part of Postulate 4, $b \cdot z = z \cdot b = z$, can be proved from the other postulates.

EXAMPLE 8

Prove that $b \cdot z = z \cdot b = z$ from the remaining postulates using z as the zero element and $(b + z)^*$ as the additive inverse of the element $(b + z)$

$$b \cdot z = b \cdot (z + z) \qquad \text{by Postulate 4}$$

$$= (b \cdot z) + (b \cdot z) \qquad \text{by Postulate 8}$$

Now by Postulates 1 and 6 there exists an element $(b \cdot z)^*$ which is the additive inverse of the element $(b \cdot z)$. Adding this to each member we have

$$(b \cdot z) + (b \cdot z)^* = [(b \cdot z) + (b \cdot z)] + (b \cdot z)^*$$

$$= (b \cdot z) + [(b \cdot z) + (b \cdot z)^*] \qquad \text{by Postulate 3}$$

$$z = b \cdot z \qquad \text{by Postulates 6 and 4a}$$

$$= z \cdot b \qquad \text{by Postulate 2}$$

Thus we have proved the last sentence of Postulate 4, namely $b \cdot z = z \cdot b = z$, from the remaining postulates. Although it is possible to prove portions of certain postulates from the remaining postulates, a clearer picture of the true nature of an integral domain is obtained by considering the given set of postulates, rather than by attempting to find a minimal set and proving the others as theorems. Furthermore, as certain postulates are relaxed ($a \cdot b = b \cdot a$, for example), it is desirable to consider all of the remaining properties listed in our postulates.

Problem Set 1.7

It is important that you read the text with care *before* you start to work problems. You *can't* be more than 35 pages behind now. Perhaps you should start at the beginning and reread the material presented this far if you are having difficulties.

1. Rework any problems of Set 1.4 with which you had difficulty before.

2. Let S be the set of numbers of the form $\dfrac{a}{2} + \dfrac{b}{2}\sqrt{-3}$ where a and b are integers. Is S an integral domain?

3. Let T be the set of numbers of the form $\dfrac{a}{2} + \dfrac{b}{2}\sqrt{-3}$ where a and b are integers such that $(a - b)$ is even. Is T an integral domain?

4. The postulates for an integral domain assert that there exists an element z such that for every b in D, $b + z = z + b = b$. Prove that there is only one such element in a given domain; i.e., prove the uniqueness of the additive identity. [HINT: Assume there are two elements z and z', each of which is an additive identity.

> Then: $z + z' = z$, since z' is an additive identity.
> Also, $z + z' = z'$, since z is an additive identity.
> Hence, $z = z'$ (Why?)]

5. Prove that the multiplicative identity, u, of an integral domain is unique. [HINT: See Problem 4.]

6. Let D be an integral domain, and S be a subset of the elements of D (that is, $S \subset D$) such that S is closed with respect to addition, $+$, and with respect to multiplication, $\cdot$. Is S necessarily an integral domain? Justify your answer with a proof or a counterexample, as appropriate.

7. If D is an integral domain *prove* from the domain postulates that $(a^*)\cdot(b^*) = (a\cdot b)$ and that $z^* = z$. Do not just say that "minus times minus is plus" or some such remark but *prove* it from the integral domain postulates.

1.8 CLASSIFICATION OF THE INTEGERS

Since the integers form an integral domain, Postulate 6 states that each integer N has an inverse with respect to addition. That is, that the equation $N + x = 0$ has an integral solution. However, it is not true that each integer N has an inverse *with respect to multiplication* (that is, that $N\cdot x = 1$ has an integral solution). If such a multiplicative inverse (reciprocal) exists for a number N, then N is called a *unit*. Do not confuse "unit" with the "unity" of Postulate 5. The unity is *always* a unit. (Prove this.) There is only one unity, while there may be many units. A unit is a number which has a reciprocal in the system. The units in the system of integers are $+1$ and -1. In the system of rational numbers, every rational number except zero is a unit. (Why?)

An integer D is said to be a *divisor* of an integer N (symbolically written $D \mid N$) if there exists an integer K such that $D\cdot K = N$. A positive divisor of N, less than N itself, is called a *proper divisor* of N.

An integer N is *prime* if N is neither zero nor a unit ($N \neq 0, 1, -1$), and if $A\cdot B = N$ implies that either A or B must be a unit. In other words,[6] an integer N is prime if $N \neq 0, 1, -1$, and if the only divisors of N are ± 1 and $\pm N$.

If C is an integer which is not zero, not a unit, and not a prime, then C is called *composite*. It can be shown that every composite integer can be expressed as the product of a unit (± 1) times a product of positive primes. Thus, $-24 = -1\cdot 2\cdot 2\cdot 2\cdot 3$.

The integers may be divided into four classifications:

(1) Zero
(2) The units ($+1$ and -1)
(3) Primes
(4) Composite numbers.

Examples of prime integers are $\pm 2, \pm 3, \pm 5, \pm 7, \pm 11, \pm 13, \pm 17, \cdots,$ $\pm 307, \pm 311, \pm 313, \cdots, \pm 1973, \pm 1979, \cdots, \pm 10006721, \cdots.$

Handbooks, such as the *Handbook of Chemistry and Physics*, C. D.

---

[6] If, as in this case, a definition is stated in two forms, *the reader is expected to show that the forms are equivalent*. In showing that each definition implies the other, you will gain facility both with the definitions and with the techniques of modern algebra.

Hodgeman, Ed. (Cleveland: Chemical Rubber Pub. Co., 1962) contain tables of prime integers and of factors of composite integers. More extensive lists are available in D. N. Lehmer's *List of Prime Numbers from 1 to 10,006,721* and his *Factor Tables for the First Tem Millions* (New York: Hafner, 1909) and in *British Association for the Advancement of Science V, Factor Tables*. A more recent volume is *The First Six Million Prime Numbers* by C. L. Baker and F. J. Gruenberger published on Microcards (Madison, Wisconsin).

There is no complete list containing all prime integers. Indeed, the following theorem shows that no such list can ever be published.

THEOREM 1.3   *The number of positive primes is infinite* (that is, is not finite).

Proof: It is sufficient to show that, for every prime P_k, there exists a larger prime. (Why?) Use an indirect proof, as follows: Assume there exists a largest prime, say P_k. Add 1 to the product of all positive primes

$$N = 1 + (2 \cdot 3 \cdot 5 \cdot 7 \cdot 11 \cdot 13 \cdot \cdots P_k)$$

Since N is greater than 1, N is not a unit and must be either prime or composite.

The integer N has no positive prime factor $\leq P_k$, since every such division leaves a remainder of 1. Hence, either N itself is a prime $> P_k$, or N is composite, in which case each of its prime factors must be greater than P_k. In either case, the existence of a prime larger than the supposed largest prime has been demonstrated. Hence, the number of (positive) primes must be infinite. (Why?)

This proof is credited to Euclid, of geometry fame (*Elements*, Book 9). It is interesting to note that examples in which N is prime, such as $N = 1 + 2 \cdot 3$ or $N = 1 + 2 \cdot 3 \cdot 5$, are known, and that examples in which N is composite may also be found. You are asked to find one such example in the following problem set.

EXAMPLE 9

Much is known about prime numbers, but much still remains to be discovered. It is known that, if $x > 7/2$, then there exists at least one prime p such that $x < p < 2x - 2$. In 1937, Ingham proved that there exists a positive constant k such that for each x there exists at least one prime p between x and $x + kx^{5/8}$; i.e., such that $x < p < x + kx^{5/8}$. It is interesting that, although it has been shown that such a k exists, no one has yet found an actual value of k.

Many years ago (1742), Goldbach made two conjectures:

(1) Every even integer $N \geq 6$ is the sum of two odd primes.
(2) Every odd integer $N \geq 9$ is the sum of three odd primes.

If the first conjecture is true, then the second conjecture can easily be proved. (Can you do it?—Try!) However, no one has yet been able to prove the first conjecture. N. Pipping has verified the first conjecture for all $N \leq 100,000$ by producing examples, but this is of little assistance in proving that it is valid for *all* N. In 1937, Vinogradov proved that there exists an integer K such that, for all odd $N > K$, the second conjecture is valid. However, again, this is an existence proof and no one knows how large K actually is. Recent results show that K is less than $10^{400,000}$, thus it might seem possible to examine all integers less than $10^{400,000}$ on a computer to complete the proof—however $10^{400,000}$ is a large number and the task is formidable. Viggo Brun proved that every positive even integer N can be written as the sum of two positive odd integers, each of which is the product of nine or fewer prime factors. Recently, it has been possible to reduce the "nine" in this result to "four," but this is still far short of Goldbach's conjecture.

It is well known that the infinite series known as the harmonic series,

$$\sum \frac{1}{N} = 1 + \frac{1}{2} + \frac{1}{3} + \frac{1}{4} + \frac{1}{5} + \frac{1}{6} + \cdots, \text{ is divergent. It is true, but not}$$

as well known, that, if every term of the harmonic series in which the digit nine appears is deleted, the resulting series is convergent. Perhaps you would be interested in making a guess as to the convergence or divergence of

the series $\sum \dfrac{1}{P_n}$, where P_n is the nth prime number. The series is divergent.

(Show that the divergence of this series implies Theorem 1.3 as a corollary.)

The subject of "twin primes" holds many secrets. Two consecutive odd integers, both of which are prime, are called *twin primes*. Examples are 5, 7 or 17, 19 or 1997, 1999 or 78862601, 78862603. It is still an open problem whether or not there exist infinitely many pairs of twin primes; i.e., if there is a largest pair of twin primes. Interestingly enough, it has been shown that the series of reciprocals of twin primes, namely

$$\left(\frac{1}{3} + \frac{1}{5}\right) + \left(\frac{1}{5} + \frac{1}{7}\right) + \left(\frac{1}{11} + \frac{1}{13}\right) + \left(\frac{1}{17} + \frac{1}{19}\right) + \left(\frac{1}{29} + \frac{1}{31}\right) + \cdots$$

is convergent, or perhaps finite.

A perusal of the *Mathematical Reviews* under "Number Theory" will convince you that much is currently being discovered about primes.

Problem Set 1.8

1. Find the divisors of (a) 52 (b) 91 (c) 103 (d) 231. Underline the divisors which are *not* proper divisors. Is 1 a proper divisor of 1? of 6?

2. Find the proper divisors of (a) 194 (b) 64 (c) 210 (d) 223.

3. In each case justify your answer:

   (a) Is 3 a unit in the system of integers?
   (b) Is 3 a unit in the system of rational numbers?
   (c) Is $2 + 3i$ a unit in the system of rational numbers? Why?
   (d) Is $2 + 3i$ a unit in the system of complex integers $a + bi$, where a and b are integers?
   (e) Is $2 + 3i$ a unit in the system of rational complex numbers (all numbers of the form $a + bi$, where $i^2 = -1$ and a and b are rational numbers)?

4. Which of the following numbers are prime integers? If an integer is *not* prime, explain why it is not: $-31, -27, -14/3, -3, -1, 0, 1, 2, 7/3, 3, 7, 10, 16, 22, 103, 231, 497, (2^{273} - 8), (4^{51} - 2)$.

5. It is not always an easy task to determine whether a large number is prime or composite. Show that, if $|N| > 1$ is *not* divisible by any positive prime $\leq \sqrt{|N|}$, then N is prime.

6. The largest prime number listed in the *Handbook of Chemistry and Physics* is 2003. Consult other tables and journals in your library and find four larger prime numbers. The list of prime numbers compiled by Lehmer may be of some help, or the microcard list of *The First Six Million Prime Numbers* by Baker and Gruenberger.

7. Consult the tables and journals available in your library and bring to class the largest prime number you can find. Several gigantic prime numbers (Mersenne primes) have been discovered since 1960.

8. Prove that, if $a \mid b$ and $a \mid c$, then $a \mid (b + c)$. [HINT: There exist integers k_1 and k_2 such that $ak_1 = b$ and $ak_2 = c$.]

9. (a) Find a method of determining all positive primes less than a given number $N > 2$. [HINT: List the positive integers less than

N. Discard 1. Two is prime. Discard all other multiples of 2. The next remaining number is a prime. Discard all multiples of this prime. Continue this process for integers less than $\sqrt{N}$. Use Problem 5 to complete the proof.]

(b) List all positive primes less than 100 by discarding multiples of 2, 3, 5, 7 $\cdots$ from the integers less than 100. This method is known as the sieve of Eratosthenes.

10. If $b \neq 0$ and $a \mid b$, show that $\mid a \mid \leq \mid b \mid$, for integers a and b.

11. Show, by example, that the N of Theorem 1.3 may be composite.

12. Prove, from the definition of division, "$D \mid N$, if and only if, the equation $Dx = N$ has a solution in the system under consideration," that every nonzero integer divides zero. Discuss further the observation that, "If one does not demand a unique solution, then one may say that zero divides zero, but in any case zero does not divide any other integer."

13. Prove: If N is an integer, then the product $N(N + 1)(N + 2)$ is divisible by 6.

14. Prove: If N is a positive integer, then the product $N(N - 1)(2N - 1)$ is divisible by 6.

15. Show that the product of any five consecutive positive integers is divisible by 30. Can you find a larger integer which always divides such a product?

16. (a) 12 is divisible by *each* positive integer less than or equal to its square root. Find two other positive integers having this property.
(b) Since $6 < \sqrt{37}$ and 6 does not divide 37, the integer 37 does not have the property mentioned in part (a). Find two other positive integers which do not have the property.

*17.[7] Prove that 24 is the largest number which is divisible by each positive integer not exceeding its square root.

18. Prove that there are infinitely many primes of the form $4x - 1$. [HINT: Consider the number $N = -1 + 4(3 \cdot 7 \cdot 11 \cdot \cdots P_k)$, where P_k is the kth prime of the form $4x - 1$. Note also that the product of primes all of the form $4x + 1$ is itself of the form $4x + 1$.]

---

[7] Sections and problems preceded by a star $\star$ are optional and may be omitted without loss of continuity, but your author hopes you will at least read them.

19. A receipted bill was found in an old book. The bill states

    72 lambs *67.9*

    The first and last digits of the total price, replaced here by *'s, were so faded they were illegible. What are the two faded digits, and what was the price paid for one lamb? (This problem requires common sense as well as a knowledge of multiplication.)

20. Two consecutive odd prime numbers, such as 11 and 13, or 29 and 31, are called twin primes. The existence or nonexistence of an infinitude of such "prime pairs" is still (in 1971) one of the unsolved problems of number theory. The question of the number of "prime triplets," such as 3, 5, 7 is, however, completely solved. Prove that the number of prime triplets is finite and, if possible, determine the exact number of such prime triplets which exist.

21. Although the question of how many pairs of twin primes exist has not yet been settled, it is known that there do exist *arbitrarily long* strings of consecutive composite numbers! Prove this. [HINT: Consider the set $(n! + 2)$, $(n! + 3)$, $\cdots$, $(n! + n - 1)$, where $n \geq k + 2$. Show that this sequence contains k consecutive composite numbers.]

22. Prove, or disprove: If n is an odd positive integer, then 8 divides $(n^2 - 1)$.

23. Prove that *if* Goldbach's first conjecture is true, then the second conjecture is also valid. (See p. 34.)

24. Is the number 46069 prime or composite?

25. Problem 20 defined twin primes. The numbers 13001, 13003, 13007, 13009 are each prime. The first pair are twin primes and the last pair are also twin primes. Can you find two other sets of twin primes such that the difference between the largest and the smallest of the four members is only 8 units? (Could it be less than eight units?). Why is 1, 3, 5, 7 *not* an example of two sets of twin primes whose difference between the largest and smallest members is less than eight?

1.9 GREATEST COMMON DIVISOR

If $N \mid A$ and $N \mid B$, then N is a *common divisor* of A and B. A number G is called *greatest common divisor* of two or more numbers if both of the following conditions hold:

(1) G is a common divisor of each of these numbers.
(2)   Every common divisor of these numbers also divides G.

Thus, the greatest common divisors of 12 and 30 are -6 and $+6$. Note that the "greatest" part of "greatest common divisor" has been generalized, since both -6 and 6 are greatest common divisors of 12 and 30. The "greatest common divisor" is also applied to systems *in which no order* ("less than") *relationship is defined*. Conditions (1) and (2) do not require an order relationship.

In work with integers, the symbol (A, B) is used to denote the *positive* greatest common divisor of A and B. Thus, $(12, 30) = 6$ and $(-14, 21) = 7$. A similar restriction is made in elementary mathematics, where the equation $x^2 = 4$ has two solutions, $x = 2$ and $x = -2$, but the symbol $\sqrt{4}$ is used to denote the *positive* square root of 4. The reader should show, from the definition, that $(45, 60) = 15$, and that $(0, A) = A$ for $A \neq 0$. (Show that A satisfies the definition of greatest common divisor given above.) An interesting note discussing $(0, 0)$ will be found in the *American Mathematical Monthly*, **51,** p. 345.

1.10 ARCHIMEDES' AXIOM AND EUCLID'S ALGORITHM

The reader who is interested in examining a simple proof of Archimedes' Axiom will find one on page 9 of C. MacDuffee's *Introduction to Abstract Algebra* (New York: Dover, 1956) and in other texts. Our purposes in this course are better served by assuming it, as Euclid did. It is quoted here to help emphasize the difference between the ideas of "positive" and "nonnegative."

ARCHIMEDES' DIVISION AXIOM   Let A be a non-negative integer and B a positive integer. Then there exist two non-negative integers q and r such that $A = Bq + r$, where $0 \leq r < B$.

EXAMPLE 10

If $A = 37$ and $B = 5$,

$$
\begin{array}{r}
7 \\
5\overline{)37} \\
35 \\
\hline
2
\end{array}
\qquad \text{or} \quad 37 = 5(7) + 2
$$

so that $q = 7$ and $r = 2$.

In general:

$$B \overline{)A}^{\,q} \quad \text{or} \quad A = Bq + r \quad \text{with} \quad 0 \leq r < B$$

Using this process of division, a method of finding the positive greatest common divisor of two numbers A and B may be established. This process, or algorithm, is known as Euclid's Algorithm, after the Euclid of geometry fame. The 1945–49 translations of the Plimpton Babylonian tablets (see *O. U. Mathematics Letter*, May 1953, p. 4, or *Mathematics Magazine*, **27,** No. 1, p. 39) make it seem likely that this algorithm is actually of much earlier origin. We now develop Euclid's Algorithm for computing (A, B) by a series (chain) of divisions.

Let us first consider a numerical example:

Compute the greatest common divisor (30,252). We begin by dividing 252 by 30 to obtain

$$252 = 8 \cdot 30 + 12$$

Then we divide the former divisor (30) by the remainder (12) to obtain

$$30 = 2 \cdot 12 + 6$$

Continuing to divide the divisor (12) by the remainder (6) we obtain

$$12 = 2 \cdot 6 + 0$$

The zero remainder is significant and the previous remainder (6) is the desired *gcd* of (30, 252).

It is not expected that it will yet be obvious to the reader that this procedure is a valid algorithm for producing the *gcd* of two integers. We examine the given example further.

Note that from next to the last equation we may obtain:

$$6 = 30 - 2 \cdot 12$$

If the previous equation is then solved for its remainder one obtains $12 = 252 - 8 \cdot 30$ and by substituting this expression for 12 into the equation $6 = \mathbf{30} - 2\cdot\mathbf{12}$ one obtains

$$6 = \mathbf{30} - 2(\mathbf{252} - 8\cdot\mathbf{30})$$

or

$$6 = 17\cdot\mathbf{30} - 2\cdot\mathbf{252}$$

which expresses the supposed *gcd* (6) as a linear combination of the two original numbers 30 and 252.

Clearly any common divisor of 30 and 252 must also divide 6, since $6 = 17\cdot\mathbf{30} - 2\cdot\mathbf{252}$. By working backward in the equations

$$\mathbf{252} = 8\cdot\mathbf{30} + 12$$

$$\mathbf{30} = 2\cdot\mathbf{12} + 6$$

$$\mathbf{12} = 2\cdot\mathbf{6} + 0$$

we reason that 6 divides 12 (third equation) and hence 6 divides 30 (second equation); then, since 6 divides both 12 and 30, 6 divides 252 (first equation). Thus 6 divides both 30 and 252, and furthermore any integer that divides both 30 and 252 must also divide 6. Thus 6 satisfies the definition of *gcd* given in Section 1.9.

The alert student will feel it would be easier to factor 30 and 252 and pick out the common prime factors to determine the *gcd*. He will be correct, but the Euclidean Algorithm we are discussing is of far greater importance than as a tool to determine the *gcd* of two integers. The Euclidean Algorithm applies to algebraic expressions and to even more general, nonfactorable systems. We now turn our attention to a statement of this algorithm in general symbols rather than for a specific example. Don't get lost in the symbols, we are going to do the same thing we just did with (252, 30) for (A, B).

[Side problem: Can you show that (A, B) and (B, A) are always the same number? It may be "obvious," but can you *prove* it?]

If $A > 0$ and $B > 0$ and $A > B$, then by Archimedes' Axiom (ordinary division) there exist q and r with $0 < q$

such that $A = Bq + r$ with $0 \le r < B.$

Similarly, there exist r_1 and q_1

such that $B = rq_1 + r_1$ with $0 \le r_1 < r.$

Similarly,

$$r = r_1 q_2 + r_2 \qquad\qquad 0 \le r_2 < r_1$$

$$r_1 = r_2 q_3 + r_3 \qquad\qquad 0 \le r_3 < r_2$$

$$\begin{array}{c} \cdot \quad \cdot\;\cdot \quad \cdot \\ \cdot \quad \cdot\;\cdot \quad \cdot \end{array} \qquad\qquad \begin{array}{c} \cdot \\ \cdot \end{array}$$

$$r_{k-3} = r_{k-2} q_{k-1} + r_{k-1} \qquad\qquad 0 \le r_{k-1} < r_{k-2}$$

$$r_{k-2} = r_{k-1} q_k + r_k \qquad\qquad 0 \le r_k < r_{k-1}$$

$$r_{k-1} = r_k q_{k+1} + 0$$

Eventually, the final remainder must be zero, since

$$B > r > r_1 > r_2 > r_3 > \cdots \ge 0$$

and there exist only a finite number of positive integers less than a given B.

Euclid's Algorithm yields $(A, B) = r_k$, the remainder just before the zero remainder is obtained. To prove this, solve the next to the last equation for r_k, obtaining

$$r_k = r_{k-2} - r_{k-1} q_k$$

Then substitute successively from the chain of equations, the remainder from the previous equation, thus:

$$r_k = r_{k-2} - r_{k-1} q_k$$

$$= r_{k-2} - (r_{k-3} - r_{k-2} q_{k-1}) q_k = r_{k-2}(1 + q_{k-1} q_k) - r_{k-3} q_k$$

$$= \text{substitute again, this time eliminating } r_{k-2}$$

Eventually, by eliminating $r_{k-1}, r_{k-2}, r_{k-3}, \cdots, r_3, r_2, r_1$, and finally r, in succession, one obtains a t and an s such that

$$r_k = tA + sB$$

To show that $r_k = (A, B)$, we must show that r_k fulfills the two conditions given in Section 1.9:

    (1) r_k is a common divisor of A and of B.
    (2) Every common divisor of A and B also divides r_k.

The reader is asked to show this in Problems 11 and 12 of the next set.

EXAMPLE 11

Find $(1596, 96)$.

$$A = Bq + r \quad \text{with} \quad 0 \le r < B$$

$$1596 = 96(q) + r \qquad 0 \le r < 96$$

which, by ordinary division gives

$$1596 = 96(16) + 60$$

$$96 = 60(1) + 36$$

$$60 = 36(1) + 24$$

$$36 = 24(1) + 12$$

$$24 = 12(2) + 0$$

Hence, $(1596, 96) = 12$.

To obtain a t and an s such that $12 = t\cdot1596 + s\cdot96$, eliminate the remainders, in succession, by substitution into the following equation, thus:

$$12 = 36 - 24(1)$$

$$= 36 - (60 - 36\cdot1)(1) = -60 + 36(2)$$

$$= -60 + (96 - 60\cdot1)(2) = 96(2) - 60(3)$$

$$= 96(2) - (1596 - 96(16))(3) = 1596(-3) + 96(50)$$

or

$$12 = t\cdot1596 + s\cdot96$$

where
$$t = -3 \quad \text{and} \quad s = 50.$$

Thus $\qquad 12 = (-3)1596 + (50)96$ as desired.

The fact that it is *always* possible to find t and s such that $(A, B) = tA + sB$ is of considerable importance in further theory.

Problem Set 1.10

In Problems 1–9 find the indicated greatest common divisor and also find t and s such that $(A, B) = tA + sB$ for each problem.

1. $(1500, 570)$ 4. $(365, 146)$ 7. $(203, 343)$

2. $(875, 651)$ 5. $(506, 33)$ 8. $(21, 796)$

3. $(352, 221)$ 6. $(11, 3)$ 9. $(418, 1376)$

10. Geometrize Archimedes' Axiom by using a number axis. "Lay off" the B q times between 0 and A with remainder r, where $0 \leq r < B$.

11. (a) Show that, if $A = Bq + r$, then every common divisor of A and B is also a common divisor of B and r.
    (b) Show that the r_k obtained in Euclid's Algorithm is a common divisor of A and B.

12. (a) Show that, since there exist a p and s such that $r_k = pA + sB$, every common divisor of A and B also divides r_k.
    (b) Use Problems 11(b) and 12(a) to show that $r_k = (A, B)$ in Euclid's Algorithm.

13. Prove that $(ma, mb) = m(a, b)$. Do not guess and say that it looks likely, or obvious. Instead, show that the two conditions of (ma, mb) are both satisfied.

*14. Show that, if $(a, b) = 1$, then $(a + b, a - b)$ is either 1 or 2. [HINT: Assume $g = (a + b, a - b)$ and hence that

$$g \mid (a + b)$$
$$g \mid (a - b).$$

Then show $g \mid 2a$ and $g \mid 2b$. Then continue until g is either 1 or 2.]

15. Extend the concept of greatest common divisor to more than two integers and show that $(405, 285, 495, 675) = 15$.

16. (a) Show that, if a prime, P, divides AB, then either $P \mid A$ or $P \mid B$. [HINT: If $P \mid A$, then $(P, A) = 1$ and there exist integers s and t, such that $sP + tA = 1$. Then $sPB + tAB = B$. Complete the proof.]
    (b) Find a counterexample to show that, if P is not prime, the corresponding statement is not valid.

17. Prove or disprove: If the product of two integers divides an integer C, then C is divisible by the *square* of the greatest common divisor of

the two given integers. In symbols:

$$\text{If } a \cdot b \mid c, \quad \text{then} \quad [(a, b)]^2 \mid c.$$

18. Prove or disprove: If $N = 2k + 1$, where k is an integer which is *not* of the form $\left(p \cdot m + \dfrac{p - 1}{2}\right)$ for any odd prime p and integer m, then N is prime. Example: $k = 6$ is *not* of the form $\left(p \cdot m + \dfrac{p - 1}{2}\right)$, since if $p = 3, m = 1$, then $p \cdot m + \dfrac{p - 1}{2} = 4$ and for all other choices of p and m, $p \cdot m + \dfrac{p - 1}{2}$ is greater than 6. It is true that for $k = 6$ $N = 2 \cdot 6 + 1 = 13$ is prime.

19. (A puzzle problem—just for fun.) In the following integer multiplication problem, each even digit has been replaced with the letter E, while each odd digit has been replaced by the letter O. The problem was worked in base 10 notation. Determine the three possible solutions and show that no other solutions exist.

$$
\begin{array}{r}
E\ E\ O \\
O\ O \\
\hline
E\ O\ E\ O \\
E\ E\ O \\
\hline
O\ O\ O\ O\ O
\end{array}
$$

20. In about 1769, while discussing the Fermat problem, the mathematician Euler declared that he thought it was probably impossible to find positive integers (I, J, K, L, M) that satisfy the equation $I^5 + J^5 + K^5 + L^5 = M^5$. He also felt it would be impossible to determine positive integers (X, Y, Z, W) such that $X^4 + Y^4 + Z^4 = W^4$; but was unable to find a proof. He conjectured that if $K > 2$, then at least K positive integers would be required before their Kth powers would sum to a perfect Kth power.

Almost two centuries later (1966) Lander and Parkin proved, using a CDC 6600 computer, that the first conjecture is false since

$$27^5 + 84^5 + 110^5 + 133^5 = 144^5$$

(a) Consult your library to see if you can find the paper by Lander and Parkin in the Bulletin of The American Mathematical Society, Vol. **72,** p. 1079 (1966) which discusses this result. Also consult L. E. Dickson's *History of the Theory of Numbers,* (New York: Chelsea, 1950) Vol. **1,** p. 6048 for a discussion of Euler's conjecture.

(b) Your author still does not know (1971) whether or not the conjecture about the sum of three fourth powers is true or false.

See if you can find any recent results on the solution of $X^4 + Y^4 + Z^4 = W^4$ in positive integers by consulting journals, including *Mathematical Reviews* and *Computing Reviews.*

21. Rework Problem 26, Set 1.4 page 17 using your current knowledge.

★1.11[8] PERFECT NUMBERS

If a positive integer R divides a positive integer N, then R is called a positive divisor of N. For example, the positive divisors of 18 are 1, 2, 3, 6, 9, and 18. A number is said to be an improper divisor of itself. All other positive divisors are called proper divisors. The proper divisors of 18 are 1, 2, 3, 6, and 9 (not 18).

A positive integer N is said to be perfect if the sum of the proper divisors of N is N. For example, the proper divisors of 6 are 1, 2, and 3, and since $1 + 2 + 3 = 6$, 6 is a perfect number. The number 28 is also a perfect number, since the proper divisors of 28 are 1, 2, 4, 7, and 14; with sum $1 + 2 + 4 + 7 + 14 = 28$. The four smallest perfect numbers are 6, 28, 496, and 8128. The fifth perfect number is 33,550,336. The eighth perfect number contains 19 digits.

Perfect numbers were studied by the ancient Greeks and are still of interest today. They hold their secrets well.

No one knows whether or not an odd number can be perfect! None has been found, yet no one has proved they do not exist. Since 1949 it has been shown that, if an odd perfect number exists, it must be greater than 10 billion and must be equal to either $12m + 1$ or to $38m + 9$ for some integer m. Still, no odd perfect number has been discovered.

More is known about even perfect numbers. For example, every even perfect number must end in either 28 or 6. Another interesting fact is that the sum of the reciprocals of *all* the divisors of an even perfect number must equal 2. The perfect number 6 has divisors 1, 2, 3, and 6, and their

---

[8] The remainder of Chapter 1 is included for those interested in optional subject matter. It is not essential to the further development of this course. You should glance at Section 1.16 before starting Chapter 2.

reciprocals total $1 + \frac{1}{2} + \frac{1}{3} + \frac{1}{6} = 2$. In a similar fashion, the divisors of 28 are 1, 2, 4, 7, 14, 28, and $1 + \frac{1}{2} + \frac{1}{4} + \frac{1}{7} + \frac{1}{14} + \frac{1}{28} = 2$. These facts have been proved true for every even perfect number—those which were known before 1950, those which were discovered between 1950 and the present, and those which may be discovered in years to come.

An important theorem on perfect numbers states that $2^{p-1}(2^p - 1)$ is an even perfect number if, and only if, $(2^p - 1)$ is prime, and that every even perfect number is of this form. If, for example, $p = 2$, then $(2^2 - 1) = 3$ is prime. Thus $2^{p-1}(2^p - 1) = 2(3) = 6$ is perfect. If $p = 3$, then $(2^3 - 1) = 7$ is prime, and $2^{p-1}(2^p - 1) = 2^2(7) = 28$ is perfect. If $p = 4$, $(2^p - 1) = (2^4 - 1) = 15$ is not prime, and hence $p = 4$ does not lead to a perfect number. If the number $(2^p - 1)$ is prime, it is called a Mersenne prime. Every Mersenne prime determines a perfect number and every *even* perfect number contains a factor which is a Mersenne prime. Euclid, of geometry fame, proved that if $(2^p - 1)$ is prime then $2^{p-1}(2^p - 1)$ is perfect, but it was not until 2000 years later that Euler proved that every even perfect number is of the form $2^{p-1}(2^p - 1)$ with $(2^p - 1)$ prime. Be sure you comprehend the distinction.

The first four perfect numbers were discovered by the end of the first century. By 1870, only four more had been found. Between 1870 and 1950, four additional even perfect numbers were discovered. Considering all the facts and formulae known about perfect numbers, it may surprise you to learn that, in the 2000 years prior to 1951, only 12 perfect numbers had been discovered. Between 1951 and 1960, five more perfect numbers were found, using the SWAC (electronic) computing machine at the National Bureau of Standards Institute for Numerical Analysis at U.C.L.A. They are: $2^{520}(2^{521} - 1)$, $2^{606}(2^{607} - 1)$, $2^{1278}(2^{1279} - 1)$, $2^{2202}(2^{2203} - 1)$, $2^{2280}(2^{2281} - 1)$. Work sponsored by the United States Office of Naval Research, 1960–1962, considered $2300 < p < 6000$ and revealed three more Mersenne primes. They result from $p = 3217, 4253, 4423$.

By 1964 D. B. Gillies had found $2^p - 1$ is prime for $p = 9689, 9941$, and 11213. By 1970 it had been shown that $N = 2^{p-1}(2^p - 1)$ is perfect for the 23 values $p = 2, 3, 5, 7, 13, 17, 19, 31, 61, 89, 107, 127, 521, 607, 1279,$ 2203, 2281, 3217, 4253, 4423, 9689, 9914 and 11213. However, the principal problems, namely, "How many perfect numbers are there?" and, "Do odd perfect numbers exist?" are still unsolved mysteries which await your research or that of one of your contemporaries.

Problem Set 1.11

1. Show that 496 is perfect.

2. Show that 8128 is perfect.

3. Show that 762 is *not* perfect.

4. Show, by enumerating the divisors, that $2^{p-1}(2^p - 1)$ is perfect if $(2^p - 1)$ is prime.

5. Read the account of perfect numbers given in Chapter IV, Sec. 10 of J. Uspensky and M. Heaslet, *Elementary Number Theory* (see Selected Reading List at end of Chapter 1) and prepare a report on it for the class. Note, in particular, the gaps in our knowledge about perfect numbers which existed when that book was published *and which have since been filled*.

6. Reread Uspensky and Heaslet's last sentence mentioned in Problem 5. By actual experiment, try to determine whether or not $2^p - 1$ is prime for some of the six values listed. Do not spend more than 10 or 15 minutes on this problem.

7. Read the article in the November 1949 issue of the *American Mathematical Monthly* which deals with odd perfect numbers and prepare a report on this article. Also, read the review of the article given in the *Mathematical Reviews*. Part of this problem is using the index in the reviews to find where it was reviewed. [HINT: It is unlikely that the article was reviewed before it appeared in print.]

8. Consult the article on perfect numbers in *Scripta Mathematica*, Vol. XIX, p. 128, or Vol. XIX, p. 38, or Vol. XVIII, p. 122, and prepare a report.

9. Prepare a talk or paper on perfect numbers for your speech or English class.

*10. Prove that the sum of the reciprocals of all the divisors of an even perfect number is 2.

*11. Prove that an integer N that is a perfect square cannot also be a perfect number. (This is easy to prove if N is even, but since no one has yet proved that odd perfect numbers do not exist, the proof for even N is not sufficient.) [HINT: If $N = p_1{}^{a_1}p_2{}^{a_2}\cdots p_k{}^{a_k}$, where the p_i are the distinct prime factors of N, then N has

$$(a_1 + 1)(a_2 + 1) \cdots (a_k + 1) - 1$$

proper divisors. (Why?)]

12. Read Martin Gardner's article beginning on page 121 of the March 1968 issue of *Scientific American* (**218**, No. 3) and write a brief report on it.

13. The pair of integers 284 and 220 are called *amicable* (friendly) numbers because the sum of the proper divisors of each is the other. Consult your library and prepare a brief report on amicable numbers. If you have a computer available, write a program to find pairs of amicable numbers. There are 13 known pairs of amicable numbers having five or fewer digits, and hundreds of larger pairs.

★1.12 NUMBER SYSTEMS

You have probably heard that our number system is "positional, based on 10." You may even know what it means. The idea is simple. The number **476** really means $4 \cdot 10^2 + 7 \cdot 10 + 6$, while

$$20395 = 2 \cdot 10^4 + 0 \cdot 10^3 + 3 \cdot 10^2 + 9 \cdot 10 + 5$$

In general, only the *coefficients* of the powers of 10 are listed, and the *position* tells to what power of 10 the coefficient belongs. It may be of momentary interest to note why 10 was chosen as a base. The answer lies in the fact that primitive man had 10 fingers, just as we do today. He said, "5 double hands and 3," meaning 53. Some of the primitive tribes still use this system, and in English the word "digit" means both "a finger" and "one of the fundamental blocks on which our counting system is based."

It is instructive to consider how man might count if he had used one hand, rather than two hands, as a base of his number system.

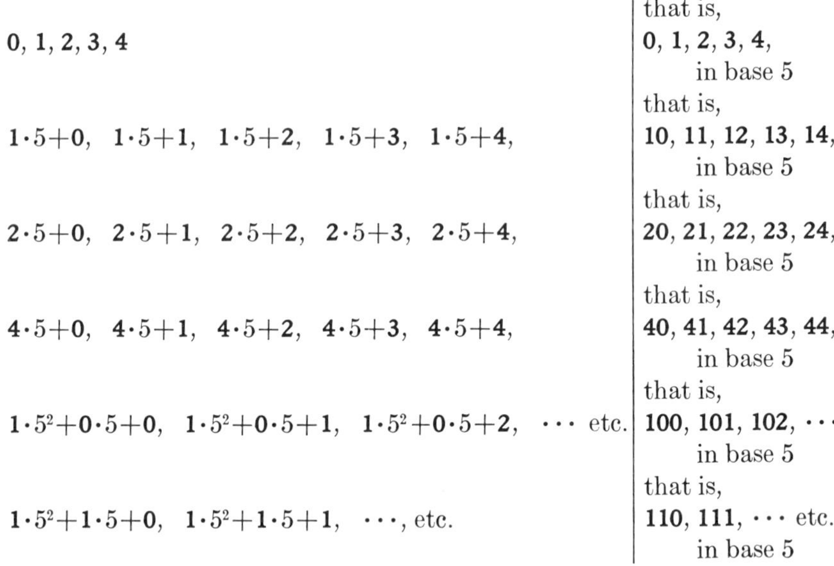

	that is,
0, 1, 2, 3, 4	**0, 1, 2, 3, 4,** in base 5
	that is,
$1 \cdot 5 + 0$, $\ 1 \cdot 5 + 1$, $\ 1 \cdot 5 + 2$, $\ 1 \cdot 5 + 3$, $\ 1 \cdot 5 + 4$,	**10, 11, 12, 13, 14,** in base 5
	that is,
$2 \cdot 5 + 0$, $\ 2 \cdot 5 + 1$, $\ 2 \cdot 5 + 2$, $\ 2 \cdot 5 + 3$, $\ 2 \cdot 5 + 4$,	**20, 21, 22, 23, 24,** in base 5
	that is,
$4 \cdot 5 + 0$, $\ 4 \cdot 5 + 1$, $\ 4 \cdot 5 + 2$, $\ 4 \cdot 5 + 3$, $\ 4 \cdot 5 + 4$,	**40, 41, 42, 43, 44,** in base 5
	that is,
$1 \cdot 5^2 + 0 \cdot 5 + 0$, $\ 1 \cdot 5^2 + 0 \cdot 5 + 1$, $\ 1 \cdot 5^2 + 0 \cdot 5 + 2$, $\cdots$ etc.	**100, 101, 102,** $\cdots$ in base 5
	that is,
$1 \cdot 5^2 + 1 \cdot 5 + 0$, $\ 1 \cdot 5^2 + 1 \cdot 5 + 1$, $\cdots$, etc.	**110, 111,** $\cdots$ etc., in base 5

If a subscript in parentheses is used to indicate the base, then the number 4123 in base 5 means

$$4123_{(5)} = 4 \cdot 5^3 + 1 \cdot 5^2 + 2 \cdot 5 + 3 = 538_{(10)}$$

Number systems with bases other than 10 provide many interesting puzzle problems. However, the study of other bases can also be justified on grounds of severe practicality. Modern computers are usually constructed using bases 2, 8, or 16. The Marchant Company makes a desk calculator which calculates in base 8. This should suggest that research today actually uses base 8 and base 2 arithmetic. It does.

★1.13 THE BINARY SYSTEM

One of the easiest and most important bases is the base 2. It is easy, because the laws of binary arithmetic are so simple.

Addition base 2			*Multiplication base 2*		
$+$	0	1	$\times$	0	1
0	0	1	0	0	0
1	1	$10_{(2)}$	1	0	1

Note that $10_{(2)} = 1 \times 2 + 0 = 2_{(10)}$ and $746_{(10)} = 1011101010_{(2)}$.

This system is of great importance today, since many of the new high-speed computing machines use base 2 for all computations. An electrical switch has only two positions (open and closed) and, similarly, the binary system needs only two digits (0 and 1). Thus, they are well suited to work together.

See page 9 of Courant and Robbins *What Is Mathematics?* for an interesting problem.

★1.14 NIM

The ancient game of Nim (probably Chinese) is played by two people as follows:

Matches are placed in three piles containing 3, 5, and 7 matches, or A, B, C matches, respectively. The contestants play alternately, and may pick as many matches as they wish at one time from *one* pile, but they must not take matches from more than one pile on a given turn, and at least one match must be taken on each turn. On the next turn, matches may be selected from a different pile or, if it still remains, from the same pile. The player who takes the last match is the winner. (Sometimes an alternate game is played in which the player who is forced to take the last match is the loser.) The game and its generalization, in which the number of piles of matches and the number of matches per pile may be arbitrary, can be completely analyzed through the use of the binary number system. Definite rules may be established such that, if one knows these rules, and has a favorable opening move, he can always win the game. Briefly, the number of matches in each pile may be expressed in the binary notation. If the sum of the coefficients of each power of 2 (column digits) is *even*, the position is a safe one. An unsafe position always results when a play is made from a safe position. If a play is made from an unsafe position it is always possible to arrive at a safe position. A player who always maintains a safe position will certainly win. (This is *not* obvious; you are asked to prove it in the next problem set.) A more complete discussion of the game of Nim will be found on pages 15 to 19 of *Elementary Number Theory* by Uspensky and Heaslet.

EXAMPLE 12

Find a move which will yield a safe position on a game of Nim which now stands at three piles containing 7, 9, and 15 matches, respectively.

$$15_{(10)} = 1 \quad 1 \quad 1 \quad 1_{(2)}$$
$$9_{(10)} = 1 \quad 0 \quad 0 \quad 1_{(2)}$$
$$7_{(10)} = \quad\quad 1 \quad 1 \quad 1_{(2)}$$

Column Totals   2   2   2   3

In order to win, the column totals must be kept even. This may here be accomplished by taking one match from any one of the piles.

EXAMPLE 13

You select one match from the second pile of Example 1, and your opponent then selects 6 matches from the first pile, leaving 9, 8, 7 matches. Find a safe position for your next move.

$$9_{(10)} = 1 \quad 0 \quad 0 \quad 1_{(2)}$$
$$8_{(10)} = 1 \quad 0 \quad 0 \quad 0_{(2)}$$
$$7_{(10)} = \quad\quad 1 \quad 1 \quad 1_{(2)}$$

Column Totals   2   1   1   2

You may put yourself in a safe position by taking 6 matches from the third pile. (Why?) The University of Oklahoma has a machine which will play Nim with all comers. Could you design one?

★Problem Set 1.14

1. Express the following numbers in the base 10 system: (a) $202_{(3)}$ (b) $2304_{(5)}$ (c) $11010_{(2)}$ (d) $19342_{(12)}$ (e) $1492_{(12)}$ (f) $1254_{(7)}$.

2. Express the numbers $297_{(10)}$ and $34_{(10)}$ in each of the following bases: (a) 2, (b) 3, (c) 7, (d) 9, (e) 4, (f) 8, (g) 6.

3. Make up a set of rules for addition and multiplication in the base 6 system, and carry out the following:

   (a) $43105_{(6)} + 24014_{(6)} + 51235_{(6)}$
   (b) $1342_{(6)} + 5541_{(6)} + 15_{(6)}$
   (c) $31243_{(6)} - 1454_{(6)}$
   (d) $[4215_{(6)}] \times [4_{(6)}]$
   (e) $[13124_{(6)}] \times [125_{(6)}]$
   (f) $[4312_{(6)}] \div [3_{(6)}]$

4. Make up a set of rules for addition and multiplication in the base 2 system, and carry out the following:

   (a) $1101_{(2)} + 101101_{(2)} + 10110_{(2)}$
   (b) $1011_{(2)} + 10001_{(2)} + 10011_{(2)}$
   (c) $1011_{(2)} - 110_{(2)}$
   (d) $[10011_{(2)}] \times [10_{(2)}]$
   (e) $[101001_{(2)}] \times [101_{(2)}]$
   (f) $[10011_{(2)}] \times [10_{(2)}]$

5. If the opponent of Example 2 selects 2 matches from the first pile, leaving 7, 8, 1, what possible plays are open to you which will leave you in a safe position?

*6. (a) Show that a Nim player in an unsafe position cannot leave his opponent in an unsafe position.
   (b) Show that the process described for playing the game of Nim will inevitably produce a winning combination.

7. If the game of Nim is played with piles of 5, 14, and 19 matches, and your opponent draws 7 matches from the third pile, can you put yourself in a safe position?

8. Is 23, 14, 19 a safe position in a game of Nim? Is 12, 9, 3 safe? Is 3, 5, 7 safe?

9. Show that your opponent can be beaten as soon as he leaves you with only two unequal piles remaining.

10. Consider the alternate game of Nim in which the person who is forced to take the last match loses. Can you make up a set of rules, based on the binary system, which will enable you to win such a game?

11. Determine all safe positions for a game of Nim beginning with 3, 5, 7 matches.

12. The following articles in *The American Mathematical Monthly* deal with the game of Nim. Read one of them and write a short report on your findings.

    (a) Machines for playing Nim: **49,** p. 330; **55,** p. 343.
    (b) Other articles on Nim: **25,** p. 139; **49,** p. 44;
                                      50, p. 435; **52,** p. 441.

13. Discuss: For what integral values of m is $(m + 3) \cdot (m + 2) \cdot (m - 1)$ the square of an integer?

14. Show that, from any sequence of n integers, a block of adjacent integers may be selected such that their sum is divisible by n. [HINT: See solution to Problem 4300 in *American Mathematical Monthly*.]

15. Find all positive integers a, b, c such that $a^2 + b^2 + 3c^2 = (a + b + c)^2$. [HINT: See Problem E941 in *American Mathematical Monthly*.]

★1.15 TWO *wee* RESEARCH PROBLEMS

Can you do mathematical research at your present level of mathematical maturity? Of course you can. You must not be disappointed if you later discover that someone else has also done the same research—if your work is independent and is carried out in a workmanlike manner it is a credit to you and to your teachers. The fact that someone else beat you to the publication of a particular piece of research may well mean only that they were born before you were. Below you will find two brief research projects which may interest you.

We turn our attention to a rather interesting subset of the primes which we call *superprimes*.

The integer 7331 is a prime. So is any integer obtained by deleting digits from the right edge of 7331, since 733, 73 and 7 are each prime.

A prime integer which has the property that every integer obtained by deleting an arbitrary number of its right-most digits is again prime, is called a *superprime*. The integers 317 and 2399 are also superprimes. (You should verify this—it is part of reading mathematics.)

The superprimes 7331 and 317 are rather special, even among superprimes and are called *superprime leaders* since there is no digit X which will make either $7331X$ or $317X$ into a superprime. (If you wish to do some mathematical research get out your pencil and paper and *prove* this last statement before you continue reading. It isn't very difficult.)

A *superprime leader* is a superprime which cannot be obtained by deleting digits from a larger superprime. The superprime 2399 is *not* a superprime leader since 23993 is also a superprime. Actually 23993 is not a superprime leader either.

Ascertain which four of the following seven integers are superprime leaders:

$$59; \quad 23339; \quad 59393339; \quad 7323; \quad 7331; \quad 239933; \quad 739397$$

You may save considerable time and effort if you are observant enough to make and prove the validity of a few observations (that is, theorems) along the way. The types of theorems which could be helpful might include:

THEOREM 1.4   If a superprime contains either of the digits 2 or 5 then 2 or 5 can only be the *left-most* digit of the superprime.

THEOREM 1.5   The digits 4, 6, 8, 0 will not appear in a superprime.

THEOREM 1.6   The digit 1 can*not* appear as the left-most digit of a superprime.

These and other theorems which are not difficult to prove can be most helpful in any search for superprime leaders.

Research Proposition I

Select an integer $K \geq 4$ and determine *all* of the superprime leaders having K or fewer digits. If you have a computer available you may wish to use it. If not, use your library prime tables.

For those who would like some assurance that you are progressing satisfactorily the following partial table is presented.

Number of digits	2	3	4	5	6	7	8
Number of known superprime leaders having the above number of digits	1	3	4	6	5	3	5

Research Proposition II

Determine whether the number of superprime leaders is finite or infinite. Don't guess. This requires a *proof*, but the proof is within your ability if you have solved Research Proposition I for sufficiently large K.

★1.16 A PROBLEM FOR FURTHER INVESTIGATION

The problem given here is one that has not as yet (1971) been investigated very fully. Perhaps you or some of your classmates may be able to add to our current knowledge of this little problem which arose while creating an example of an integral recursive function for a computer class being taught to high school students on Saturdays on the campus of The University of Oklahoma.

The problem is simple.

Let N_0 be any integer.

We now obtain a sequence of integers starting with N_0 using the recursive rule.

$$N_{k+1} = \begin{cases} N_k/2, & \text{if } N_k \text{ is even} \\ 3N_k + 1, & \text{if } N_k \text{ is odd} \end{cases}$$

If $N_0 = 11$, the corresponding sequence is **11** 34 17 52 26 13 40 20 10 5 16 8 4 2 1. If a chain contains the digit 1, it repeats the digits 1 4 2 1 4 2 1 4 2 $\cdots$ in a periodic fashion thereafter. In this case we terminate the chain at the first 1 and say the chain converged to 1. Furthermore, we define the number of steps from N_0 to the first 1 to be the *length* of the chain; that is, if the integer 1 appears in the sequence for the first time as N_α, then the length of the chain is α.

Several typical chains are given below.

25 76 38 19 58 29 88 44 22 11 34 17 52 26 13 40 20 10 5 16 8 4 2 1

26 13 40 20 10 5 16 8 4 2 1

27 82 41 124 62 31 94 47 142 71 214 107 322 161 484 242 121 364 182 91 274 137 412 206 103 310 155 466 233 700 350 175 526 263 790 395 1186 593 1780 890 445 1336 668 334 167 502 251 754 377 1132 566 283 850 425 1276 638 319 958 479 1438 719 2158 1079 3238 1619 4858 2429 7288 3644 1822 911 2734 1367 4102 2051 6154 3077 9232 4616 2308 1154 577 1732 866 433 1300 650 325 976 488 244 122 61 184 92 46 23 70 35 106 53 160 80 40 20 10 5 16 8 4 2 1

28 14 7 22 11 34 17 52 26 13 40 20 10 5 16 8 4 2 1

29 88 44 22 11 34 17 52 26 13 40 20 10 5 16 8 4 2 1

30 15 46 23 70 35 106 53 160 80 40 20 10 5 16 8 4 2 1

It will be noted that each of the given chains terminates with the integers 16, 8, 4, 2, 1. This makes us wonder if *every* chain of length 4 or greater will end with this sequence. We do not know the answer to this question. However, it is easy to prove the following theorem:

THEOREM 1.7   If a chain of the sequence

$$
N_{k+1} = \begin{cases} N_k/2, & \text{if } N_k \text{ is even} \\[2ex] 3N_k + 1, & \text{if } N_k \text{ is odd} \end{cases}
$$

does converge to 1, then its terminal elements will be 16, 8, 4, 2, 1 if it has length 4 or greater.

The proof is by working backward from 1, and is left for the reader. The rub, of course, is that we do *not* yet know that for every starting value N_0, the chain will converge to 1. Actual experiment (on a computer, of course) shows that all $N_0 < 10,000$ do converge to 1, but this does not prove the conjecture that all chains eventually converge to 1.

Another theorem that is important sounding, but easy to prove is:

THEOREM 1.8   Given an integer $\alpha > 0$, there exists a starting value N_0 such that the length of the chain from N_0 to 1 is α.

It is left for the reader to discover a simple formula which yields a starting value $N_0(\alpha)$ which will produce a chain of length α.

Unfortunately no formula that will enable one to look at N_0 and easily determine the length of the chain from N_0 to 1 is known *for general* N_0. Perhaps you or one of your classmates will eventually solve this problem, but let us put it aside for now and look at some other problems.

If there is a starting value N_0 which does not converge to 1 (and there could well be either a value that "cycles" somewhere else, or one that increases without bound as far as our knowledge given here shows) then there must be a *smallest* N_0 which does not converge to 1. It is almost obvious that such a "smallest N_0 that does not converge to 1" can*not* be even. (Why not?)

Your research problem is to investigate the recursive function

$$N_{k+1} = \begin{cases} N_k/2, & \text{if } N_k \text{ is even} \\[2ex] 3N_k + 1, & \text{if } N_k \text{ is odd} \end{cases}$$

and to prove as many theorems about it as possible. As far as your author knows no one has yet (1971) proved that for every integral starting value $N_0 > 0$, the sequence always converges to 1, nor has anyone yet (1971) devised a formula to determine the length α of a chain from N_0 to 1 that works for an arbitrary starting value N_0. Still there are a number of theorems that can be proved. Thinking up possible theorems (call them conjectures) and then trying to either prove or disprove them is an important phase of mathematical research.

⋆1.17 SUGGESTED TOPICS FOR FURTHER INDEPENDENT STUDY

I hope you have enjoyed the first chapter. It is always difficult to begin a voyage of mathematical discovery. Chapter 1 has provided you with a number of topics on which you may make future independent investigations and discoveries. I have sincerely tried to guide you past treacherous spots, and to point out various interesting sights along the way. In addition to providing a guide and prod to your thinking, groundwork was laid for excursions yet to come.

One of the primary purposes of this book is to help develop the so-called "mathematical maturity" which many authors deem the prime prerequisite for the study of advanced mathematics. One requisite of such mathematical sophistication is the ability to use the library and to study topics that interest you without the constant prodding of a classroom teacher. It is your author's hope that you will take the time *today* to read at least one reference in the list below. Perhaps you can find material to start a term project or paper in one of your courses, or to prepare a lecture for an interested group of students. If not, you should at least get acquainted with the mathematics section of your library.

Problem Set 1.17

In obtaining answers to these problems you will become acquainted with your mathematics library. This is important.

1. What is the color of the cover of last month's copy of *Mathematical Reviews*? Of month before last's issue?

2. Who wrote the first article in last month's issue of the *American Mathematical Monthly*?

3. Consult the index for the *Mathematical Reviews* of last year. Is there an author whose initials are the same as yours?

4. What is the call number used for books in Number Theory? How many books on Number Theory does your library have?

5. In what year did volume 50 of the *American Mathematical Monthly* appear?

6. Name six mathematical journals to which your library subscribes.

7. Examine the various Mathematical Tables available in your library. What is the largest prime number listed? Is it larger or smaller than the largest prime discussed in Section 1.11?

8. Many university libraries do *not* give shelf space to texts in current use. If your library had this text, under what call number would it be shelved? What other books are available under that call number in your library?

9. What is the oldest mathematical journal you can find in your library? Do the names of any of the authors sound familiar?

10. Follow up at least one reference in the bibliographic suggestions given below:

    Further material on integral domains will be found in almost any book on Modern Algebra or Abstract Algebra in your library (512.8 in Dewey system, QA266 in Library of Congress System). Work on modular systems, Euclid's algorithm, divisibility, prime and composite numbers and related topics will be found in books on number theory which are usually catalogued under 512.18 in the Dewey system or QA 241–248 in the Library of Congress system. The book, *Solved and Unsolved Problems in Number Theory* by D. Shanks (New York: Spartan Books, 1962), is of more than usual interest. Work on the nature of proof is apt to be scattered through many books. Two favorites are

> P. Suppes and S. Hill, *First Course in Mathematical Logic* (Waltham, Mass.: Blaisdell, 1964), especially Chap. 2.
>
> R. L. Wilder, *Introduction to the Foundations of Mathematics*, (New York: Wiley, 1965) Chap. 1.

A student interested in perfect numbers should check the journal *Mathematics of Computation* for articles on Mersenne primes as well as on perfect numbers. (Why?) Donald Gilles' article, *Three New Mersenne Primes and a Statistical Theory*, which appeared in January 1964 *Mathematics of Computation* (**18**, No. 85, p. 93) will give you a start. Note also page 87 and page 176 in the same issue for related articles. Martin Gardner's excellent article in the March 1968 issue of *Scientific American* also merits perusal.

11. Read and write a brief report (one page or less) on an article that has appeared during the past six months in one of the journals listed below:

    > *American Mathematical Monthly*
    > *Mathematics Teacher*
    > *Pi Mu Epsilon Journal*
    > *Mathematics Magazine*
    > *Journal in Undergraduate Mathematics*
    > *Computer Surveys*
    > *Mathematical Log*

12. Let $\oplus$ be an operation on the set of integers defined by

$$x \oplus y = x + y - 1$$

   (a) Is there an identity element under the operation $\oplus$, that is does there exist an integer i such that $i \oplus x = x \oplus i = x$ for all integers x?
   (b) What is the inverse of the integer k under the operation $\oplus$?
   (c) Do the integers form an integral domain under the operations $\oplus$ and $\cdot$?

*13. Prove that if D is an integral domain having only a finite number of elements, then for each element $x \in D$ there exists an multiplicative inverse $x^\dagger$ such that $x \cdot x^\dagger = x^\dagger \cdot x = u$ where u is the multiplicative identity of D.

Selected Reading List, Chapter 1
NUMBER THEORY AND PROOF

The following books and journals will provide additional reading for students interested in term projects or independent study.

Books

Hardy, G. and E. Wright, *Introduction to the Theory of Numbers.* London: Oxford 1960.

Jones, B. W., *The Theory of Numbers.* New York: Holt, Rinehart, and Winston 1955.

LeVeque, W. J., *Topics in Number Theory.* Reading, Mass.: Addison–Wesley Vol. 1 1962.

Ore, O., *Number Theory and Its History.* New York: McGraw-Hill 1948.

Uspensky, J. and M. Heaslet, *Elementary Number Theory.* New York: McGraw-Hill, 1939.

Andree, R. V. and others. *Table of Indices and Power Residues for Primes and Prime Powers Less than 2000.* New York: Norton, 1962.

Articles

Botts, T., "A Chain Reaction Process in Number Theory," *Mathematics Magazine,* **40,** No. 2 (Mar. 1967), pp. 55–65.

Chiang, P. -S. and A. J. MacIntyre, "Integers, No Three in Arithmetic Progression," *Mathematics Magazine,* **41,** No. 3 (May 1968), pp. 128–130.

Grimm, C. A., "A Conjecture on Consecutive Composite Numbers," *American Mathematical Monthly,* **76,** No. 10 (Dec. 1969), p. 1126.

Holladay, J. C., "Matrix Nim," *The American Mathematical Monthly,* **65,** No. 2 (Feb. 1958), p. 107.

Maxfield, J. E., "A Note on N!," *Mathematics Magazine,* **43,** No. 2 (Mar. 1970), pp. 64–67.

Moser, L., "On the Series $\sum 1/p$," *The American Mathematical Monthly,* **65,** No. 2, p. 104.

Reich, S., "On the Rational Positive Solutions of $m^n = n^m$, $m > n$," *The American Mathematical Monthly,* **75,** No. 10 (Dec. 1968), p. 1104.

Vaidya, A. M., "On Primes in A. P.," *Mathematics Magazine,* **40,** No. 1 (Jan. 1967), pp. 29–30.

"An Application of the Fibonacci Numbers," Problem E 2022, *The American Mathematical Monthly,* **75,** No. 10 (Dec. 1968), p. 1117.

"A Super-number," Problem E 2024, *The American Mathematical Monthly,* **75,** No. 10 (Dec. 1968), p. 1119.

"Matrix with Prescribed First Row," Problem E 1911, *The American Mathematical Monthly,* **75,** No. 1 (Jan. 1968), p. 81.

"n Distinct Prime Divisors," Problem E 2014, *The American Mathematical Monthly,* **75,** No. 9 (Nov. 1968), p. 1016.

"Representable Integers," Problem E 1967, *The American Mathematical Monthly,* **75,** No. 6 (June–July 1968), p. 675.

Notes on Chapter 1

Notes on Chapter 1

EQUIVALENCE AND CONGRUENCE

2

2.1 EQUIVALENCE RELATION

The study of mathematics almost always involves the use of some sort of an *equals relation* or *equivalence relation*. This relationship constitutes one of the most fundamental ideas in mathematics, as well as outside of it. Arithmetic is meaningless if one cannot tell when two numbers are equal; geometry demands the ability to determine when figures are congruent or similar. The equivalence relation was mentioned briefly in Chapter 1; here we shall discuss it at greater length. Do you know what an *equals* or *equivalence relation* is? Can you tell whether or not a given relation (for example, "is a divisor of, for integers") is a proper equals (equivalence) relation? There is a set of three simple postulates which determine an equals (equivalence) relation.

Originally, the concept of equality meant identity. Two things were equal if, and only if, they were identical (exactly the same). Soon it became necessary to extend the idea of equality to nonidentical objects. For example,

$\dfrac{47}{94}, \dfrac{13}{26}$, and $\dfrac{i(2 + 3i)}{-6 + 4i}$ are each said to be equal to $\frac{1}{2}$, although the expressions are *not* identical. This is an acceptable generalization of the concept of equality.

Fundamentally, the judgement as to whether or not two items are, or are not, equivalent depends upon what attributes are important for the particular situation. From one point of view (the view of the father's bank) two cars are "equal" if their prices are the same. From another view (the son's, perhaps), they may be "equal" if they have the same horsepower. Certain car owners feel that cars of the same make, year, and model need not be equivalent. In geometry, if "shape" is the criterion of interest, then similar triangles are the same; while if "area" is the point at issue, then equivalent triangles are the same. In general, any relationship, E, is an equivalence (or equals) relation if it satisfies the three postulates given below.

The symbol E, "is equivalent to," will be used to represent a general equivalence relationship. The relationship must be applied to some specific set of elements. The set may be quite general, but it must be specified. Furthermore, given any two elements a, b in the set, it must be true that either $a\ E\ b$, or $a\ \not\!E\ b$ (but not both), where $\not\!E$ means "is *not* equivalent to."

The relation E is then an equivalence or equals relation if, and only if, the following three postulates are satisfied for all elements a, b, c of the set.

1. *Reflexive Law:*   for each a, $a\ E\ a$.
2. *Symmetric Law:* if $a\ E\ b$, then $b\ E\ a$.
3. *Transitive Law:* if $a\ E\ b$ and $b\ E\ c$, then $a\ E\ c$.

The transitive law is familiar to many as, "Things equal to the same thing are equal to each other."

Any relation which satisfies these postulates is apt to have interesting mathematical (and nonmathematical) properties.

We examine several relations to see whether or not they satisfy these postulates.

EXAMPLE 1

Identity, $\equiv$, for the set of algebraic expressions in one variable, x: Given any two algebraic expressions, $f(x)$, $g(x)$, then either $f(x) \equiv g(x)$, or $f(x) \not\equiv g(x)$.

(1) "For each $a(x)$, $a(x) \equiv a(x)$" is valid, since an algebraic expression is identical to itself.
(2) "If $a(x) \equiv b(x)$, then $b(x) \equiv a(x)$" is valid.
(3) "If $a(x) \equiv b(x)$ and $b(x) \equiv c(x)$, then $a(x) \equiv c(x)$" is valid.

Hence, identity, $\equiv$, is an equivalence relation for the set of algebraic expressions in one variable, x, since it satisfies the required postulates.

EXAMPLE 2

Equality, $=$, for the set of real numbers:

(1) "For each a, $a = a$" is valid.
(2) "If $a = b$, then $b = a$" is valid.
(3) "If $a = b$ and $b = c$, then $a = c$" is valid.

Hence, equality of numbers, $=$, is an equivalence relation. (Why?)

EXAMPLE 3

Less than or equal to, $\leq$, for the set of rational fractions:

(1) "For each a, $a \leq a$" is valid, since $a = a$.
(2) "If $a \leq b$, then $b \leq a$" is *not* valid, since $3 \leq 5$, but $5 \not\leq 3$.[1]
(3) "If $a \leq b$ and $b \leq c$, then $a \leq c$" is valid.

The relationship $\leq$ is *not* an equivalence relation, since it fails to satisfy Postulate 2, the symmetric law.

EXAMPLE 4

Geometric congruence, $\cong$, for the set of plane triangles

(1) "For each $\triangle AXY$, $\triangle AXY \cong \triangle AXY$" is valid.
(2) "If $\triangle AXY \cong \triangle BST$, then $\triangle BST \cong \triangle AXY$" is valid.
(3) "If $\triangle AXY \cong \triangle BST$ and $\triangle BST \cong \triangle CUV$, then

$$\triangle AXY \cong \triangle CUV"$$

is valid.

Hence, congruence of triangles, $\cong$, is an equivalence relationship.

EXAMPLE 5

"Has the same color hair as," for girls

(1) "For each girl G, G 'has the same color hair as' G" is satisfied.
(2) "If A 'has the same color as' B, then B 'has the same color hair as' A" is valid.
(3) "If A 'has the same color hair as' B and B 'has the same color hair as' C, then A 'has the same color hair as' C" is also valid.

---

[1] It is customary to indicate that a relationship does *not* hold by making a slash through the symbol for the relationship, for example $5 \not\leq 3$, or $7 \neq 1$, or $\triangle ABC \not\cong \triangle AEB$.

Hence, "has the same color hair as" is a valid equivalence relationship in the set of all girls. (Would it be valid for the set of all animals having hair?)

EXAMPLE 6

"Is a sibling of," for the set of all people

(1) "For all A, A 'is a sibling of' A" is *not* satisfied.
(2) "If A 'is a sibling of' B, then B 'is a sibling of' A" is valid.
(3) "If A 'is a sibling of' B and B 'is a sibling of' C, then A 'is a sibling of' C" is not satisfied, since A and C may be identical.

Hence, "is a sibling of" is *not* an equivalence relationship for the given set, since (1) and (3) are not satisfied. (Would "has the same parents as" be an equivalence relationship?)

EXAMPLE 7

"Weighs within half a pound of," for the set of all mathematics students in the United States

(1) "For all A, A 'weighs within half a pound of' A" is valid.
(2) "If A 'weighs within half a pound of' B, then B 'weighs within half a pound of' A" is also valid.
(3) "If A 'weighs within half a pound of' B and B 'weighs within half a pound of' C, then A 'weighs within half a pound of' C" is *not* valid. (Why not? Give a counterexample.)

Hence, "weighs within half a pound of" is not an equivalence relationship for the set given. (Why not?)

EXAMPLE 8

Consider the set of all rational numbers: $a \mathrel{E} b$ shall mean a and b have reciprocals (in lowest terms), each of which has the same denominator. Thus, $3 \mathrel{E} \frac{3}{7}$, since $\frac{1}{3}$ and $\frac{7}{3}$ are reciprocals having the same denominator. Also, $\frac{17}{4} \mathrel{E} \frac{17}{121}$ and $8 \mathrel{E} \frac{8}{5}$.

(1) "For all a, $a \mathrel{\not{E}} a$" does *not* hold, since 0 has no reciprocal, and hence $0 \mathrel{\not{E}} 0$.
(2) "If $a \mathrel{E} b$, then $b \mathrel{E} a$," since the denominators of the reciprocals are unchanged.
(3) "If $a \mathrel{E} b$ and $b \mathrel{E} c$, then $a \mathrel{E} c$," since the denominators of the reciprocals are unchanged.

Students familiar with elementary logic will note that Examples 3, 7, and 8 provide instances in which each of the three postulates, in turn, is violated while the other two hold. This proves our postulates are *independent*; that is, no one of the postulates could be proved from the other two.

Problem Set 2.1

In Problems 1–16, test each of the given relationships to see which postulates of an equivalence relationship are satisfied. If a postulate is not satisfied, give a counterexample.

1. The relation *similar*, $\sim$, for triangles.

2. The relation *is not equal to*, $\neq$, for whole numbers (integers).

3. The relation *divides*, $|$, for whole numbers (integers).

4. "Is a brother of" (a) for the set of living persons
   (b) for the set of living men and boys.

5. "Is a descendant of" for the set of all people.

6. "Has the same parents as" for (a) the set of all living people
   (b) all people.

7. "Is the same age as" for the set of all clocks.

8. "Is approximately equal to" for the set of all positive decimals. (You yourself must decide upon a meaning for "approximately equal," and then use it. Two possible meanings would be "their difference is less than 1 percent of the larger value" or "their difference is less than 0.001.")

9. "Is married to" for the set of all people.

10. "Has the same marital status (that is, single, married, divorced, or spouse dead) as" for the set of all living persons.

11. Has the same "oddness or evenness as" for the set of integers.

12. "Has the same number of sides as" for the set of all polygons.

13. "Has a larger integral divisor than" for the set of all positive integers (positive whole numbers).

14. "Is perpendicular to" for the set of all lines in a plane.

15. "Is parallel to" for the set of all lines in space.

16. "Has the same denominator as" for the set of rational fractions.

★17. Illustrative Example 8 shows that the relation "has a lowest term reciprocal having the same denominator as" on the set of rational numbers satisfies the symmetric and the transitive laws, but fails to be reflexive $(0 \not\mathrel{E} 0)$. The following argument purports to show that the reflexive law is a consequence of the symmetric and transitive laws. This is a paradox.

$$a \mathrel{E} b \text{ implies } b \mathrel{E} a \text{ by the symmetric law.}$$

Then $a \mathrel{E} b$ and $b \mathrel{E} a$ implies $a \mathrel{E} a$ by the transitive law, with a used in place of c. Hence $a \mathrel{E} a$. Where is the error?

18. Is congruence mod 6 an equivalence relation? Justify your answer.

19. Same as Problem 18 for mod 7.

2.2 EQUIVALENCE CLASSES

In general, an equivalence relation separates the objects of a set into *equivalence classes* (baskets or heaps) of equivalent objects.

Mathematics uses the various *equivalence classes* of objects as elements, not the objects themselves. An operation is said to be *well defined* with respect to a given relation if each object in a class may be used to represent the entire class under that operation. A simple example may help:

EXAMPLE 9

The equals relation, $=$, separates the rational numbers into classes such that all fractions ka/kb with $k \neq 0$, $b \neq 0$ belong in the same class. In adding $\frac{1}{3} + \frac{1}{5}$, $\frac{1}{3}$ is replaced by another member, $\frac{5}{15}$, of its equivalence class, while $\frac{1}{5}$ is replaced by $\frac{3}{15}$. Thus,

$$\tfrac{1}{3} + \tfrac{1}{5} = \tfrac{5}{15} + \tfrac{3}{15} = 5 \cdot \tfrac{1}{15} + 3 \cdot \tfrac{1}{15} = (5 + 3)\tfrac{1}{15} = 8 \cdot \tfrac{1}{15} = \tfrac{8}{15}$$

Which of the equalities in the above string depends upon the distributative law for its justification?

EXAMPLE 10

Let the original set of objects be all the positive integers (whole numbers)

$$1, 2, 3, 4, \cdots, 1769, \cdots.^2$$

---

[2] In mathematics, as in literature, the three dots indicate that something has been omitted. In mathematics, the reader is expected to understand what has been omitted.

Problem 11, Set 2.1, showed that "has the same oddness or evenness as" was a valid equivalence relation.

Separate the positive integers into two classes, Class 1 (*Odd*), containing the odd integers, and Class 2 (*Even*), the even integers.

Class 1, *Odd*: $[1, 3, 5, 7, 9, \cdots, 171, \cdots, 1943277, \cdots, 2n + 1, \cdots]$.
Class 2, *Even*: $[2, 4, 6, 8, 10, 12, \cdots, 218, \cdots, 17694218, \cdots, 2n, \cdots]$.

The following properties are apparent:

(1) Each integer in Class 1 (*Odd*) is equivalent to every other integer in Class 1 under the relation "has the same oddness or evenness as." (Why?)
(2) Each integer in Class 2 (*Even*) is equivalent to every other integer in Class 2 under the relation "has the same oddness or evenness as."
(3) Each positive integer is in one, and only one, of Class 1 or 2.
(4) No integer in Class 1 (*Odd*) is equivalent to an integer in Class 2 (*Even*), and no integer in Class 2 (*Even*) is equivalent to an integer in Class 1 (*Odd*) under the relationship "has the same evenness or oddness as."

Therefore, the positive integers have been separated into two classes with respect to the equivalence relationship "has the same evenness or oddness as." The new mathematical system, having as its elements the classes *Odd*, *Even*, has the arithmetic given below.

Tabular presentations

$Odd + Odd = Even$	$+$	O	E		$+$	(1)	(2)
$Even + Even = Even$	O	E	O	(1)	(2)	(1)
$Odd + Even = Odd$	E	O	E	(2)	(1)	(2)
$Even + Odd = Odd$						
$Odd \cdot Odd = Odd$	$\cdot$	O	E		$\cdot$	(1)	(2)
$Even \cdot Even = Even$	O	O	E	(1)	(1)	(2)
$Odd \cdot Even = Even$	E	E	E	(2)	(2)	(2)
$Even \cdot Odd = Even$						

In the first tabular presentation, we used O to represent Class 1, *Odd*, and E to represent Class 2, *Even*. In the second tabular presentation, Class 1, *Odd*, is represented by (1) and Class 2, *Even*, by (2). Note that, since (2) represents an entire class $(2, 4, 6, \cdots, 2m, \cdots)$ of even numbers, not merely the one integer, 2, it is correct and reasonable to say $(2) + (2) = (2)$ and $(2) \cdot (2) = (2)$.

It is in a similar fashion that one treats $\dfrac{1}{2}, \dfrac{47}{94}, \dfrac{13}{36}$, and $\dfrac{i(2 + 3i)}{-6 + 4i}$ as members of the same class in ordinary arithmetic.

In more advanced mathematics, the rational numbers are developed as sets of equivalence classes of integers [ordered pairs of integers (ka, kb) where a is the integer corresponding to the numerator, and b to the denominator, of the rational number a/b]. The real numbers are developed as sets of equivalent classes of sequences of rational numbers. The complex numbers are developed as sets of equivalent classes of pairs of real numbers. The quaternions (to be studied later in this text) are developed as sets of equivalent classes of complex numbers.

The heart of the concept of equivalence classes is, of course, the equivalence relation. Every equivalence relation determines equivalence classes. (Why?) Furthermore, every classification of a set into equivalence classes determines an equivalence relation.

Problem Set 2.2

Please read the text carefully before you attempt a problem set.

1. Example 5, Section 2.1, gives an equivalence relation. Find the equivalence classes so determined. List the names of three elements of G which belong to the same equivalence class.

2. Some of the relations given in Problems 1–16 of Problem Set 2.1 are equivalence relations. For each valid equivalence relation given there, determine three members of the set in question which do belong to the same equivalence class, and one member which does not.

3. Let F be the set of all men who played college football last season. Divide F into three sets:

   $F_1 = $ those men in F who scored more than 50 points.
   $F_2 = $ those men in F who scored from 1 to 50 points, inclusive.
   $F_3 = $ those men in F who did not score.

   Are the three sets in question equivalence classes? If so, *what is the equivalence relation*? If not, why not?

4. Let the F of Problem 3 be divided into three sets as follows:

   F_1 = those men of F who played line positions only.
   F_2 = those men of F who played backfield positions only.
   F_3 = those men of F who played more than one position.

   Is this a partition of F into equivalence classes? Why not?

5. Make up an equivalence relationship not mentioned in the text. Be sure to specify the set, S, on which it operates. Prove that it is an equivalence relation and describe the equivalence classes into which S is separated.

6. Does the mod 7 congruence relation, $\equiv$ (mod 7), separate the set of integers into equivalence classes? Demonstrate.

7. Same as Problem 6 for $\equiv$ (mod 6).

8. (a) Let the set, N, consist of the following integers: 1, 2, 4, 6, 10. Let $a \boxtimes b$ mean that $a + b \equiv 0$ (mod 4). Is the described relation an equivalence relation for the set N?

   (b) Let $N = [0, 2, 4, 5, 8, 10]$ and $a \boxtimes b$ mean $2a + b \equiv 0$ mod 6. Is the relation an equivalence relation for the set N? Why or why not?

2.3 CONGRUENCES

Let a, b, and m be integers. Define $a \equiv b$ (mod m) (read "a is congruent to b, modulo m") to mean that $m \mid (a - b)$. It would be reasonable to say that $a \equiv b$ (mod m) means that there exists an integer k such that $a = b + km$. (The reader is expected to show that these two statements are equivalent; that is, that each statement implies the other.) The mod 7 and mod 6 systems discussed in Sections 1.2 and 1.3 are examples of particular modular systems.

In the introductory material of Chapter 1, the mod 7 and mod 6 systems were considered as having only 7 and 6 elements, respectively. This is a useful and valid interpretation. However, another viewpoint is also beneficial at times. Consider mod m arithmetic, not as a system composed of only m elements, but as a system having all the integers as elements, and a new type of equals relationship (congruence) as given above. It is the equals relationship which classifies the integers into m distinct sets (equivalence classes). It would be wise to review Problems 14 and 15 of Set 1.2 at this point. With the generalized equals relationship in mind, it will be reasonable to say $5 \equiv 17$ (mod 12) and to solve congruences like $13x \equiv 7$ (mod 5). (Can you find a solution?)

EXAMPLE 11

Find $x \equiv 3 + 7 - 4 + 6 - 2 + 8 + 14$, (mod 11)

$$x \equiv 32 \equiv 10 \; (\text{mod } 11)$$

[Note that $x \equiv -1$ is also a valid solution, since $-1 \equiv 10 \; (\text{mod } 11)$.]

EXAMPLE 12

Find a value of x such that $5x \equiv 4 \; (\text{mod } 6)$.

By actual substitutions of $x = 0$, 1, 2, 3, 4, 5 into $5x \equiv 4 \; (\text{mod } 6)$, one finds that $x \equiv 2$ is the only solution.

The astute reader may wonder if the phrase "*is* the only solution" should not be replaced by the phrase "*are* the only solutions." His point is well taken, since both

$$x = 2, 8, 14, 20, 26, \cdots, 2 + 6k, \quad \text{and} \quad x = -4, -10, \cdots, 2 - 6k, \cdots$$

all are solutions of $5x \equiv 4 \; (\text{mod } 6)$. Each of these solutions is congruent to 2 (mod 6). Many times in mathematics it is desirable to lump together objects which have certain properties in common. One does not hesitate to say that $\frac{6}{12}$, $14\pi/28\pi$, $\frac{1}{2}$, and $(3\sqrt{11})/(6\sqrt{11})$ are all the same number, and after a little computation one agrees that $\dfrac{(1 - \sqrt{2})}{-2(1 + \sqrt{2})(3 - 2\sqrt{2})}$ is also $\frac{1}{2}$.

In fact, there is a whole equivalence class of number representations (infinitely many) which are all equal to $\frac{1}{2}$. It is usual to think of the entire class as being the same number, $\frac{1}{2}$, and one says that the *unique* solution of the equation $2x = 1$ is $x = \frac{1}{2}$. Actually, the solution is unique only because we agree to consider the $\frac{1}{2}$ as a symbol for the entire equivalence class which contains all the numbers equal to $\frac{1}{2}$. In advanced calculus you will group together all sequences which have the same limit, and consider this huge class of sequences to be one object. It is usual to pick out one of the sequences to represent the entire class, just as it is usual to pick out the fraction $\frac{1}{2}$ to represent the entire, infinite class of numbers of the form $x/2x$, where x is any nonzero number (integer, real number, or perhaps even complex number).

The same technique is used in the solution of congruences. If one speaks of congruences mod 6, then all the integers are separated into six equivalence classes, $0 + 6k$, $1 + 6k$, $2 + 6k$, $3 + 6k$, $4 + 6k$, $5 + 6k$. Each equivalence class may be represented by any member of that class. It is often convenient, but not vital, to pick the smallest nonnegative number in each class to represent it. It is in this usage that one says $x \equiv 2 \pmod 6$ *is the only* solution of $5x \equiv 4 \pmod 6$.

It may never have occurred to you that it is surprising that the same answer is obtained in adding $\frac{1}{4} + \frac{1}{12} + \frac{1}{16}$ no matter whether one uses $4 \cdot 3 \cdot 4 = 48$ or $4 \cdot 12 \cdot 16 = 768$ as the common denominator.

$$\frac{1}{4} + \frac{1}{12} + \frac{1}{16} = \frac{12}{4 \cdot 3 \cdot 4} + \frac{4}{4 \cdot 3 \cdot 4} + \frac{3}{4 \cdot 3 \cdot 4} = \frac{19}{4 \cdot 3 \cdot 4} = \frac{19}{48}$$

$$\frac{1}{4} + \frac{1}{12} + \frac{1}{16} = \frac{192}{4 \cdot 12 \cdot 16} + \frac{64}{4 \cdot 12 \cdot 16} + \frac{48}{4 \cdot 12 \cdot 16} = \frac{304}{4 \cdot 12 \cdot 16} = \frac{19}{48}$$

You have used this principle ever since you learned to add fractions in grade school, and familiarity has bred a complacency. Nevertheless, it is really quite a remarkable thing that, no matter which representatives of the classes of fractions equal to $\frac{1}{4}$, $\frac{1}{12}$, and $\frac{1}{16}$ are chosen, the result is always one of the representatives of the class of fractions equal to 19/48. This is, of course, a fundamental property of equivalence classes. These proofs can wait until all students have developed enough mathematical sophistication to see that they are required. Now, however, we shall prove similar theorems for the theory of congruences, which you may be more willing to admit need proof and which, incidentally, are also easier to prove, since the similar theorems in the system of integers will be assumed.

THEOREM 2.1 If $a \equiv b \pmod m$, then *for every integer N, it follows that* $a + N \equiv b + N \pmod m$, *and* $aN \equiv bN \pmod m$.

The hypothesis states that there exists an integer k such that $a = b + km$. The theorem then follows by applying the definition of congruence to obtain $a + N = b + N + km$ and $aN = bN + (kN)m$, which are immediate consequences of the hypothesis. The reader is expected to write out the steps of this proof and supply a reason for each step. The theorem states that congruence is well defined with respect to addition and multiplication.

The following corollary is a special case of Theorem 2.1.

Corollary: If $a \equiv b \pmod m$, then $-a \equiv -b \pmod m$.

The reader will find it instructive to prove this corollary directly from the definition of congruence.

THEOREM 2.2   If $a \equiv b \pmod{m}$ and $c \equiv d \pmod{m}$, then
$$a + c \equiv b + d \pmod{m} \quad \text{and} \quad ac \equiv bd \pmod{m}.$$

By hypothesis, there exist integers j and k such that $a = b + jm$ and $c = d + km$. Therefore, $a + c = b + d + (j + k)m$ and $a + c \equiv b + d$. The reader should supply the proof that $ac \equiv bd \pmod{m}$.

THEOREM 2.3   If $a \equiv b \pmod{m}$, and *if $P(x)$ is a polynomial with integral coefficients*, then $P(a) \equiv P(b) \pmod{m}$.

Theorem 2.3 follows from Theorems 2.1 and 2.2, since the polynomial $P(x)$ is a sum of products of integers and powers of x.

Problem Set 2.3

In each of Problems 1–10, find the value of x by selecting the smallest non-negative representative of the equivalence class into which the given modulus separates the integers.

1. $x \equiv 4 + 16 - 37 + 11 \pmod{15}$.

2. $x \equiv 3 + 11 - 15 + 76 - 1 + 41 + 5 \pmod{7}$.

3. $x \equiv 14 + 27 + 35 - 3672 \pmod{5}$.

4. $x \equiv 1 + 2 + 3 + 4 + 5 + 6 + 7 + 8 \pmod{31}$.

5. $x \equiv 9 + 10 + 11 + 12 + 13 + 14 + 15 \pmod{31}$.

6. $x \equiv 17 + 9 + 6 - 5 + 11 - 2 \pmod{13}$.

7. $x \equiv 4 - 7 - 16 + 5 - 37 + 1 - 16 \pmod{24}$.

8. $x \equiv 21 - 16 + 7 - 39 - 52 + 5 \pmod{18}$.

9. $x \equiv 29 - 17 + 16 + 5 + 37 \pmod{11}$.

10. $x \equiv 1 + 2 - 7 + 6 - 8 + 11 \pmod{2}$.

11.–20. Find two other representatives of the equivalence class of integers congruent to x in Problems 1–10, respectively.

21. (a) Show that congruence mod -15 is equivalent to congruence mod 15.
    (b) Show, in general, that congruence mod $-|m|$ is equivalent to congruence mod $|m|$.

22. Show that congruence mod 0 is ordinary equality.

23. Prove the corollary of Theorem 2.1 directly from the definition of congruence.

24. (a) Find the first ten powers of 5 (mod 12) with a minimum of labor, by noting that $5^2 = 25 \equiv 1$ (mod 12) and hence,

$$5^3 = 5^2 \cdot 5 \equiv 1 \cdot 5 \equiv 5, \quad \text{etc.}$$

(b) Find 5^{237} and 5^{238} (mod 12) by generalizing the results of part (a) of this problem.

25. (a) Find the first ten powers of 5 (mod 9) with a minimum of labor noting that $5^2 = 25 \equiv 7$ (mod 9) and hence, $5^3 = 5^2 \cdot 5 \equiv 35 \equiv 8$, and $5^4 = 5^3 \cdot 5 \equiv 40 \equiv 4$, etc.

(b) If possible, generalize the result of part (a) to find 5^{237} and 5^{238} (mod 9).

(c) After working part (b), find a general rule for 5^n (mod 9). Can you express this result using congruences?

26. The rule known as "casting out nines" states: "If a number, N, is divided by 9, the remainder is the same as the remainder when the sum of the digits of N is divided by 9." Use congruences to prove this rule. [HINT: Every number can be written $N = P(10)$, where $P(x) = a_n x^n + a_{n-1} x^{n-1} + \cdots + a_1 x + a_0$. In this case, the sum of the digits of N is $P(1)$.]

27. Prove: The remainder after division of N by 3 is the same as the remainder after division of the sum of the digits of N by 3.

28. Obtain tests for divisibility by 3 and 9 as corollaries of Problems 26 and 27.

29. Show that a number N is divisible by 11 if, and only if, K is divisible by 11, where K is formed by taking the unit's digit minus the ten's digit plus the hundred's digit minus the thousand's digit plus $\cdots$ etc. [that is, $P(-1)$ in the notation of Problem 26].

30. Make up and prove a test for divisibility by some number other than 2, 3, 5, 9, 10, or 11. [NOTE: It may turn out to be a bit impractical to use, but should be instructive to manufacture.]

31. Show that congruence mod m satisfies the three postulates for an equivalence relationship.

In Problems 32 to 44, solve the stated congruences. Be sure to obtain all solutions (i.e., a representative of *each* equivalence class of congruent numbers which satisfies the original congruence).

32. $3x \equiv 7 \pmod{13}$.

33. $4x \equiv 5 \pmod{10}$.

34. $5x \equiv 7 \pmod{12}$.

35. $6x \equiv 9 \pmod{7}$.

36. $3x \equiv 3 \pmod{9}$.

37. $9x \equiv 5 \pmod{11}$.

38. $17x \equiv 2 \pmod{13}$.

39. $41x \equiv 1 \pmod{14}$.

40. $36x \equiv 17 \pmod{15}$.

41. $29x \equiv 13 + 4x \pmod{60}$.

42. $14x + 11 \equiv 3 \pmod{19}$.

43. $23x + 19 \equiv 2x \pmod{35}$.

44. $4x + 6 \equiv 18 \pmod{12}$.

45. Prove the second statement of Theorem 2.2.

46. Does congruence mod p, where p is a prime, satisfy the cancellation law $c \cdot a = c \cdot b$ and $c \neq 0$ imply $a = b$? Is it an integral domain?

47. Does congruence mod m, where m is an integer, satisfy the cancellation law?

48. Describe the equivalence classes into which congruence mod m separates the integers.

*49. Congruence mod 24 fails to satisfy the cancellation law, since $6 \cdot 5 \equiv 6 \cdot 1$, but $5 \not\equiv 1 \pmod{24}$. Can you determine a stronger condition than $c \not\equiv 0$ which would imply that, if c satisfies this condition, then $ca \equiv cb$ would imply $a \equiv b \pmod{24}$.

*50. Generalize Problem 49 to mod m, where m is any composite modulus.

51. Which of the elements of the modulo 10 system do *not* have multiplicative inverses?

52. How many distinct equivalence classes are there: (a) modulo 10? (b) modulo 11? (c) modulo N?

53. Is E an equivalence relation for the set of real numbers if $a \; E \; b$ means that $(a - b)$ is rational?

54. Is E an equivalence relation for the set of all lines in the x, y-plane if $a \; E \; b$ means that line a is parallel to line b?

2.4 LINEAR CONGRUENCES

A linear real equation in one unknown has a unique solution. The corresponding example $3x \equiv 5 \pmod{6}$ has no solution at all, while

$4x \equiv 2 \pmod{6}$ has two solutions, and $5x \equiv 2 \pmod{6}$ has only one solution. (The reader is expected to verify *each* of these statements. Verifying statements and supplying missing steps is an important part of *reading* mathematics.)

THEOREM 2.4   *The congruence $Ax \equiv B \pmod{m}$ has exactly $g = (A, m)$ incongruent solutions mod m if $g \mid B$, and no solutions if $g \nmid B$.*

The proof is presented in five parts:

- I.   If $g \mid B$, there is no solution.
- II. If $g \mid B$, there is at least one solution.
- III. If x_0 is a solution, then so is $x_0 + k(m/g)$ for every integer k.
- IV. Every solution is obtained in Part III.
- V.   If solutions exist, then there exist exactly $g = (A, m)$ incongruent solutions.

Proof:

I. If $g \nmid B$, there is no solution.

If there exists a solution x_0 of the congruence $Ax \equiv B \pmod{m}$, then there exists a constant k such that $Ax_0 = B + km$. Since $g = (A, m)$, it follows that $g \mid A$ and $g \mid m$. Hence, if a solution x_0 exists, then we also have that $g \mid B$. (Why?)

Thus, if $g \nmid B$, no solution exists.

II. If $g \mid B$, then at least one solution exists.

Since $g \mid B$, and $g = (A, m)$, there exist integers B_1, A_1, and m_1 such that

$$gB_1 = B, \quad gA_1 = A, \quad gm_1 = m$$

where $(A_1, m_1) = 1$. (Why?)

Hence, there exist integers p and s, such that

$$1 = pA_1 + sm_1 \qquad \text{(Why?)}$$
$$B = BpA_1 + Bsm_1$$
$$\text{or} \qquad B = gB_1pA_1 + gB_1sm_1$$
$$B = pB_1A + sB_1m$$

Thus, $x_0 = pB_1$ satisfies the original congruence. (Why?)

III. If x_0 is a solution, then so is $x_0 + k(m/g)$ for every integer k.

Each solution of $Ax \equiv B \pmod{m}$ is an integer x such that $Ax = B + km$ for some integer k.

Hence, there exists an integer k_1 such that

$$Ax_0 = B + k_1 m$$

Note that
$$A[x_0 + n(m/g)] = Ax_0 + An(m/g)$$
$$= Ax_0 + A_1 gn(m/g)$$
$$= Ax_0 + (A_1 n)m$$

Since $Ax_0 = B + k_1 m$, it follows that

$$A[x_0 + n(m/g)] = B + k_1 m + (A_1 n)m$$
$$= B + (k_1 + A_1 n)m$$

Hence, $A[x_0 + n(m/g)] \equiv B \pmod{m}$, if $Ax_0 \equiv B \pmod{m}$.

IV. Every solution is obtained in Part III.

Let x_0 and x_1 be two solutions, that is $Ax_0 \equiv B \pmod{m}$ and $Ax_1 \equiv B \pmod{m}$. Then $Ax_0 \equiv Ax_1 \pmod{m}$ (by Problem 31, Set 2.3)

or
$$Ax_1 = Ax_0 + km$$

Setting $gA_1 = A$ and $gm_1 = m$ as before, one obtains

$$gA_1 x_1 = gA_1 x_0 + kgm_1$$
or
$$A_1 x_1 = A_1 x_0 + km_1$$
$$A_1(x_1 - x_0) = km_1$$

Hence, $A_1 \mid km_1$.

Since $(A_1, m_1) = 1$ (Why?), thus A_1 must divide k (Why?) and $k = k_1 A_1$ giving

$$A_1(x_1 - x_0) = k_1 A_1 m_1$$
$$x_1 - x_0 = k_1 m_1.$$

Hence,
$$x_1 = x_0 + k_1 m_1$$
$$= x_0 + k_1(m/g)$$

Thus, every solution is obtained in Part III.

V. If solutions exist, then there exist exactly $g = (A, m)$ incongruent solutions mod m.

If x_0 is a solution, all numbers of the form $x_0 + k_1(m/g)$ are solutions by III. By IV, only such numbers are solutions. Hence, all that remains is to determine into how many sets which are incongruent mod m the numbers

$$\cdots, x_0 - 2(m/g), x_0 - (m/g), x_0, x_0 + (m/g), x_0 + 2(m/g),$$

$$x_0 + 3(m/g), \cdots, x_0 + g(m/g), \cdots$$

may be divided. By actual count there are g such sets.

This completes the proof of Theorem 2.4.

In solving linear congruences having more than one solution, it is often desirable to divide each member of the congruence and the modulus by $g = (A, m)$. This may be done, since

$$Ax = B + km$$

$$gA_1x = gB_1 + kgm_1$$

$$A_1x = B_1 + km_1$$

$$A_1x \equiv B_1 \ (\text{mod } m_1)$$

In general practice, it is unusual to write out these intermediate steps. Rather, one proceeds directly from $Ax \equiv B \ (\text{mod } m)$ to $A_1x \equiv B_1 \ (\text{mod } m_1)$ after verifying that $(A, m) = g$ actually divides B.

EXAMPLE 13

Solve $6x \equiv 8 \ (\text{mod } 20)$.

Since $g = (A, m) = (6, 20) = 2$, and $2 \mid 8$, there will exist exactly $g = 2$ solutions of $6x \equiv 8 \ (\text{mod } 20)$, which are incongruent $(\text{mod } 20)$.

$$6x \equiv 8 \ (\text{mod } 20)$$

$$3x \equiv 4 \ (\text{mod } 10) \qquad (\text{Note change of modulus.})$$

Since $7 \cdot 3 = 21 \equiv 1 \bmod 10$, we multiply each member by 7 (that is, multiply by the inverse of 3) to obtain:

$$7 \cdot 3x \equiv 28 \ (\mathrm{mod}\ 10)$$

$$x \equiv 8 \ (\mathrm{mod}\ 10)$$

then

$$x \equiv 8, 18 \ (\mathrm{mod}\ 20) \qquad (\text{Note change of modulus.})$$

EXAMPLE 14

Solve $7x \equiv 10 \ (\mathrm{mod}\ 14)$.
Since $g = (A, m) = (7, 14) = 7$ and $7 \nmid 10$, there are no solutions.

EXAMPLE 15

Solve $8x \equiv 16 \ (\mathrm{mod}\ 12)$.
Since $g = (8, 12) = 4$ and $4 \mid 16$, there exist exactly $g = 4$ incongruent solutions mod 12.

$$8x \equiv 16 \ (\mathrm{mod}\ 12)$$

$$2x \equiv 4 \ (\mathrm{mod}\ 3) \qquad (\text{Note change of modulus.})$$

$$x \equiv 2 \ (\mathrm{mod}\ 3)$$

$$x \equiv 2, 5, 8, 11 \ (\mathrm{mod}\ 12).$$

The integers are classified into m equivalence classes, modulo m, by considering the classes

$$(0 + k \cdot m) = [\text{all integers} \equiv 0 \ (\mathrm{mod}\ m)]$$

$$(1 + k \cdot m) = [\text{all integers} \equiv 1 \ (\mathrm{mod}\ m)]$$

$$(2 + k \cdot m) = [\text{all integers} \equiv 2 \ (\mathrm{mod}\ m)]$$

$$\vdots \qquad\qquad\qquad \vdots$$

$$(m - 1 + k \cdot m) = [\text{all integers} \equiv (m - 1) \ (\mathrm{mod}\ m)].$$

Every integer lies in one of these *residue classes*, and no integer is contained in more than one class. Thus, in writing $x \equiv 2, 5, 8, 11 \ (\mathrm{mod}\ 12)$ as solutions of $8x \equiv 16 \ (\mathrm{mod}\ 12)$, we refer to the classes $(2 + k \cdot 12)$, $(5 + k \cdot 12)$, $(8 + k \cdot 12)$, $(11 + k \cdot 12)$. These classes are, of course, equivalence classes of the mod 12 equivalence relation.

Problem Set 2.4

In Problems 1–15, determine the number of incongruent solutions of the given congruences and, if solutions exist, find them.

1. $3x + 7 \equiv 5 \pmod 9$.
2. $4x + 2 \equiv 11 \pmod 5$.
3. $7x + 5 \equiv 0 \pmod{12}$.
4. $9z \equiv 17 \pmod{12}$.
5. $4x \equiv 5 + x \pmod 6$.
6. $3x \equiv 2 + 5x \pmod{28}$.
7. $5w + 3 \equiv 2w - 7 \pmod{35}$.
8. $4x + 2 \equiv 0 \pmod{12}$.
9. $6y \equiv 20 \pmod{15}$.
10. $6y \equiv 20 \pmod{16}$.
11. $6y \equiv 20 \pmod{17}$.
12. $9x + 2 \equiv 4x \pmod{15}$.
13. $3x \equiv 7 \pmod{15}$.
14. $7w \equiv 3 \pmod{15}$.
15. $21z \equiv 17 + 4z \pmod{285}$.

16. Show that $3^{2n+2} \equiv 8n + 9 \pmod{64}$, where n is any integer ≥ 1. [HINT: Write 3^{2n+2} as $(1 + 8)$ with the proper exponent.]

17. If $p \neq 2$ is a prime and $(p, b) = 1$, prove that $1^2 b^2$, $2^2 b^2$, $3^2 b^2$, $\cdots$, $[(p - 1)/2]^2 b^2$ yield distinct remainders when divided by p.

18. Prove that, if $(a, k) = 1$, then $(b, k) = 1$ for any b such that $a \equiv b \pmod k$.

19. If p_n is the nth prime number, how large must k be so that a complete residue system modulo 6 is obtained among the primes $p_1, p_2, p_3, \cdots, p_k$?

20. Same as Problem 19, but modulo 7.

21. In the mod 11 system, divide $4x^3 + 7x^2 + 3x - 5$ by $5x - 3 \pmod{11}$, until a constant remainder is obtained.

    [HINT:
$$
\begin{array}{r}
Qx^2 + ? + ? \pmod{11} \\
\hline
5x - 3 \;\big|\; 4x^3 + 7x^2 + 3x - 5
\end{array}
$$

    where Q is a solution of the congruence $5Q \equiv 4 \pmod{11}$.]

22. Divide $4x^5 - 3x^4 + 3x^3 + 5x^2 - 2x + 3$ by $3x + 2 \pmod 7$.

23. Divide $14x^3 - 7x^2 + 5$ by $3x^2 + x - 2 \pmod{23}$.

24. Divide $7x^5 - 4x^3 + 3x^2 + 2x - 5$ by $2x^2 + 3x - 1 \pmod{11}$.

2.5 A MORE SOPHISTICATED VERSION

It is not essential that you understand how the concepts of Chapters 1 and 2 are expressed as *Cartesian products* in order to complete the material in this text, but it is worthy of consideration, since more advanced books in abstract algebra make use of this powerful concept.

Cartesian Product of Two Sets

The set of all ordered pairs (a, b) with $a \in A$ and $b \in B$ is called the *Cartesian product*, $A \times B$ of sets A and B. Two ordered pairs (a, b) and (a', b') are equal in $A \times B$ if and only if $a = a'$ in A and $b = b'$ in B. It is not essential that A and B of $A \times B$ be distinct. The familiar "Cartesian coordinates" of analytic geometry is a Cartesian product $R \times R$, where R is the set of real numbers.

Let the set S consist of that subset of the ordered pairs of $R \times R$ in which the second element is the cube of the first element, that is S consists of those ordered pairs of the form (x, x^3) for x a real number. Then S is the mapping or function of R into R which is often designated as $y = x^3$.

Equivalence

We now turn our attention to the notions of equivalence relations and equivalence classes. These concepts may be stated in terms of ordered pairs.

Let E be a subset of the set of ordered pairs $A \times A$. E defines an *equivalence relation* on A if

Reflexivity: $(a, a) \in E$ for all $a \in A$
Symmetry: $(a, b) \in E$ implies $(b, a) \in E$
Transitivity: $(a, b) \in E$ and $(b, c) \in E$ imply $(a, c) \in E$.

If E is an equivalence relation on A, then A is separated into subsets called equivalence classes. Every element $a \in A$ determines an equivalence class $\{a\}$ consisting of all elements of A which are equivalent to a, clearly $a \in \{a\}$. It is also true that if the equivalence classes $\{a\}$ and $\{b\}$ have even one element in common, then the two equivalence classes are identical and every element of either is also an element of the other (that is, $\{a\} = \{b\}$).

If we define the set operations of addition and multiplication of equivalence classes as $\{a\} + \{b\} = \{a + b\}$ and $\{a\} \cdot \{b\} = \{a \cdot b\}$, a consistent arithmetic of equivalence classes is obtained. Congruences modulo m are obtained as the subset E_m of $I \times I$, where I is the set of integers and E_m consists of those pairs (a, b) such that m divides $a - b$. The reader should try to relate this to the ideas of congruence presented in Chapters 1 and 2.

Functions and Mappings

The concept of a *mapping* or a *function* may be the most basic useful concept in mathematics. It is found in every branch of mathematics and should seem familiar from your high school mathematics. It may take the additional sophistication of a college student to see that the two definitions given below really describe the same ideas:

(I) A *function* (or mapping) of a set A into a set B is a correspondence such that for each $a \in A$, there corresponds one and only one element b of the set B. We often write $b = f(a)$.

(II) A *mapping* (or function) of a set A into a set B is a set S of ordered pairs (a, b), where $a \in A$ and $b \in B$, such that S contains *one and only one pair with the first element a* for each $a \in A$. We often write $a \to aS$ to indicate the mapping (function or correspondence) of the element $a \in A$ into the element $aS = b \in B$ and refer to aS as the image of a. We may write a representative ordered pair either as $[a, f(a)]$ or (a, aS).

The subset of B consisting of those elements $b \in B$ which are actually images of elements $a \in A$ is called the image set of A (or the graph of the function $[a, f(a)]$) and may be designated as AS. Clearly $AS \subseteq B$. If $AS = B$ (i.e. if $AS \subseteq B$ and $B \subseteq AS$) we may say that S is a mapping of A *onto* B rather than a mapping *into* B. An *onto* mapping may be further specialized as a *one-to-one* mapping of A onto B if each element b of B is the image of exactly one element of A. In this case, there also exists an *inverse mapping* S^{-1} of B onto A, namely $b \to bS^{-1} = a$ or in our former notation $aS \to a$, where $b = aS \in B$ and $a \in A$.

If R is the set of real numbers, the real function $y = x^2$ is represented by the mapping (x, x^2), that is by the "points" in $R \times R$ for which the second element is the square of the first. This function is merely "into," not "onto" and is not "one to one." The reader should verify these two statements. The function $y = x^3$ represented by the mapping (x, x^3) on the other hand is both "onto" and "one to one." Please verify these statements too; that is part of reading a mathematics text.

The thoughtful reader will discover that it is perfectly feasible to construct an algebra of mappings (functions). Two mappings S_1 and S_2 from A into B can be considered as equal if (and only if) $xS_1 = xS_2$ for every $x \in A$. This means that S_1 and S_2 have the same graph (image set) in $A \times B$. If M_1 is a mapping of A into B and M_2 is a mapping of B into some third set C, then the mapping M_3 which maps A directly into C such that $a \in A$ maps into the corresponding $(aM_1)M_2 \in C$ may be called the *product* of the two mappings M_1 and M_2 and written as $M_1 \otimes M_2 = M_3$. In this fashion we have a mathematical system in which the elements are mappings and which uses the definitions of equality of mappings and products discussed in this section.

If this seems difficult, stop for a moment and consider the technical vocabulary of advanced zoology or medicine or aviation. With experience you will become familiar with the mathematical vocabulary and eventually understand the ideas it expresses so well. A student often experiences confusion the first time he sees the ϵ, δ definition of

$$\lim_{x \to a} f(x) = L$$

but as he becomes more sophisticated he agrees that this is a perfectly natural and reasonable way of expressing the concept. *It is not essential that you understand the way in which ideas were expressed in the above section to continue in this text.* The purpose of this text is to get you ready to study from texts that use such vocabulary by introducing you *gradually* to the important concepts of abstract algebra. Indeed, if the above "explanation" seems the natural way to express the concepts involved, you should probably be studying from a more sophisticated text at this juncture. Your author recommends I. N. Herstein's *Topics in Algebra* (Waltham, Mass.: Blaisdell, 1964); S. MacLane's and G. Birkhoff's *Algebra* (New York: Macmillan, 1967); N. Jacobson's *Lectures in Abstract Algebra* (Princeton: Van Nostrand, Vol. I 1951, Vol. II 1953); and Bourbaki's *Algèbre* (Hermann, 1964 revised) for such sophisticated students. For those of us who are not quite that advanced yet, but who found Chapters 1 and 2 interesting, Section 2.6 contains several suggestions for further independent study which should, without undue strain, challenge and extend your developing mathematical maturity.

★2.6 SELECTED TOPICS FOR INDEPENDENT STUDY

It would defeat the purpose of this book to spend more time on the theory of numbers. Instead, brief suggestions of topics for outside reading and reports, along with a list of suitable books, are presented. The indexes of these, or other texts on number theory or modern abstract algebra, may be consulted. Your teacher may wish to let you study and report on one of these topics as a term project.

Subjects for Further Study

1. *The Euler ϕ function*, or *totient*, or *indicator function* is defined as: $\phi(N) = $ (the number of positive integers $\leq N$, which are relatively prime to N).

    Hence, $\phi(7) = 6, \quad \phi(8) = 4, \quad \phi(15) = 8.$

    The ϕ function is of considerable importance. The student may be interested in making and proving the validity of a conjecture concerning the existence or nonexistence of an *unbounded monotone increasing function* $F(N) < \phi(N)$. That is, does there exist a function $F(N)$ satisfying the three conditions:

    (1) If $K < L$, then $F(K) < F(L)$.
    (2) $\lim_{N\to\infty} F(N) = \infty$; that is, $F(N)$ may be made as large as desired.

    (3) For all positive integers N, $F(N) < \phi(N)$.
    You can have the thrill of doing some *original mathematical research* by solving this problem for yourself. It is not excessively difficult.

2. *The Fermat–Euler Theorem*
    "If $(a, m) = 1$, then $a^{\phi(m)} \equiv 1 \ (\mathrm{mod}\ m)$."
    The Fermat form states that, "if p is a prime and $(a, p) = 1$, then $a^{p-1} \equiv 1 \ (\mathrm{mod}\ p)$." The converse of Fermat's Theorem is not valid, since $3^{90} \equiv 1 \ (\mathrm{mod}\ 91)$, but $91 = 7 \cdot 13$ is not prime. Much has been written both on the converse of Fermat's Theorem and on Euler's Theorem. Either should provide an interesting report.

3. *The Chinese Remainder Theorem* is a special case of the following theorem.
    "A necessary and sufficient condition that n linear congruences

$$x \equiv a_1 \ (\mathrm{mod}\ m_1), \quad x \equiv a_2 \ (\mathrm{mod}\ m_2), \cdots, \quad x \equiv a_n \ \mathrm{mod}\ m_n$$

have a common solution is that

$$(m_i, m_j) \mid a_i - a_j \quad \text{whenever} \quad i \neq j$$

In this case, there is one common solution mod M where M is the least common multiple of all the m_i's."

This theorem determines the existence and number of solutions of a set of n linear congruences in n moduli and one unknown. Can you prove the stated theorem, without consulting any reference material, for the special case in which the modulii are relatively prime in pairs?

4. *Indices modulo p* are the theoretical counterpart, in congruences, of the ordinary arithmetic logarithm. Like logarithms, their most important role is in theory, not in computation, although both display computational advantages as well. For example, like $\log (ab) = \log a + \log b$, we have Ind $(ab) = $ Ind $a + $ Ind $b \pmod{p-1}$, or like $\log a^n = n \log a$, we have Ind $a^n = n$ Ind $a \pmod{p-1}$.

An excellent historical survey of the use of indices is presented in the foreword (by H. S. Vandiver) to R. V. Andree and others *Table of Indices and Power Residues for Primes and Prime Powers Less Than 2000*. This book should be available through your college library.

5. *Quadratic Residues and the Legendre Symbol*

If $(b, m) = 1$ and $x^2 \equiv b \pmod{m}$ has a solution, then b is called a quadratic residue; if $x^2 \equiv b \pmod{m}$ has no solution, then b is called a quadratic nonresidue. If $(b, m) \neq 1$, then b is neither a residue nor a nonresidue mod m. Indices are helpful in the theory of quadratic residues.

The Legendre Symbol $\left(\dfrac{b}{p}\right)$, where p is an odd prime, is defined to be

1 if b is a quadratic residue mod p; -1 if b a quadratic nonresidue mod p; and 0 otherwise. It leads to the quadratic reciprocity law.

6. *Quadratic Reciprocity Law*

Gauss is said to have claimed that, "Mathematics is the queen of the sciences; number theory is the crown upon her head; and the quadratic reciprocity law is the jewel in this crown." Some students may wish to have a closer look at this particular jewel.

Selected Reading List, Chapter 2
EQUIVALENCE AND CONGRUENCE

The following books and journals will provide additional reading for students interested in term projects or independent study.

Books

Andree, R. V. and others, *Table of Indices and Power Residues for Primes and Prime Powers Less than 2000*. New York: Norton, 1962.

Fraleigh, J. B., *A First Course in Abstract Algebra*. Reading, Mass.: Addison-Wesley, 1967.

Jones, B. W., *The Theory of Numbers*. New York: Holt, Rinehart and Winston, 1955.

Niven, I., and H. Zuckerman, *An Introduction to the Theory of Numbers*. New York: Wiley, 1960.

Maxfield, J. and M., *Discovering Number Theory*. Phila.: Saunders, 1972.

Shanks, D., *Solved and Unsolved Problems in Number Theory*. New York: Spartan, 1962.

Articles

Bachman, A. L., "Patterns in Algorithms for Determining Whether Large Numbers are Prime," *The Mathematics Teacher*, **63,** No. 1 (Jan. 1970), p. 30.

Cohen, E., "A Generalized Euler ϕ-Function," *Mathematics Magazine*, **41,** No. 5 (Nov. 1968), pp. 276–279.

Fraser, O. and B. Gordon, "On Representing a Square as the Sum of Three Squares," *The American Mathematical Monthly*, **76,** No. 8 (Oct. 1969), p. 922.

Klee, V., "Is There an n for Which $\phi(x) = n$ has a Unique Solution?," *The American Mathematical Monthly*, **76,** No. 3 (Mar. 1969), p. 288.

Perisastri, M., "On Fermat's Last Theorem," *The American Mathematical Monthly*, **75,** No. 6 (June–July 1969), p. 671.

Schwartz, B. L., "Mathematical Theory of Think-A-Dot," *Mathematics Magazine*, **40,** No. 4 (Sept. 1967), pp. 187–193.

"A Prime That can be put in Many Forms," Problem E 1922, *The American Mathematical Monthly*, **75,** No. 2 (Feb. 1968), p. 193.

"100th Power Residues," Problem E 2146, *The American Mathematical Monthly*, **76,** No. 9 (Nov. 1969), p. 1071.

Notes on Chapter 2

BOOLEAN ALGEBRA

3

3.1 INTRODUCTION

We now turn our attention to the subject of Boolean algebra—the algebra of switching circuits and the algebra of logic. You may already have studied Venn diagrams or Euler circles in your earlier mathematics.

If the pattern  or looks familiar to you, so much the better. See how quickly you can associate your knowledge with the work presented in this chapter. Much of the original work in this highly practical area of mathematics was done by the Irish mathematician George Boole (1815–1864) as part of his study of mathematical logic.

3.2 DUALITY

The *principle of duality* is of considerable importance in mathematics. It is not, strictly speaking, a theorem of algebra or geometry in the usual sense; instead, it is a theorem about theorems. It belongs to "metamathematics" or "logical syntax," a study of prime importance in modern symbolic logic.

The *dual* of a statement is that statement obtained from the original by interchanging certain pairs of words or phrases.

The interchange red $\leftrightarrow$ white means, "Whenever the word *red* appears, substitute the word *white*, and whenever the word *white* appears in the original, substitute the word *red*." Thus:

The girl in the *red* and blue dress has a corsage of *white* roses

has as its dual, under the interchange red $\leftrightarrow$ white:

The girl in the *white* and blue dress has a corsage of *red* roses.

In algebra, theorems on inequalities (deduced from the postulates of partially ordered systems) remain provable from the postulates if the symbols $\leq$ and $\geq$ are interchanged throughout the statement of the theorem. For example:

$$\text{If} \quad a \leq b, \quad \text{then} \quad a + c \leq b + c$$

has as its dual under the interchange $\leq \leftrightarrow \geq$

$$\text{If} \quad a \geq b, \quad \text{then} \quad a + c \geq b + c.$$

If the words "line" and "point" are interchanged, the geometric postulate

Two distinct points determine a unique line.

has as its dual under line $\leftrightarrow$ point,

Two distinct lines determine a unique point.

In this case the "dual" is *not* a postulate of *ordinary Euclidean geometry*, since distinct lines may be parallel. However, both are usual postulates in projective geometry, in which each postulate and each theorem has a valid dual obtained by interchanging the words "line" and "point." The dual of each provable theorem is also provable, using duals in the same proof structure.

The reader should notice that, with respect to a given set of interchanges, the dual of a dual is the original theorem. (Why?) In a mathematical system in which the dual of each postulate is also a postulate or a provable theorem, much time can be saved by employing the principle of duality. *If every postulate of the system has its dual in the system, then the dual of each theorem is also valid.* (The same proof structure, using duals, yields the dual theorem.) The dual of *each* postulate must also be a postulate or be provable from the postulates, before the principle of duality is used.

If "0" and "1" are interchanged, and at the same time "+" and "·" are interchanged, then

Postulate		*Dual*
$1 \cdot 1 = 1$	becomes	$0 + 0 = 0$
$1 \cdot 0 = 0 \cdot 1 = 0$	becomes	$0 + 1 = 1 + 0 = 1$

The duality interchange $[0 \leftrightarrow 1, + \leftrightarrow \cdot]$ will be employed in this chapter.

3.3 BINARY BOOLEAN ARITHMETIC

The familiar variables of ordinary algebra may take many different values. Binary Boolean algebra is a much simpler algebra in which the variables may assume only two possible values, namely 0 or 1. Two operations, addition $(+)$ and multiplication $(\cdot)$, will be permitted, and the following laws of arithmetic will be used:

(1) $0 \cdot 0 = 0$
$1 + 1 = 1$ dual postulates.

(2) $1 \cdot 1 = 1$
$0 + 0 = 0$ dual postulates.

(3) $1 \cdot 0 = 0 \cdot 1 = 0$
$0 + 1 = 1 + 0 = 1$ dual postulates.

These laws all "look normal," with the possible exception of the second part of (1), namely $1 + 1 = 1$. However, whether or not they "look normal" is unimportant. These are the laws of arithmetic which were postulated. The problem is to see what theorems can be derived for an algebra whose variables are governed by this arithmetic.

It may help you to feel at ease with these postulates if you think of 1 as representing the universe (that is, "everything" or "all"), and 0 as representing a void (that is, "nothing"). Then it seems more reasonable to say,

$$(\text{everything}) + (\text{everything}) = (\text{everything})$$

$$1 \quad + \quad 1 \quad = \quad 1$$

since there is no class larger than "everything."

Similarily, in an electrical circuit containing two switches in parallel

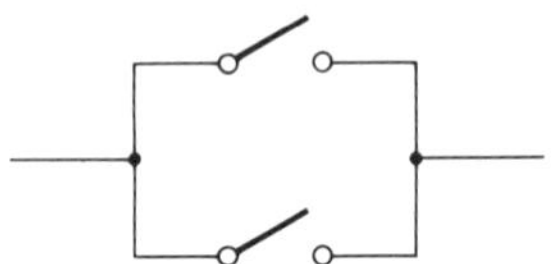

we again have

$$ON + ON = ON$$

$$1 + 1 = 1$$

since the current does *not* double if both switches are closed. In many applications the postulate $1 + 1 = 1$ is indeed meaningful.

3.4 BOOLEAN FUNCTIONS

In ordinary algebra, it is common practice to evaluate a function, say $f(x, y) = x^2 - 3x + 2y - 7$, by direct substitution, obtaining $f(1, 3) = 1 - 3 + 6 - 7 = -3$. However, it is not feasible to "prove" an algebraic or trigonometric identity by substituting *all possible* values of the unknowns, since there may be infinitely many values. In binary Boolean arithmetic and algebra, each variable may take on only two possible values. A function $B(x, y)$ of two variables has only four possible choices for the variables x and y, namely

x	y
0	0
0	1
1	0
1	1

Examining the Boolean polynomial $B(x, y) = x + (x \cdot y)$ in the four possible cases, one obtains

x	y	$x + (x \cdot y)$
0	0	$0 + (0 \cdot 0) = 0$
0	1	$0 + (0 \cdot 1) = 0$
1	0	$1 + (1 \cdot 0) = 1$
1	1	$1 + (1 \cdot 1) = 1$

Since the last column is identical with the first column, the unlikely looking theorem

$$x + (x \cdot y) = x$$

is valid in Boolean algebra.

3.5 ISOMORPHIC SYSTEMS

Let us examine two switching arrangements common in electrical circuits.

Two switches in series. Output only if *both x and y* are closed.

Two switches in parallel. Output if either *x or y or both* are closed.

Four tables are presented here in summary:

Series Switches

x	y	circuit
open	open	open
closed	closed	closed
closed	open	open
open	closed	open

Parallel Switches

x	y	circuit
open	open	open
closed	closed	closed
closed	open	closed
open	closed	closed

Arithmetic under $\cdot$

x	y	$x \cdot y$
0	0	0
1	1	1
1	0	0
0	1	0

Arithmetic under $+$

x	y	$x + y$
0	0	0
1	1	1
1	0	1
0	1	1

Please check all four of the above tables to see that they *do* represent the stated situations. Note the similarities between the switching tables and the arithmetic tables. When you have done this, you will discover that, if 0 is substituted for "open" and 1 for "closed," then the *switches in series table* becomes the *multiplication table*, while the *parallel switches* may be represented by *addition*. This suggests an important application of the Boolean algebra. There are other equally important applications in set theory, logic, and in language (analysis of contracts and laws).

The mathematician says that, under the following correspondence, an *isomorphism* exists between the binary Boolean algebra and the electrical circuits:

Switches		*Arithmetic*
open	$\leftrightarrow$	0
closed	$\leftrightarrow$	1
series	$\leftrightarrow$	$\cdot$
parallel	$\leftrightarrow$	$+$

This means that the systems are abstractly identical.

A mechanical system isomorphic to Boolean algebra is obtained by considering networks of pipes containing stopcocks and flowing liquid or gas.

An isomorphism also exists between Boolean algebra and certain very general types of statements involving "and" and "or." The expression $(A \cdot B)$ means "both A and B." The "or" involved in $(A + B)$ is the *in*clusive "A or B or both" type. An *exclusive* "or" is indicated by $(A + B) \cdot [\text{not } (A \cdot B)]$. Later, when a "not function" is introduced, important consequences in the analysis of written contracts and laws will become evident. A simple illustration involves two insurance contracts:

(1) Policy will pay if insured is killed in an accident involving a car *and* a train. $(C \cdot T)$

(2) Policy will pay if insured is killed in an accident involving a car *or* a train (or both). $(C + T)$

$$\text{"and"} \leftrightarrow \, \cdot \, \leftrightarrow \text{series circuits}$$

$$\text{"or"} \leftrightarrow + \leftrightarrow \text{parallel circuits}$$

In the first case, money will "flow" only if both a car *and* a train are involved. In the second case, money will "flow" if *either* a car *or* a train (or both) is involved. The analogy between this and the electrical or mechanical systems already mentioned should be clear. Each is represented by one of the two Boolean polynomials $x \cdot y$ and $x + y$.

In ordinary algebra it is usual to save notation by adopting the convention that multiplication shall take precedence over addition. Thus,

$$x + y \cdot z = x + (y \cdot z), \quad \text{not} \quad (x + y) \cdot z$$

We adopt the same convention here for similar reasons.

Problem Set 3.5

Evaluate the following Boolean functions for (a) $x = 0, y = 0, z = 1$; (b) $x = 0, y = 1, z = 0$; (c) $x = 1, y = 0, z = 1$; (d) $x = 1, y = 1, z = 1$. The results should be either 0 or 1 in each case.

1. $x + y$.   2. $x \cdot y$.   3. $(x + y) + y \cdot z$.

4. $x \cdot y + x \cdot z$.   5. $(x + y) \cdot (x + z)$.   6. $x + y \cdot z$.

7. $x \cdot (x + x \cdot x) + x \cdot y$.   8[1]. $z \cdot (\overline{x + y + z})$.   9. z.

---

[1] The American reader is already familiar with the use of a bar (vinculum) over quantities as a method of grouping in the square root symbol. Thus

$$\sqrt{A^2 + B^2} = (A^2 + B^2)^{1/2}.$$

10. Show that the polynomials given in Problems 5 and 6 determine equivalent Boolean functions.

11. Show that the polynomials given in Problems 8 and 9 determine equivalent Boolean functions.

12. Construct a diagram for a simple "electrical brain" which will tell you when the following insurance policy will pay off:

    Three partners, George, Bob, and Sam, take out an insurance policy which is to pay \$50,000 into the partnership treasury if either George or Bob dies, providing that either George or Sam also dies. [HINT: $(G + B) \cdot (G + S)$.]

13. Are the conditions for payment given in Problem 12 the same as any of the following?

    (a) Pays if, and only if, two of the three die.
    (b) Pays if, and only if, George dies.
    (c) Pays if George dies, or if both Bob and Sam die.
    (d) Pays if two of the three die, except that if both Bob and Sam die it does not pay off.

14. Show that the postulates of Section 3.3 occur in dual pairs.

15. Construct electrical network diagrams for $x + x \cdot y$, and show that the unlikely looking theorem proved in Section 3.4 is applicable to electrical networks.

16. An insurance company agrees to pay a \$10,000 claim either if both the insured and his wife die, or if the insured dies. Use this to illustrate the theorem $x \cdot y + x = x$ of Section 3.4.

17. How much higher a premium should be charged for the policy of Problem 16 than for a straight policy on the life of the insured man?

3.6 BINARY BOOLEAN ALGEBRA

If you will examine the postulates of Boolean arithmetic, you will note that each lettered postulate contains two parts which may be obtained from one another by interchanging 0 and 1 and at the same time interchanging $\cdot$ and $+$.

Under the interchange

$$0 \leftrightarrow 1$$

$$+ \leftrightarrow \cdot$$

(1) $0 \cdot 0 = 0$ becomes $1 + 1 = 1$
(3) $1 \cdot 1 = 1$ becomes $0 + 0 = 0$
(3) $1 \cdot 0 = 0 \cdot 1 = 0$ becomes $0 + 1 = 1 + 0 = 1$.

If, conversely, the suggested substitution is made in the second expression, the first will result. Since the dual of each postulate is also a postulate, it follows that if a theorem is proved from these postulates, then the dual of the theorem is also proved.

Postulates of Binary Boolean Arithmetic

(1) $0 \cdot 0 = 0$
    $1 + 1 = 1$ dual postulates.
(2) $1 \cdot 1 = 1$
    $0 + 0 = 0$ dual postulates.
(3) $1 \cdot 0 = 0 \cdot 1 = 0$
    $0 + 1 = 1 + 0 = 1$ dual postulates.

We are now ready to consider an algebra based on the above arithmetic.

The variables in binary Boolean algebra are restricted to the values 0 and 1 of the Boolean arithmetic. It is actually possible to prove, by exhaustion, certain theorems which it would be impossible to prove if the arithmetic were infinite. These would need to be postulated in an infinite system.

THEOREM 3.1a $x + y = y + x$.

Commutative Laws

THEOREM 3.1b $x \cdot y = y \cdot x$.

The proof of Theorem 3.1a consists of listing all possible cases and showing that in each case $x + y = y + x$.

x	y	$x + y$	$y + x$
0	0	0	0
0	1	1	1
1	0	1	1
1	1	1	1

Theorem 3.1b is obtained from Theorem 3.1a by the interchange $\cdot \leftrightarrow +$; i.e., 3.1b is the dual of 3.1a. Hence, Theorem 3.1b also is proved. Since algebraic theorems hold for all values of x and y for which the members are defined, it is unnecessary to use the interchange $0 \leftrightarrow 1$; or rather, when the interchange $0 \leftrightarrow 1$ is used, no change is apparent in the statement of the theorem. (This is an interesting arrangement; be sure you understand it.)

THEOREM 3.2a $(x + y) + z = x + (y + z)$.

Associative Laws

THEOREM 3.2b $(x \cdot y) \cdot z = x \cdot (y \cdot z)$.

Theorem 3.2a may be proved by listing the eight possible values of (x, y, z) and showing that the polynomials are indeed identical. The dual of Theorem 3.2a is obtained by the interchange $+ \leftrightarrow \cdot$, giving $(x \cdot y) \cdot z = x \cdot (y \cdot z)$, as desired.

If you prefer, 3.2b can be proved by considering eight possible cases.

The expression $(x + y + z)$ is defined as either one of the equal expressions $(x + y) + z$ or $x + (y + z)$. This expression was undefined prior to this. (Why?) In a similar fashion, $(x \cdot y \cdot z)$ is given meaning. These definitions may be extended, by induction, to $(x + y + z + w + \cdots + t)$ and $(x \cdot y \cdot z \cdot w \cdot \cdots \cdot t)$, where the number of terms or factors involved is arbitrary, but finite.

Now that the commutative and associative laws of ordinary algebra have been established as valid in Boolean algebra, let us examine another formative rule—the distributive law.

THEOREM 3.3a $x \cdot (y + z) = (x \cdot y) + (x \cdot z)$.

Surely this old friend should also hold in Boolean algebra. Before we go about a proof, let us examine the dual of the above statement, namely

THEOREM 3.3b $x + (y \cdot z) = (x + y) \cdot (x + z)$.

The expression $x + (y \cdot z)$ could be written as $x + y \cdot z$ in accord with the convention discussed just before Problem Set 3.5.

Theorem 3.3b may appear unlikely. Certainly it is not true in ordinary algebra. For this reason, Theorem 3.3b will be proved directly, and Theorem 3.3a obtained as its dual. The alternate procedure is equally valid. The proof is again by exhaustion of cases, a method which is not available in ordinary algebra. (Why not?)

x	y	z	$y \cdot z$	$(x + y)$	$(x + z)$	$x + (y \cdot z)$	$(x + y) \cdot (x + z)$
0	0	0	0	0	0	0	0
0	0	1	0	0	1	0	0
0	1	0	0	1	0	0	0
0	1	1	1	1	1	1	1
1	0	0	0	1	1	1	1
1	0	1	0	1	1	1	1
1	1	0	0	1	1	1	1
1	1	1	1	1	1	1	1

The last two columns are identical. Hence, Theorem 3.3b is proved, and Theorem 3.3a, its dual, follows at once.

Consider the meaning of Theorem 3.3b in terms of isomorphic electrical networks.

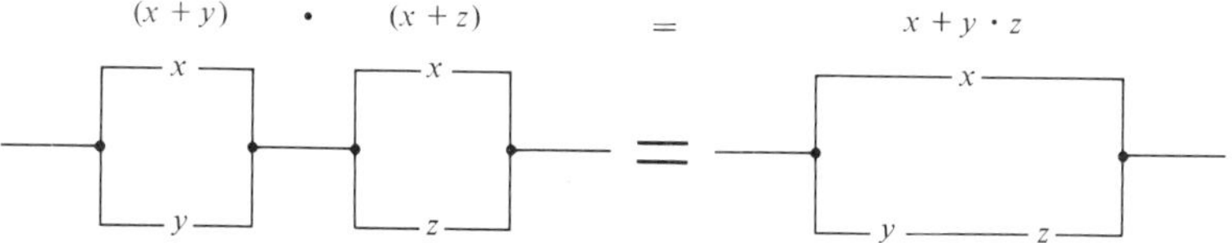

The second circuit is mechanically simpler than the first, and hence offers some advantage in electrical circuit theory.

By examining the two networks for all possible conditions of the three switches x, y, and z, it will be found that the left network is closed (or open) for exactly the same conditions as the right network, and hence the two networks are equivalent. That is, they perform the same switching operations. It is not necessary to actually check this statement. The check is the same as the algebraic proof by cases. It may aid your understanding to work out three or four of the eight possible situations. Problem 12, Set 3.5 provides an example of a language application of Theorem 3.3b.

Let us examine Theorem 3.3b in more detail:

$$x + y{\cdot}z = (x + y) \cdot (x + z)$$

In ordinary algebra the right member $(x + y) \cdot (x + z)$ equals $x{\cdot}x + x{\cdot}z + y{\cdot}x + y{\cdot}z$. *We do not know that this is valid in Boolean algebra.* However, since we are searching for possible additional theorems to prove or disprove, let us see what theorems the above equality might suggest— after all, one must get ideas for possible theorems to prove from some- where—they don't grow on theorem bushes.

If it were true that (Note that Theorem 3.1b has already been used.)

$$x + y{\cdot}z \overset{?}{=} x{\cdot}x + x{\cdot}z + x{\cdot}y + y{\cdot}z$$

it *might* (we haven't proved it yet, we are just looking for propitious hunches)—it just might be possible that

$$\text{Guess 1:} \qquad x = x{\cdot}x$$

and

$$\text{Guess 2:} \qquad y{\cdot}z = x{\cdot}z + x{\cdot}y + y{\cdot}z$$

Certainly if these equalities were true, Theorem 3.3b would follow, but just because we know Theorem 3.3b to be true does *not* guarantee that either Guess 1 or Guess 2 is valid. Let us examine Guess 1, $x{\cdot}x = x$, and its dual, $x + x = x$. Neither really looks very likely from our experience in ordinary algebra, but it won't take long to examine them.

x	$x{\cdot}x$	$x + x$
0	0	0
1	1	1

Thus both Guess 1, $x = x{\cdot}x$, and its dual, $x = x + x$, have been verified by direct examination of all possible cases. It was, of course, unnecessary to prove both Guess 1 and its dual (why?), but we didn't waste much time. We state the theorems just proved as

THEOREM 3.4a $x + x = x.$

THEOREM 3.4b $x \cdot x = x.$

Theorem 3.4b follows as the dual of 3.4a. You will be asked to set up the equivalent electrical networks in the next problem set—try it now. The interested reader should note how naturally these theorems arise in the language interpretation. Let x be the statement, "John will come" $(x \text{ or } x) = x \leftrightarrow$ Either John will come or John will come = John will come. $(x \text{ and } x) = x \leftrightarrow$ John will come and John will come = John will come.

Theorems 3.4 explain why exponents and coefficients other than 0 and 1 are unnecessary in Boolean algebra.

Inspired by our success we hasten to examine Guess 2.

$$\text{Guess 2:} \quad y \cdot z = x \cdot z + x \cdot y + y \cdot z$$

Since each member contains the terms $y \cdot z$, we may hypothesize at once that

$$\text{Guess 3:} \quad 0 = x \cdot z + x \cdot y$$

Unfortunately, Guess 3 is invalid. A moment's reflection will suggest several counterexamples, such as $(x, y, z) = (1, 0, 1)$ or $(x, y, z) = (1, 1, 1)$, either of which eliminates Guess 3 from the realm of possible theorems of Boolean algebra. If Guess 3 had been valid, then guess 2 would also, but the fact that Guess 3 is invalid does *not* imply that Guess 2 is also invalid. (Why not?)

What about Guess 2 then? Actually Guess 2 is valid and can be proved. We leave it as an exercise for the reader (Set 3.6, Problem 10).

Let us turn our attention instead to the original tentative equality from which these guesses originated.

$$x + y \cdot z \overset{?}{=} x \cdot x + x \cdot z + x \cdot y + y \cdot z$$

Possibly another grouping of the right hand members would be fruitful. Using Theorems 3.3a, 3.2a, 3.1a, and 3.4b we may rearrange the right hand member to obtain

$$x + y \cdot z = x \cdot (x + y + z) + y \cdot z$$

Possibly the entire underlined portion $x \cdot (x + y + z)$ might equal x. Indeed it does! If $x = 1$, then $x \cdot (x + y + z) = 1$ no matter what values y and z may have. Similarly if $x = 0$, so does $x \cdot (x + y + z)$ irrespective of the values of y and z. The result is apparently independent of y and z therefore the simpler expression

$$x \cdot (x + y) = x$$

conveys the same meaning as the one just discussed. We present it, and its dual as Theorems 3.5a and 3.5b. The important thing is not so much the proofs of the theorems, which are fairly easy, but rather to notice how experience and intuition played a vital role in helping us discover new *possible* theorems to be proved or disproved. Just because your intuition says something looks good or bad is *not* sufficient reason to accept or discard it without a proof or counterexample. Guess 3 seems as likely a consequence of our original hypothesis as Guess 2, but one is valid while the other is not.

THEOREM 3.5a $x + x \cdot y = x.$

THEOREM 3.5b $x \cdot (x + y) = x.$

The proof is left for the reader. The electrical circuit interpretation of Theorem 3.5a makes the algebraic theorem seem less unusual. (Compare last paragraph of Section 3–2.)

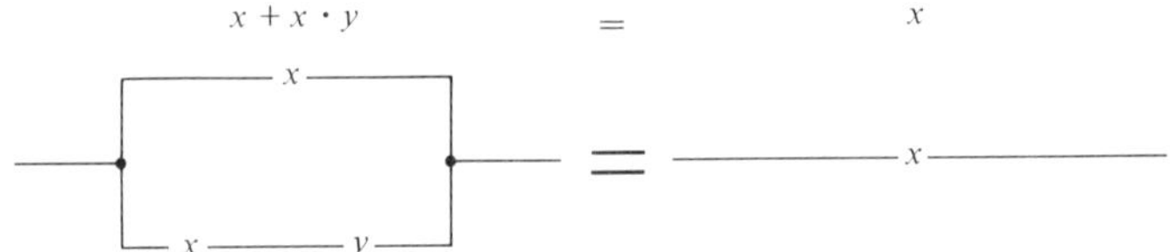

Problem Set 3.6

1. Prove Theorem 3.1b directly by examining all possible cases.

2. Prove Theorem 3.2a by direct substitution and obtain Theorem 3.2b as its dual.

3. Prove either Theorem 3.5a or Theorem 3.5b, and show that the other is its dual.

4. Express the following as Boolean polynomials:

    (a) A and $[B$ or $(C$ and $A)]$ and $(C$ or $B)$.
    (b) $[(A$ and $B)$ or $\overline{(C \text{ and } A) \text{ and } C}]$ or B.
    (c) A and $(B$ or $C)$.

5. Are any of the statements of Problem 4 equivalent? Prove your assertion using Boolean algebra.

6. Set up electrical networks representing Theorems 3.5a and 3.5b.

7. Which theorem is represented by:

    (a) "David kissed Helen" and "Either David kissed Helen or Peter kissed Helen" imply "David kissed Helen."
    (b) "Bob will eat ice cream" and "Bob or Mary will eat ice cream" imply "Bob will eat ice cream."

8. Try to prove Theorems 3.5a and 3.5b by the use of postulates and previously proved theorems, rather than by enumeration of cases.

9. Which of the integral domain postulates are satisfied in Boolean algebra?

10. Prove the validity of Guess 2, preferably by using Theorems 3.1–3.5 rather than by examination of cases.

11. Prove that the original tentative hypothesis of this section, namely that $(x + y) \cdot (x + z) = x \cdot x + x \cdot z + y \cdot x + y \cdot z$ is also valid. Be careful that you use the theorems of Boolean algebra, not of high school algebra. Place the theorem numbers used in your proof next to each step of the proof.

3.7 THE NEGATIVE OR COMPLEMENTARY OR RELAY RELATIONSHIP

In ordinary language and in legal contracts, the negative of a statement often appears. Denote the negation of an assertion by adding a prime:

$(A)'$ = (John will come)$'$ = John will *not* come.
$(B)'$ = (The explosion will occur)$'$ = The explosion will *not* occur.
$(A')'$ = not (not A) = A = John will come.
$(A + B)'$ = not (A or B) = (not A) and (not B) = $A' \cdot B'$.

Give the linguistic interpretation of $(A + B)' = A' \cdot B'$ using the A and B statements given above.

The new operator of negation introduced into our Boolean algebra will prove useful in electrical and mechanical systems as well as in linguistic analysis. It has other applications with mathematical and physical interpretations. The notations $-A$ and "complement A" are sometimes used to express A'.

A (normally closed) relay has two positions, a make contact position (x) and an open position (x'). Consider these positions to be the complement (or possibly the negative) of one another.

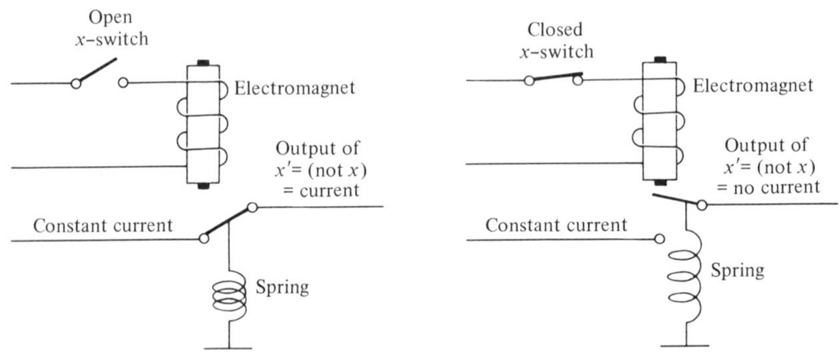

Actually, it is not essential to think of switches as carrying electricity at all. One doesn't step on the starter in his car to permit electrons to flow around a certain circuit—that is just incidental. One steps on the starter to start the motor. These circuits have been presented in terms of switches and relays, which is a valid, but not a very general, representation. Mechanical switches and relays are slow. Electronics is faster and hence less "hardware" is required to do the same job. Relays have been replaced by vacuum tubes. Vacuum tubes have been replaced by diodes and transistors, or by ferrous oxide compounds having a wide hysteresis loop (Westinghouse CYPAC, for example). These are now being replaced by deposited circuits. *The actual implementing hardware is unimportant at this time.* It will undoubtedly change again in the next decade, but the mathematics will remain valid.

THEOREM 3.6 $(x')' = x.$

The proof, in terms of the binary Boolean arithmetic, is contained in the following table:

x	x'	$(x')'$
0	1	0
1	0	1

In terms of circuits, a switch that is changed, and then changed again, has been returned to its original state. Theorem 3.6 is self-dual.

THEOREM 3.7a $(x + y + z + \cdots)' = x' \cdot y' \cdot z' \cdot \,\cdots.$

This theorem is proved by exhaustion, combined with an induction on the number of variables. In terms of circuit theory, the prime indicates that

$$(\text{circuit})' \text{ is open whenever (circuit) is closed,}$$

$$\text{and}$$

$$(\text{circuit})' \text{ is closed if (circuit) is open.}$$

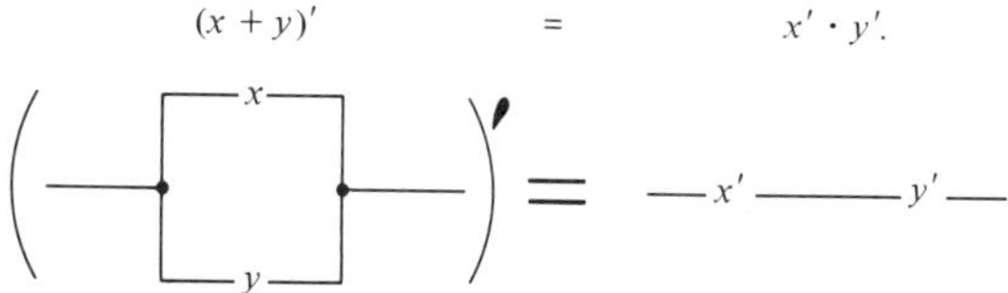

This is useful in constructing complementary circuits, and is often called De Morgan's Theorem. A language interpretation of $(x + y)' = x' \cdot y'$ was given at the beginning of Section 3.7. The dual of Theorem 3.7a is given in Theorem 3.7b.

THEOREM 3.7b $(x \cdot y \cdot z \cdot \,\cdots)' = x' + y' + z' + \cdots.$

EXAMPLE 1

$$
\begin{aligned}
[(x + y' \cdot z) \cdot (v + w')]' &= (x + y' \cdot z)' + (v + w')', \text{ by 3.7b.} \\
&= x' \cdot (y' + z)' + (v' \cdot w''), \text{ by 3.7a.} \\
&= x' \cdot (y + z') + v' \cdot w, \text{ by 3.7b and 3.6.} \\
&= x' \cdot y + x' \cdot z' + v' \cdot w, \text{ by 3.3a.}
\end{aligned}
$$

THEOREM 3.8a $x + x' = 1.$

THEOREM 3.8b $x \cdot x' = 0.$

THEOREM 3.9a $0 + x = x.$

THEOREM 3.9b $1 \cdot x = x.$

If one sets $x = 1$ in Theorem 3.5a $x + x \cdot y = x$, the statement becomes $1 + 1 \cdot y = 1$. By use of Theorem 3.9b one obtains $1 + y = 1$. Since y is a variable, this can just as well be stated as $1 + x = 1$. We present it with its dual below. Possibly the dual $0 \cdot x = 0$ has already occurred to the thoughtful reader as a likely theorem.

THEOREM 3.10a $1 + x = 1.$

THEOREM 3.10b $0 \cdot x = 0.$

Since, by Theorem 3.8b we have that $x \cdot x' = 0$, the use of this along with Theorem 3.3a and 3.9a produces 3.11a with its interesting and useful dual.

THEOREM 3.11a $x \cdot (x' + y) = x \cdot y.$

THEOREM 3.11b $x + (x' \cdot y) = x + y.$

Theorems 3.8a, 3.9a, and 3.10a may be proved directly, with the b parts obtained by the duality principle.

Theorem 3.11a may also be proved by considering the four possible cases. Instead, a proof which rests upon the previously proved Theorems 3.1–3.10 is given.

$$x \cdot (x' + y)$$

$x \cdot x' + x \cdot y$	by 3.3a
$0 \quad + x \cdot y$	by 3.8b
$x \cdot y$	by 3.9a

As usual, the b part follows by the principle of duality. The reader should be sure that he can show that each step in the above proof is actually justified *by the postulates and theorems already proved,* and that Theorem 3.11b is actually the dual of 3.11a.

In terms of network diagrams, Theorem 3.11b becomes

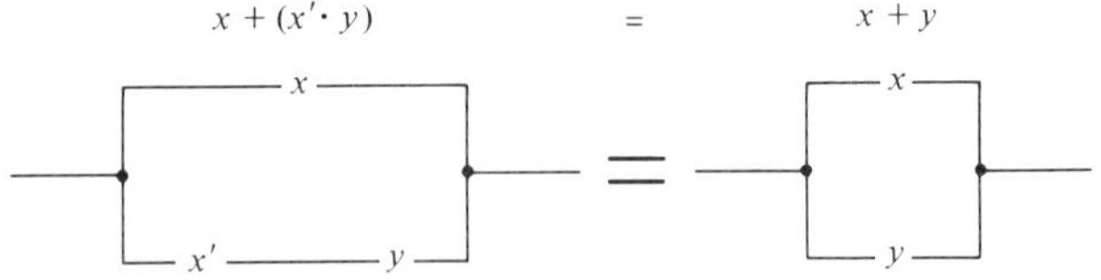

Since the existing hardware needed to accomplish x' (that is, "not x") is more complex than a simple x switch, the saving is greater than is at first apparent.

Theorems 12 and 13 are presented without suggestions concerning their origin. The reader is encouraged to speculate on how they might have been suggested if the author had not handed them to you on a silver platter.

THEOREM 3.12a $\qquad (x + y) \cdot (x' + z) \cdot (y + z) = (x + y) \cdot (x' + z).$

THEOREM 3.12b $\qquad x \cdot y + x' \cdot z + y \cdot z = x \cdot y + x' \cdot z.$

Which part of Theorem 3.12 is isomorphic to the following network?

THEOREM 3.13 $\qquad (x + y) \cdot (x' + z) = x \cdot z + x' \cdot y.$

The reader is asked to prove Theorem 3.13 by using previous theorems. It is interesting to note that the dual of Theorem 3.13 is again Theorem 3.13, in different notation.

Problem Set 3.7

1. Prove Theorem 3.5b from 3.5a and previous results rather than by enumeration of cases.

2. Prove Theorems 3.8. Give an interpretation of these theorems.

3. (a) Prove Theorems 3.9.
   (b) Prove Theorems 3.10.

4. Prove Theorems 3.11 directly, and compare the work involved by this method and the method of the text.

5. Construct electrical networks, or language statements, represented by Theorems 3.8.

6. Construct electrical networks, or language statements, represented by Theorems 3.9.

7. Construct electrical networks, or language statements, represented by Theorems 3.10.

8. Construct electrical networks, or language statements, represented by Theorems 3.11.

9. Prove Theorem 3.12a or Theorem 3.12b, and show that the other is its dual.

10. Prove Theorem 3.13 by using some of Theorems 3.1–3.12.

11. Show that Theorem 3.13 is self-dual.

12. Give an electrical network, or a language statement, representing:

   (a) Theorem 3.12a;
   (b) Theorem 3.13.

13. Give an electrical network, or a language statement, represented by:

   (a) Theorem 3.12b;
   (b) $x \cdot y = x \cdot (x' + y)$;
   (c) $(x \cdot y \cdot z) + (x \cdot y' \cdot z) + (x' \cdot y' \cdot z)$. Can you simplify this circuit?

14. Show that, since $A \cdot B = (A' + B')'$, it would be possible to construct any Boolean polynomial using the two symbols $+$ and $'$ rather than all three symbols $+$, $\cdot$, $'$.

15. Show that, since $A + B = (A' \cdot B')'$, it is possible to use only $\cdot$ and $'$ in place of $+$, $\cdot$, $'$.

16. Show that Theorem 3.12b can be obtained as a corollary of Theorem 3.13. [HINT: Expand the left member of Theorem 3.13.]

17. Theorem 3.8a, $x + x' = 1$, and Theorem 3.8b, $x \cdot x' = 0$, express the logical notions (a) "Either a statement is valid or it isn't—there is no other possibility" and, (b) "It is impossible for a statement to be both valid and not valid." Explain in a few sentences how the algebraic equations convey this interpretation.

18. Let 0 stand for a statement which is never true, while 1 is interpreted as a statement which is always true. Give a language interpretation of Theorems 3.9a and 3.9b.

19. Same as Problem 18 for Theorems 3.10.

20. The English sentence, "Either Mary will come or Mary will not come and Helen will come," is open to the following two interpretations,

$$M + (M' \cdot H)$$

$$(M + M') \cdot H$$

depending upon the insertion of a comma after the first or second appearance of the word "come." Discuss the two interpretations. Are they the same? What does each mean in terms of the number of persons expected?

21. Burton is going to a football game. He promises that he will take either Vincie or Stevie to the game with him and he promises Peggy that either he will take her to the game or he will leave Vincie home.

    (a) Make a Boolean polynomial which represents the above situation.
    (b) Find a theorem of this chapter which contains as one member, a polynomial equivalent to the one you derived in (a).
    (c) Give a verbal interpretation, in terms of this problem, of the other member of the theorem found in (b).
    (d) If Burton is true to his promises, can he take only one child to the game? Does it make a difference which one?
    (e) Burton is a bit weak on logic and Boolean algebra, and takes all three children to the game. The children are happy, but has he kept his word?

22. Problem 9, Set 3.5, asked which integral domain postulates are satisfied by Boolean algebra. You may be able to find some additional postulates by considering the not ′ relationship as an inverse. Try it and see.

3.8 APPLICATIONS TO ELECTRICAL NETWORKS

Let us examine a few circuits (Boolean polynomials) to see how these theorems are applied in actual simplifications. Remember that each letter or prime represents a relay, switch, or other piece of electrical hardware and that it is of economic interest to reduce the amount of hardware needed to do the job. The reader should draw the corresponding circuit diagrams.

EXAMPLE 2

$$(x + y) + x \cdot z =$$

$(x + x \cdot z) + y =$ by Theorems 3.1 and 3.2

$x + y$ by 3.5a

EXAMPLE 3

$$(A \cdot B + C')' + A' \cdot C + B =$$

$$(A \cdot B)' \cdot (C')' + A' \cdot C + B = \qquad \text{by 3.7a}$$

$$(A' + B') \cdot C + A' \cdot C + B = \qquad \text{by 3.7b and 3.6}$$

$$A' \cdot C + B' \cdot C + A' \cdot C + B = \qquad \text{by 3.3a and 3.1b}$$

$$A' \cdot C + B' \cdot C + B = \qquad \text{by 3.4a}$$

$$A' \cdot C + C + B = \qquad \text{by 3.11b and 3.1b}$$

$$C + B \qquad \text{by 3.5a}$$

Hence, the rather complicated network of the original is replaced by a simple parallel circuit.

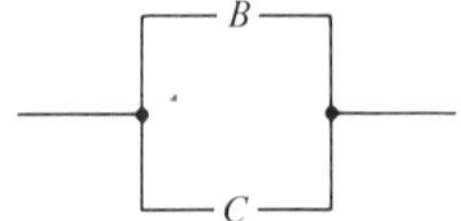

As in proving trigonometric identities, no fast rules can be given for simplification, but several useful principles can be noted. As in Example 2, it is often helpful first to remove the primes from the outside of parentheses.

EXAMPLE 4

$$A \cdot (B + C') + A' \cdot (B' \cdot C') + A \cdot B \cdot (B' + D) =$$

$$A \cdot B + A \cdot C' + A' \cdot (B' \cdot C') + (A \cdot B) \cdot B'$$
$$+ (A \cdot B) \cdot D = \text{by 3.3a}$$

$$A \cdot B + A \cdot C' + (A' \cdot B') \cdot C' + \underline{A \cdot (B \cdot B')}$$
$$+ (A \cdot B) \cdot D = \text{by 3.2b}$$

$$\underline{A \cdot B} + A \cdot C' + (A' \cdot B') \cdot C' + \underline{(A \cdot B) \cdot D} = \qquad \text{by 3.8b, 3.10b, 3.9a}$$

$$A \cdot B + A \cdot C' + (A' \cdot B') \cdot C' = \qquad \text{by 3.5a with } x = A \cdot B$$

$$A \cdot B + (A + A' \cdot B') \cdot C' = \qquad \text{by 3.3a, 3.1b}$$

$$A \cdot B + (A + B') \cdot C' = \qquad \text{by 3.11b}$$

$$A \cdot B + A \cdot C' + B' \cdot C' = \qquad \text{by 3.3a and 3.1b}$$

$$A \cdot B + B' \cdot C' \qquad \text{by 3.1 and 3.12b with}$$
$$x \to B, y \to A, z \to C'$$

The reader should show that this is also equivalent to $(A + B') \cdot (B + C')$ by use of Theorem 3.13. The circuit diagrams are included for the benefit of interested students.

$$A \cdot (B + C') + A' \cdot (B' \cdot C') + A \cdot B \cdot (B' + D)$$

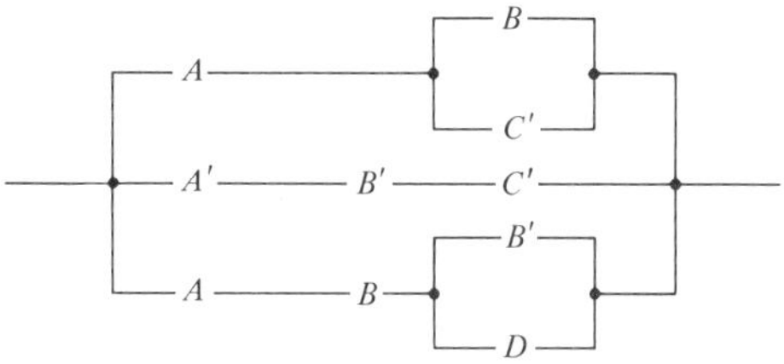

$$A \cdot B + B' \cdot C' =$$

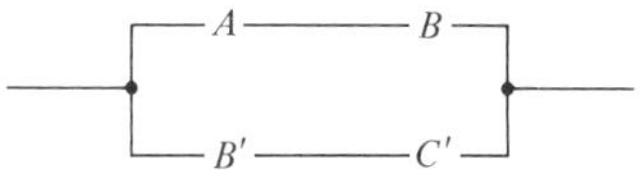

$$(A + B') \cdot (B + C') =$$

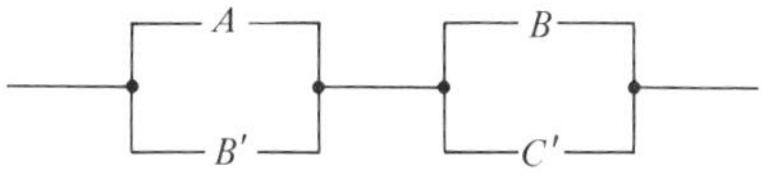

3.9 OTHER APPLICATIONS

Circuit network theory is only *one of several* important applications of Boolean algebra. If the English words "or" and "and" are identified with the Boolean symbols "$+$" and "$\cdot$", Boolean algebra may be used to examine the logical structure of laws, contracts, and conclusions drawn from data.

$$\text{or} \leftrightarrow + \leftrightarrow \text{parallel circuits}$$

$$\text{and} \leftrightarrow \cdot \leftrightarrow \text{series circuits}$$

Some of the modern logical computers make use of the isomorphism between the English connectives "or" and "and" and the circuit theory displayed above. In this way, electrical networks can be made to analyze a series of statements for consistency and logical structure. One of these interesting "machines" is so designed that military information, including rumors, may be fed into the machine, each with a weighting factor representing the supposed reliability of the information. The machine determines upper and lower limits on the reliability of any given conclusions based on the hypotheses and data given, and it will also indicate if the data is inconsistent. If additional observations or rumors are received, they may be added to those already available, and the machine will revise its credulity bounds.

Problem Set 3.9

1. Construct circuit diagrams for Example 2, Section 3.7.

2. Construct circuit diagrams for Example 3, Section 3.7.

3. Read Sections 1 and 2 of the supplement to Chapter 2 (pp. 108–114) of *What is Mathematics?* by Courant and Robbins. Prepare a written or oral report thereon.

4. Do the following tables define operations suitable for a Boolean algebra? If so, what letters play the roles of 0 and 1?

$+$	w	x	y	z
w	w	x	y	z
x	x	x	x	x
y	y	x	y	x
z	z	x	x	z

$\cdot$	w	x	y	z
w	w	w	w	w
x	w	x	y	z
y	w	y	y	w
z	w	z	w	z

5. Read and report on "Some Mathematical Aspects of Switching" by Franz Hohn, *American Mathematical Monthly*, **62** (Feb., 1955), p. 75.

6. Show that $(x + y) \cdot (x' + z) \cdot (y + z) = x' \cdot y + (x + y) \cdot z$. Compare Theorem 3.12a.

7. Prove or disprove: "If $A \cdot B = A \cdot C$ and $A + B = A + C$, then $B = C$." Show that the conclusion does *not* follow from either half of the hypothesis alone.

8. Draw circuits to represent:

   (a) $B + B' = 1$
   (b) $B \cdot B' = 0$
   (c) $1 + B = 1$
   (d) $0 + B = B$

   where 1 represents a circuit which is always closed (shorted), while 0 represents an ever-open circuit.

9. Use Boolean algebra to simplify the circuit given below:

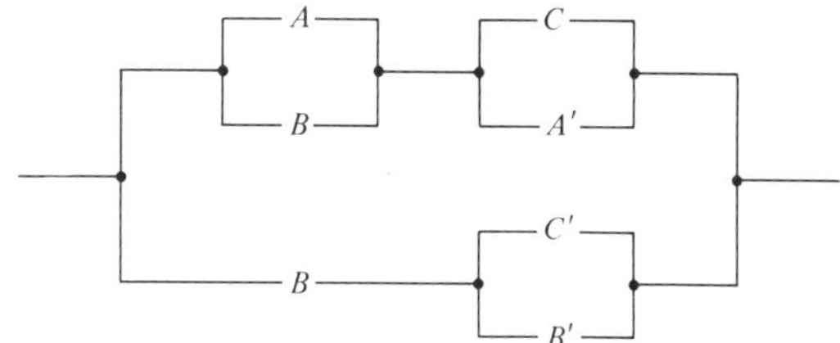

10. Use Boolean algebra to simplify the circuit given below:

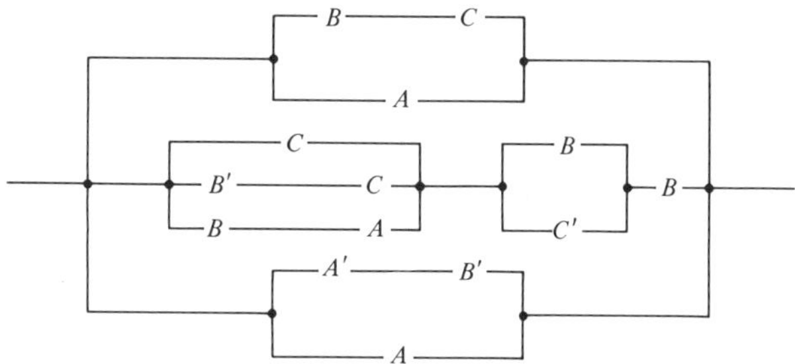

11. Prove the Boolean identity given below, by using the theorems of this section, where F and S are binary Boolean variables; that is, take on only the values 1 and 0 and obey the postulates of Section 3.2.

$$(F \cdot S) + (F' \cdot S') = (F + S') \cdot (F' + S)$$

It may be of interest to the reader to note that the above problem arises in designing an electrical circuit such that a light may be controlled independently from each of two switches.

12. It is desirable to have a stair light which may be turned on or off from either the top or bottom of the stairs, irrespective of which switch was used last. In tabular form, where "1" means "on" and "0" means "off," we seek a function having the following properties:

F (First floor switch)	S (Second floor switch)	L (Light)
1	1	1
1	0	0
0	1	0
0	0	1

(a) Show that both

$$L = (F + S') \cdot (F' + S) \ \text{ and } \ L = (F \cdot S) + (F' \cdot S')$$

have the desired properties.

(b) Give the circuits corresponding to each of the Boolean polynomials given in part (a).

13. Design a three-switch circuit which will turn on a light if at least two of the three switches are closed, but not otherwise (a simple majority secret voting machine).

14. Design a circuit containing four (double throw) switches such that, if an even number of switches are closed, the light will light and, if an odd number of switches are closed, the light will be off. Show that this permits independent control of the light from each switch station, no matter what arrangement the other switches may have.

3.10 CIRCUIT DESIGN

The question of whether or not machines think, and whether or not they deserve the name "Giant Brain"—often used as a description of large-scale computers—is one which almost invariably ends with the statement, "Well, of course, if *that* is what you mean by brain (or thinking), then ...." Actually, this is a discussion in which the author does not care to participate. The term "Supersonic Moron" is currently in more favor with people who understand high-speed computers than is the term "Giant Brain." Machines can do only things which they have been programmed to do, but they can do these with unbelievable rapidity and phenomenal accuracy. Today's fast computers can execute over 100 million instructions in one second. Who knows what tomorrow will bring?

Whether or not these machines think is unimportant and a matter of semantics. Machines can be designed to produce results, the determination of which would require thought on the part of man if the machine were not available. We shall design one such machine. Those who like the term "Giant Brain" may use the term "Midget Brain" for the simple situation which we are about to describe.

A man has a beautiful daughter, Jeraldine. Jeraldine and her father and mother are spending the summer on a small island camp. Two of Jeraldine's suitors, Willie and Charlie, accompany them. Jeraldine's father is very fond of fishing. However, he does not wish Jeraldine to be with either Willie or Charlie, nor with both of them, unless Jeraldine's mother is also present.

He finds it requires some thought, when he looks up from his boat and identifies the figures on the beach, to deduce whether or not Jeraldine is in the house (or on the beach) with one of the suitors but without her mother. The problem is to design a "black box" such that father may close switches bearing the names of those persons he can see on the beach and a red light will light at any time that Jeraldine and either of the boys are together (either in the house or on the beach) without Jeraldine's mother being present. The man wants a compact box—so he can carry it with him in the boat—containing a battery and four switches, W, J, C, M, so arranged that, when the switches of the people he can see on the beach are closed, a red light will light if a danger condition exists; namely, if Jeraldine is with either or both boys without the presence of her mother. In general, there will be trouble if the beach shows any of the following:

$$W \cdot J \cdot C \cdot M'$$

$$W \cdot J \cdot C' \cdot M'$$

$$W' \cdot J \cdot C \cdot M'$$

$$W \cdot J' \cdot C' \cdot M$$

$$W' \cdot J' \cdot C \cdot M$$

$$W' \cdot J' \cdot C' \cdot M$$

Each line of the above represents a single situation and the presences or absences occur simultaneously. Each line may be represented as a Boolean product. There is trouble if, and only if, the sum of the above products is 1, when 1 is substituted for each (unprimed) variable in the sum:

$$T = W \cdot J \cdot C \cdot M' + W \cdot J \cdot C' \cdot M' + W' \cdot J \cdot C \cdot M'$$

$$+ W \cdot J' \cdot C' \cdot M + W' \cdot J' \cdot C \cdot M + W' \cdot J' \cdot C' \cdot M$$

A circuit designed on the basis of the above polynomial will, when hooked in series with a red light and an electromotive source, light the red light when trouble is present. The use of Boolean algebra allows one, rather quickly, to reduce the circuit to a much simpler form. Problem 1 of the next problem set asks you to do this.

One day while the overzealous father was out fishing, he looked back to the beach and saw Charlie and the mother digging clams. He therefore switched C and M switches to the on position. No red light came on. The reason is simple: the light bulb had burned out! This situation was remedied by building in a complementary circuit (using Theorem 7) which lights a green light if things are O.K. Thus, we have a machine which lights a green light if things are all right and a red light if there is trouble, while no light indices a faulty circuit. The resulting circuit, which you are asked to design in Problem 2, serves to answer a problem which, were it solved each time it occurred, would certainly necessitate thinking. This is, of course, a simple problem. The resulting circuit is also simple. Please remember that it is a problem in mathematics, not in electrical engineering. If you build this machine, a minor problem in electrical engineering will occur in determining what size bulb and what size battery or e.m.f. source to use. A clerk in the ten-cent store will be glad to solve it for you.

Problem Set 3.10

1. Simplify the first circuit given in this section.

2. Construct the complementary circuit, in simplified form, for the red–green machine discussed in the latter part of this section.

3. An old puzzle states: A man is traveling with a wolf, a goat, and a basket of cabbages. They come to a river which he must cross in a boat only large enough to take him and one of the other objects. He must not leave the goat and the cabbages together (on either side of the river) when he is not present, or the goat will eat the cabbages. He must not leave the wolf and the goat together (on either side of the river) when he is not present, or the wolf will eat the goat. The problem then is to tell the man how to get across the river without losing any portion of his cargo.

   (a) Construct a circuit diagram of a machine to play the "wolf, goat, and cabbages" puzzle which will light a red light if part of the cargo is in danger and a green light if all is serene.
   (b) Make a statement concerning the problem just given which involves the word "isomorphic." [HINT: Consult the problem of Jeraldine.]

4. Look up other simple puzzles in W. R. Ball's *Mathematical Recreation and Essays* (New York: Macmillan, 1960) or M. Kraitchik's *Mathematical Recreations* (New York: Dover, 1953), and design circuits to play some of them.

5. Consult *Computer Techniques, Analysis and Mathematics* by (Andree)[3] for additional computer-related problems.

3.11 POINT SET INTERPRETATION

Many applications of Boolean algebra in logic, contracts, and language analysis stem from its use in point set theory. This is a generalization of the binary Boolean algebra discussed earlier in this chapter.

Let the universe be all the points inside a certain rectangle. Let X, Y, and Z represent the points contained in certain regions within this rectangle. For convenience, circles are used to represent these regions, but this is not essential. The circles are often called Venn circles. It is unimportant whether points on the circumference of circle X are included in X or in X' as long as you are consistent. We shall include them in X'. Thus, X consists of points inside of, but not on the circumference of, circle X.

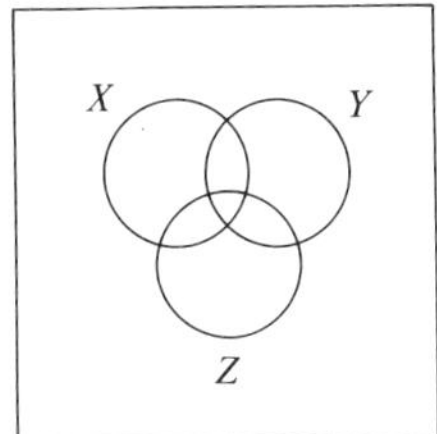

Let $X + Y$ represent all points inside of either X or Y or both. Let $X \cdot Z$ represent the points inside of both X and Z; that is, the common points. Let the prime denote the complement; that is, Z' denotes all points inside the rectangle which are outside of region Z. The symbol 1 represents the universe (all points inside the rectangle), while 0 represents the empty set (no points at all).

It is usual in point set theory to use the symbol $\cup$ in place of $+$ and the symbol $\cap$ in place of $\cdot$. The symbol $\cup$ is read "union" or "cUp" while $\cap$ is read "intersection" or "cAp."

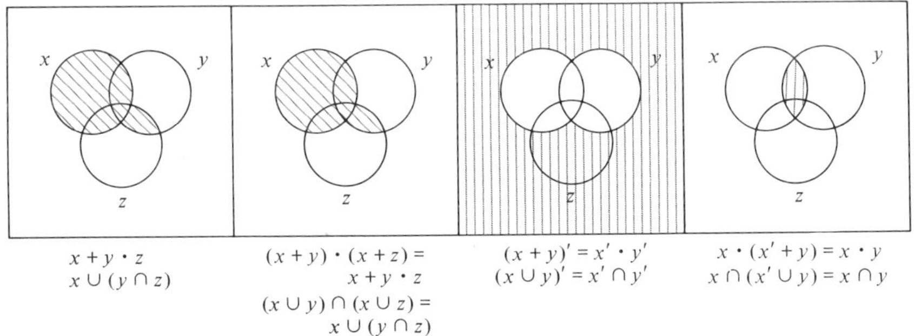

$$x + y \cdot z$$
$$x \cup (y \cap z)$$

$$(x + y) \cdot (x + z) =$$
$$x + y \cdot z$$
$$(x \cup y) \cap (x \cup z) =$$
$$x \cup (y \cap z)$$

$$(x + y)' = x' \cdot y'$$
$$(x \cup y)' = x' \cap y'$$

$$x \cdot (x' + y) = x \cdot y$$
$$x \cap (x' \cup y) = x \cap y$$

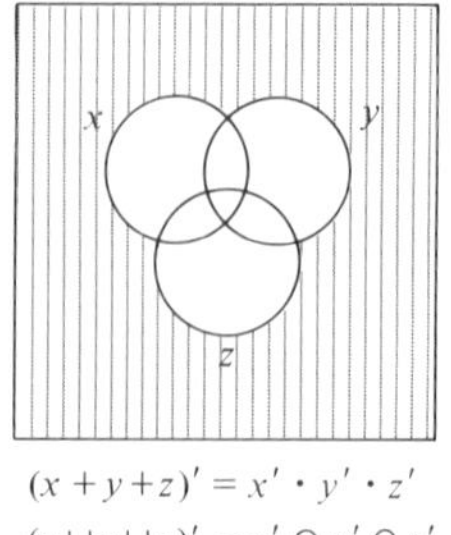

$$(x + y + z)' = x' \cdot y' \cdot z'$$
$$(x \cup y \cup z)' = x' \cap y' \cap z'$$

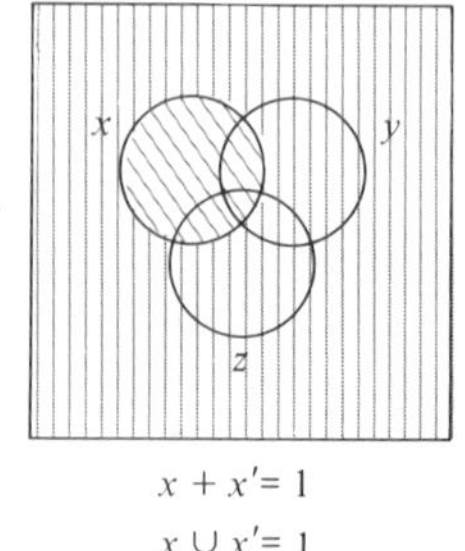

$$x + x' = 1$$
$$x \cup x' = 1$$

Examine the point set meanings of the symbols in the theorem

$$(x + y) + x \cdot z = x + y$$

$$(x \cup y) \cup (x \cap z) = x \cup y$$

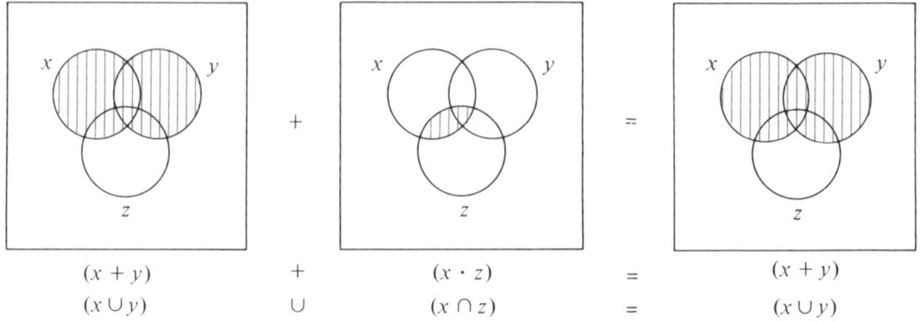

$$(x + y) \qquad + \qquad (x \cdot z) \qquad = \qquad (x + y)$$
$$(x \cup y) \qquad \cup \qquad (x \cap z) \qquad = \qquad (x \cup y)$$

One sees that $(x + y) + x \cdot z = x + y$. This result was also obtained as Example 1, binary Boolean algebra.

Most of the *proofs* given for Theorems 3.1 through 3.13 do not hold in generalized point set Boolean algebra. However, the twenty-four theorems (3.1a, 3.1b, 3.2a, $\cdots$, 3.13) are all valid. You are asked to provide intuitive proofs, using Venn diagrams, in the next problem set. Actually, these theorems may all be deduced from set theoretic postulates similar to those used in Section 3.6. To do so here would involve more repetition than is deemed desirable.

Further applications of Boolean algebra and its closely related lattice theory are studied in modern abstract algebra. One of the most important uses is in handling statistical data, where Boolean algebra is used to test the consistency of statistical data and conclusions.

Problem Set 3.11

1. Construct Venn circle diagrams illustrating each theorem of Section 3.6.

2. Construct Venn circle diagrams illustrating each theorem of Section 3.7.

3. Read and prepare a report on some of George Boole's work on logic such as his *Investigation of the Laws of Thought* (1854).

4. Show that $x \cap (x' \cup y) = x \cap y$.

5. Translate some of the theorems of this chapter into the union, $\cup$, and intersection, $\cap$, notation of point set theory.

6. Let W be a region entirely inside region X (denoted by $W \subset X$). Show that if $W \subset X$, $W + X = X$, and $W \cdot X = W$. Situations of this type do not arise in simple circuit theory applications, but they arise often in simple logic.

7. Show that $W \subset X$ (as in Problem 6) if, and only if, $W \cdot X' = 0$.

8. Let $W \subset X$, as in Problem 6. Find, by use of Venn diagrams, the regions represented by: $W' + X$, $W' \cdot X$, $X' + W$, $X' \cdot W$, $X' + W'$, $X' \cdot W'$.

9. Let Q and Y represent regions having some, but not all, points in common. Use Venn diagrams to shade the regions represented by $(Q \cdot Y)'$, $Q \cdot Y'$, $Q' \cdot Y'$, $Q' + Y'$, $(Q + Y)'$. Would you be able to forecast any equal pairs among the five regions by using Theorems 3.1a–3.13?

10. Prepare a report on "Boolean Algebra as an Introduction to Postulational Methods" by E. R. Stabler, *American Mathematical Monthly* (1943), **50**, p. 106.

11. The article by Franz Hohn mentioned in Problem 5, Set 3.9, contains an extensive bibliography on applications of Boolean algebra to electronics. Read at least one of the articles mentioned, and prepare a short report on it. If possible, arrange to present your report before an interested campus club or industrial group.

12. Read pages 108–116 of *What is Mathematics?* (London: Oxford, 1941) by R. Courant and H. Robbins, and give a brief résumé thereof.

13. Define $a + b$ to mean the least common multiple of a and b. Define $a \cdot b$ to mean the greatest common divisor of a and b. Define a' to mean $30/a$.

    (a) Do the integers $1, 2, 3, 5, 6, 10, 15, 30$ form a closed set under each of the operations $+$, $\cdot$, and $'$, as defined above?

    (b) How many of Theorems 3.1a–3.13 are satisfied by this system where $+ \leftrightarrow \cup$ and $\cdot \leftrightarrow \cap$? Which integer in this set corresponds to the Boolean 1 and which to the Boolean 0?

14. Let S be the set of all rational numbers. Interpret $x \cup y$ to mean the larger of the two rational numbers x, y, or the common value if $x = y$ [that is, $x \cup y = \max. (x, y)$]. In a similar fashion, interpret $x \cap y$ to be the minimum value of the pair x, y. Which postulates of Boolean algebra are satisfied by this system? Would your answer be changed if the set S were limited to the rational numbers between 0 and 50, inclusive?

15. Let T be the set of positive integers. Interpret $x \cup y$ to mean the greatest common divisor of x and y, while $x \cap y$ represents the least common multiple of x and y. What laws of Boolean algebra hold in this system? If an ideal element E is added to the set such that x divides E for every integer x and also E divides E, will you be able to enlarge the number of Boolean postulates which hold in this system? Is this ideal element really needed?

16. Prove or disprove: "If $A \cap B = A \cap C$ and $A \cup B = A \cup C$, then $B = C$. Would the conclusion follow from either half of the hypothesis alone? Compare Problem 7, Set 3.9.

★3.12 AN ALGEBRAIC APPROACH

In mathematics it is usual to study Boolean algebra from an algebraic approach rather than from the viewpoint of electrical switching circuits or point sets. A set of postulates somewhat akin to those used for the Integral Domain (see Section 1.7) may be set up, and theorems derived therefrom. Since the postulational system provides a reasonable *model* for both electrical circuit applications and point set applications, we may well obtain theorems which were undiscovered earlier, but which have important applications. Various sets of postulates can be used. One convenient set is presented below:

A Boolean algebra $\widetilde{\mathbf{B}}$ is a nonempty set of elements $x, y, z, \cdots$ having a well-defined equals relation and three operations, $\cup$, $\cap$, and $'$ which satisfy the following postulates:

1. *Closure*: For each pair x, y of elements of the Boolean Algebra $\tilde{\mathbf{B}}$, $x \cup y$, $x \cap y$, x' and y' are unique elements of $\tilde{\mathbf{B}}$.
2. *Commutative Laws*: For each pair x, y of elements of $\tilde{\mathbf{B}}$, $x \cup y = y \cup x$ and $x \cap y = y \cap x$.
3. *Associative Laws*: For each set x, y, z of elements of $\tilde{\mathbf{B}}$,

$$x \cup (y \cup z) = (x \cup y) \cup z \text{ and } x \cap (y \cap z) = (x \cap y) \cap z$$

4. *Zero Element*: The set $\tilde{\mathbf{B}}$ contains an element 0 such that for every element x in $\tilde{\mathbf{B}}$, $x \cup 0 = x$ and $x \cap 0 = 0$.
5. *Complement of an Element*: For each element $x \in \tilde{\mathbf{B}}$ there exists an element $x' \in \tilde{\mathbf{B}}$ such that $x \cup x' = 1$ and $x \cap x' = 0$. (If x changes, then x' may also change just as if a changes, $-a$ may also change in an integral domain.)
6. *Universal Element*: The set $\tilde{\mathbf{B}}$ contains an element $1 \neq 0$ such that for every element x in $\tilde{\mathbf{B}}$, $x \cap 1 = x$ and $x \cup 1 = 1$. (Here the same element must work for every $x \in \tilde{\mathbf{B}}$.)
7. *Distributive Laws*: For each set x, y, z of elements of $\tilde{\mathbf{B}}$

$$x \cup (y \cap z) = (x \cup y) \cap (x \cup z)$$

and

$$x \cap (y \cup z) = (x \cap y) \cup (x \cap z)$$

The above set of axioms is not the briefest possible set, but it does illustrate the structure to be expected. The reader will be asked to investigate whether certain parts of the given postulates can be deduced from the remaining postulates in the next problem set.

Problem Set 3.12

In each of the following use the *postulates of this section*, not any earlier interpretations of Boolean algebra.

1. Omit the part of Postulate 4 that says for all $x \in \tilde{\mathbf{B}}$ it is true that $x \cap 0 = 0$ and deduce this property as a theorem from the remaining postulates.

2. Omit the statement $x \cup 1 = 1$ from Postulate 5 and deduce it as a theorem from the remaining postulates.

3. Show that the elements 0 and 1 are unique elements in $\tilde{\mathbf{B}}$ (i.e., assume 0_1 and 0_2 are two elements of $\tilde{\mathbf{B}}$ such that for every $x \in \tilde{\mathbf{B}}$, $x \cup 0_1 = x$ and $x \cup 0_2 = x$, then show $0_1 = 0_2$).

4. Show that for each element $x \in \tilde{\mathbf{B}}$, the complement x' is unique.

5. Show *from the postulates* that $0' = 1$ and $1' = 0$.

6. Show that for each element $y \in \tilde{\mathbf{B}}$ that $(y')' = y$.

7. Show that for each element $z \in \tilde{\mathbf{B}}$ that $z \cup z = z$ and $z \cap z = z$.

8. Show that for all $x, y \in \tilde{\mathbf{B}}$ that $x \cup (x \cap y) = x$ and that $x \cap (x \cup y) = x$.

9. Show that for all $x, y \in \tilde{\mathbf{B}}$ that $(x \cup y)' = x' \cap y'$ and that $(x \cap y)' = x' \cup y'$.

10. Determine whether each of the Theorems 3.1a, 3.1b–3.13 given in Section 3.6 is also a theorem or postulate of postulational Boolean algebra as defined in Section 3.12.

11. Show that each of the postulates of Section 3.12 agrees with the point set (Venn diagram) interpretation of Boolean algebra; that is, that each of the seven postulates given in this section are inherent in the Venn diagram interpretation.

12. Show that each of the postulates of Section 3.12 is inherently included in the electrical circuit theory interpretation of Boolean algebra.

13. Read and prepare a brief written report on Chapter 6 (Boolean Algebras) of *Set Theory and Logic*, by R. R. Stroll (San Francisco: W. H. Freeman, 1963).

14. Show that the integers modulo 2 form a Boolean algebra. Does the operation $+$ play the role of $\cup$ or of $\cap$ in this problem, or doesn't it make any difference?

15. Read and prepare a brief report on the article "Binary Relations as Boolean Matrices" by O. Feichtinger and B. McAllister, which appeared in the January 1970 issue of *Mathematics Magazine*.

Selected Reading List, Chapter 3
BOOLEAN ALGEBRA

The following books and journals provide additional reading for students interested in term projects related to Boolean algebra.

See Problem 5, Set 2.8; Problems 3 and 10, Set 3.11; and Problems 13 and 15, Set 3.12 for suggestions in addition to the references cited below.

Books

Halmos, P. R., *Lectures on Boolean Algebra*. Princeton: Van Nostrand, 1963.

Hohn, F., *Applied Boolean Algebra*. New York: MacMillan, 1966.

Hu, S.-T., *Elements of Modern Algebra*. San Francisco: Holden-Day, 1965.

Hu, S.-T., *Mathematical Theory of Switching Circuits and Automata*. Berkeley: U. of Calif. Press, 1968.

Keister, R. and W., *The Design of Switching Circuits*. Princeton: Van Nostrand, 1955.

Rosenbloom, P., *The Elements of Mathematical Logic*. New York: Dover, 1950.

Sikorski, R., *Boolean Algebras*, Second Edition. Berlin: Springer-Verlag, 1964.

Articles

Feichtinger, O. and B. McAllister, "Binary Relations as Boolean Matrices" *Mathematics Magazine* Vol. 43 #1 p. 8–13 (1970).

Winder, R. O., "The Status of Threshold Logic" *RCA Review* Vol. 30 #1 p. 68–84 (1969).

Notes on Chapter 3

GROUPS

4

4.1 INTRODUCTION

In Chapter 1 the abstract mathematical system known as the integral domain was introduced. We now turn our attention to a much simpler abstract mathematical system. In spite of its apparent simplicity, the *group* is one of the most fruitful concepts in modern mathematics.

The *group* has important applications in heat transfer, nuclear structure, and other areas of modern science and engineering, in addition to providing basic structural concepts for use in more advanced mathematics. Basically, mathematics is the study of structure. This chapter studies one of the most simple mathematical structures—the group. It is not surprising that its applications are manifold.

4.2 MATHEMATICAL SYSTEMS

An operation, $\odot$, is said to be *well defined* with respect to an equivalence relation, $=$, if, when a and b are replaced by equivalent elements, an equivalent result is obtained.[1] This means that $x = x'$ and $y = y'$ imply

---

[1] The symbol "$\odot$" is read "operation" or "product." If the operation is commutative, the symbol "$+$," plus, is often used.

$x \odot y = x' \odot y'$. Since each element is equivalent to itself, no replacement need be apparent, as in $\frac{1}{2} + \frac{1}{4} = \frac{2}{4} + \frac{1}{4} = \frac{3}{4}$, where $\frac{2}{4}$ is equivalent to $\frac{1}{2}$, and $\frac{1}{4}$ is equivalent to $\frac{1}{4}$.

A mathematical system consists of:

(1) *A set of elements*, here denoted by letters a, b, c, $\cdots$.
(2) *An equivalence relation*, which separates the elements into classes. Each class may, conceivably, contain only one element. Often, classes contain more than one element—possibly infinitely many.
(3) *One or more well-defined operations* (multiplication and addition, for example), which associate an element c with each ordered pair of elements a, b. If the symbol $\odot$ is used to represent the operation, it is usual to write $a \odot b = c$.
(4) *A set of postulates*, which the elements and operations satisfy.

Certain mathematical systems are of such importance that special names, such as *field*, *ring*, *integral domain*, *group*, *linear algebra*, *geometry*, and *topology*, are associated with them. The integral domain, examined in Chapter 1, had two operations. We now turn our attention to a simpler mathematical system, the group, which has only one operation.

4.3 GROUP

One of the simplest, and also one of the most important, mathematical systems is the *group*. A group has only one operation, $\odot$, which may be "addition," or may be "multiplication," or may be the operation "following" or "combining", or may be any other operation. (The reader will note that *element* and *operation* are undefined terms.)

Group Postulates

If a, b, c are elements of a mathematical system, G, having a well-defined operation, $\odot$, and an equivalence relation, $=$, then G is a group if it satisfies the following postulates:

1. *Closure*: If $a, b \in G$, then $a \odot b \in G$.[2]
2. *Associative Law*: If $a, b, c \in G$, then $(a \odot b) \odot c = a \odot (b \odot c)$.
3. *Existence of an Identity (Unity)*: There exists an element $u \in G$ such that for each $b \in G$, $u \odot b = b \odot u = b$.
4. *Existence of an Inverse*: For each $b \in G$, there exists an element $b^* \in G$ such that $b \odot b^* = b^* \odot b = u$, for the u of Postulate 3.

---

[2] The symbol "$\in$" means "is an element of" or "are elements of."

It may seem that with so simple a set of postulates, no important theory could develop. However, *much of the power of modern mathematics stems from group theory.* E. P. Wigner, the physical chemist who was Director of Research at the Oak Ridge National Laboratory and the Metallurgical Laboratory at the University of Chicago, is said to have presented the essentials of the thermodynamics of heat transfer to his staff of physicists on *one sheet of paper* by using group theory.

EXAMPLE 1

The (positive, negative, and zero) integers form a group under the operation of addition, since

1. *Closure*: The sum of two integers is an integer.
2. *Associative Law*: The integers are associative under addition.
3. *Existence of an Identity (Unity)*: $u = 0$, since $b + 0 = 0 + b = b$ for all integers b.
4. *Existence of an Inverse*: The inverse of b is $b^* = -b$.

Be sure that you understand this example before continuing. What is the inverse of -11?

EXAMPLE 2

The integers do not form a group under multiplication. The first three postulates are satisfied ($u = 1$, in this case), but Postulate 4 is not, since 5 does not have an inverse in the system of integers. (Actually, no integers other than ± 1 have multiplicative inverses in the system of integers, but a single counterexample is sufficient.)

EXAMPLE 3

The rational[3] numbers, with zero excluded, form a group under the operation of multiplication.

1. *Closure*: The product of two rational numbers is a rational number.
2. *Associative Law*: The rational numbers are associative.
3. *Existence of an Identity (Unity)*: $u = 1$, since $b \cdot 1 = 1 \cdot b = b$ for all rational numbers b.
4. *Existence of an Inverse*: The inverse of $b(\neq 0)$ is $1/b$, which is rational if b is rational.

---

[3] A rational number is one which can be expressed as a quotient of two integers, a/b with $b \neq 0$.

In the above examples, the elements of the group are numbers, and the equals relationship is ordinary equality of numbers. However, a group may be more general than this.

EXAMPLE 4

Let the elements of a group be substitutions of one set of letters for another. For example, let a mean

$$a = \begin{cases} \text{for } A \text{ substitute } B \\ \text{for } B \text{ substitute } C \\ \text{for } C \text{ substitute } A \end{cases} = \begin{cases} A \to B \\ B \to C \\ C \to A \end{cases}$$

so that the substitution a carries (A, B, C) into (B, C, A), and $3A^2 - 4BC + 2A$ becomes $3B^2 - 4CA + 2B$.

$$\text{If } b = \begin{cases} A \to C \\ B \to B \\ C \to A \end{cases} \text{ is another substitution, then } b \text{ carries } (A, B, C) \text{ into}$$

(C, B, A). Also note that b carries (B, C, A) into (B, A, C).

If we make the substitution a, and then make the substitution b on the result, we obtain:

$$\begin{array}{cc} \text{Using} & \text{Using} \\ a & b \end{array}$$

(A, B, C) becomes (B, C, A) becomes (B, A, C)

However, there is a substitution which will carry (A, B, C) into (B, A, C) directly, namely

$$c = \begin{cases} A \to B \\ B \to A \\ C \to C \end{cases}$$

We write $a \odot b = c$ to express this.

There are six possible substitutions of the three letters A, B, C, including the identity substitution

$$i = \begin{cases} A \to A \\ B \to B \\ C \to C \end{cases}$$

We may let these six substitutions be elements of a mathematical system, G, in which the operation $\odot$ is the following of one substitution by another substitution, and the equals relation is identity of result, that is, $a \odot b$ means "first do 'a,' then do 'b.'" Thus, $a \odot b = c$ for the substitutions given above.

In defining a group, it is always essential that the equals relation be specified, and that it satisfy the postulates of Section 2.1.

It is usual to represent such substitutions (often called permutations) by a notation $a = \begin{pmatrix} A & B & C \\ B & C & A \end{pmatrix}$, where the notation implies that each letter in the top row is replaced by the letter just below it.

The six possible substitutions are:

$$a = \begin{cases} A \to B \\ B \to C \\ C \to A \end{cases} = \begin{pmatrix} A & B & C \\ B & C & A \end{pmatrix}; \qquad b = \begin{cases} A \to C \\ B \to B \\ C \to A \end{cases} = \begin{pmatrix} A & B & C \\ C & B & A \end{pmatrix}$$

$$c = \begin{cases} A \to B \\ B \to A \\ C \to C \end{cases} = \begin{pmatrix} A & B & C \\ B & A & C \end{pmatrix}; \qquad d = \begin{cases} A \to C \\ B \to A \\ C \to B \end{cases} = \begin{pmatrix} A & B & C \\ C & A & B \end{pmatrix}$$

$$e = \begin{cases} A \to A \\ B \to C \\ C \to B \end{cases} = \begin{pmatrix} A & B & C \\ A & C & B \end{pmatrix}; \qquad i = \begin{cases} A \to A \\ B \to B \\ C \to C \end{cases} = \begin{pmatrix} A & B & C \\ A & B & C \end{pmatrix}$$

Thus,

$$d \odot e = \begin{pmatrix} A & B & C \\ C & A & B \end{pmatrix} \odot \begin{pmatrix} A & B & C \\ A & C & B \end{pmatrix} = \begin{pmatrix} A & B & C \\ B & A & C \end{pmatrix} = c$$

while,

$$e \odot d = \begin{pmatrix} A & B & C \\ A & C & B \end{pmatrix} \odot \begin{pmatrix} A & B & C \\ C & A & B \end{pmatrix} = \begin{pmatrix} A & B & C \\ C & B & A \end{pmatrix} = b$$

Note that $d \odot e \neq e \odot d$. Hence, this system of substitutions is not commutative. It need not be to satisfy the group postulates.

The group postulates are satisfied by this system.

1. *Closure*: The "product" of two substitutions in this list is also a substitution in the list. Problem 1 of the next set asks you to complete the "multiplication" table for this group, and hence prove closure.
2. *Associative Law*: Let us examine a typical letter, under the transformation $[(a \odot b) \odot c]$ and under $[a \odot (b \odot c)]$.

$$a = \begin{pmatrix} \cdots x \cdots \cdots \\ \cdots y \cdots \cdots \end{pmatrix}$$

$$b = \begin{pmatrix} \cdots \cdots y \cdots \cdots \\ \cdots \cdots z \cdots \cdots \end{pmatrix}$$

$$c = \begin{pmatrix} \cdots \cdots z \cdots \cdots \\ \cdots \cdots w \cdots \cdots \end{pmatrix}$$

Then $(a \odot b)$ takes x into z, while $(b \odot c)$ takes y into w.

$$(a \odot b) = \begin{pmatrix} \cdots x \cdots \cdots \\ \cdots z \cdots \cdots \end{pmatrix}, \quad \text{while} \quad (b \odot c) = \begin{pmatrix} \cdots \cdots y \cdots \\ \cdots \cdots w \cdots \end{pmatrix}$$

Thus,

$$[(a \odot b) \odot c] = \begin{pmatrix} \cdots x \cdots \cdots \\ \cdots z \cdots \cdots \end{pmatrix} \odot \begin{pmatrix} \cdots \cdots z \cdots \\ \cdots \cdots w \cdots \end{pmatrix} = \begin{pmatrix} \cdots x \cdots \cdots \\ \cdots w \cdots \cdots \end{pmatrix}$$

while

$$[a \odot (b \odot c)] = \begin{pmatrix} \cdots x \cdots \cdots \\ \cdots y \cdots \cdots \end{pmatrix} \odot \begin{pmatrix} \cdots y \cdots \\ \cdots w \cdots \end{pmatrix} = \begin{pmatrix} \cdots x \cdots \cdots \\ \cdots w \cdots \cdots \end{pmatrix}$$

Thus, not only these substitutions but every set of substitutions is associative.

3. *Existence of an Identity (Unity)*: $u = i$, as already noted.
4. *Existence of an Inverse*: By forming the products $x \odot x^* = x^* \odot x = i$, show that each $x \in G$ has the inverse x^* indicated below.

$$a^* = d$$

$$b^* = b$$

$$c^* = c$$

$$d^* = a$$

$$e^* = e$$

$$i^* = i$$

EXAMPLE 5

Let the elements of a group be rotations of a plane figure through integral multiples of 60°. Let the operation be following one such rotation by another, and let the equals relation be identity of position. Rotation through $-120°$ results in the same position as rotation through $+240°$. Rotation through $(k \cdot 60° \pm n \cdot 360°)$ results in the same position as rotation through $k \cdot 60°$ for each integer n. Hence, the elements of this group may be represented as R_{60}, R_{120}, R_{180}, R_{240}, R_{300}, R_{360}. Note that each of these group elements, R_x, represents an entire equivalence class $(R_x \pm n \cdot 360°)$.

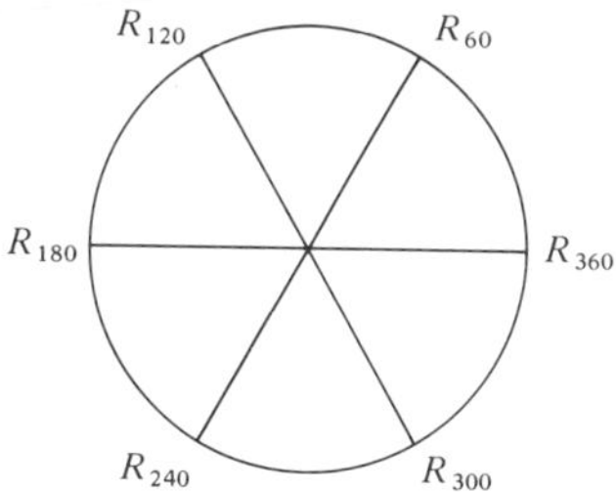

where R_x denotes a rotation through x degrees in the counterclockwise direction. Thus, $R_{120} \odot R_{180} = R_{300}$ and $R_{240} \odot R_{180} = R_{60}$. Check the group postulates for this system.

1. *Closure*: If one of the above rotations is followed by another rotation, the resulting position could have been obtained as a single rotation.
2. *Associative Law*: Valid.
3. *Existence of an Identity (Unity)*: $u = R_{360}$.
4. *Existence of an Inverse*: The inverse of R_x is $R_x{}^* = R_{360°-x}$.

It is instructive to form the multiplication tables for the groups of Examples 4 and 5. The reader is expected to fill in the entries not supplied. (See Problems 1 and 2, Problem Set 4.3.)

Example 4 (Partial multiplication table)

	i	a	b	c	d	e
i	i	a	b	c	d	e
a	a				i	
b	b		i			
c	c		a	i		
d	d	i			a	c
e	e				b	i

Example 5 (Partial multiplication table)

	R_{360}	R_{60}	R_{120}	R_{180}	R_{240}	R_{300}
R_{360}	R_{360}	R_{60}	R_{120}	R_{180}	R_{240}	R_{300}
R_{60}	R_{60}	R_{120}	R_{180}	R_{240}	R_{300}	R_{360}
R_{120}	R_{120}	R_{180}	R_{240}	R_{300}		
R_{180}	R_{180}					
R_{240}	R_{240}			R_{120}		
R_{300}	R_{300}		R_{60}			R_{240}

EXAMPLE 6

Cut out a 2-inch square of cardboard or paper. Label the vertices A, B, C, D as shown below:

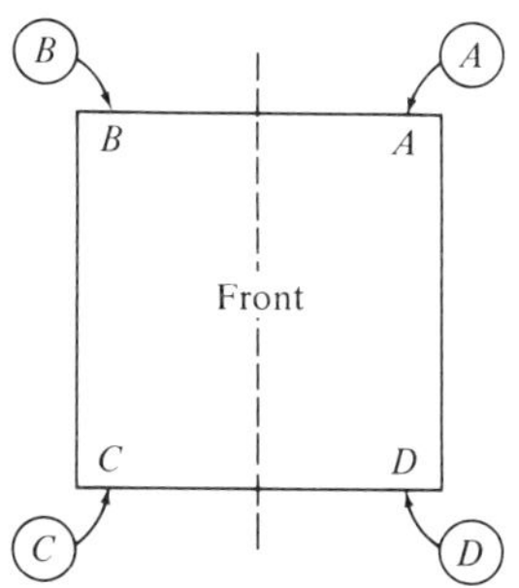
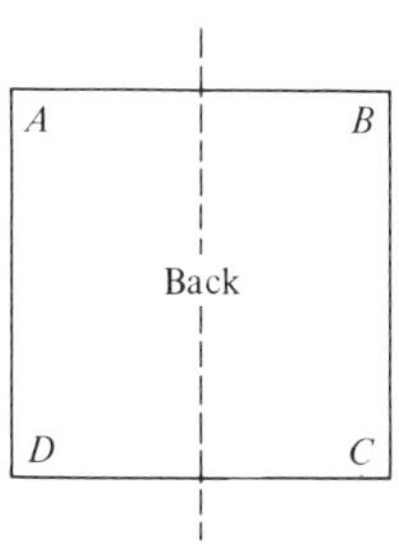

Place a letter A on the back face of the square in the same vertex as the letter A on the front. Vertex A is now uniquely identified. Repeat for vertices B, C, and D.

The elements of this group are certain movements of the square. The operation, $\odot$, consists of following one movement by another. Equality is identity of vertex positions.

The permitted movements (elements) are:

R_{180}:   A rotation (in the plane) of the square through 180° about its center.

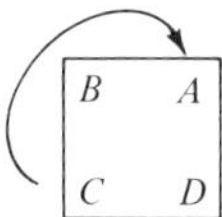
becomes
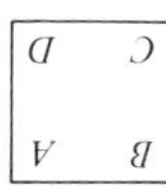

R_{360}:   A rotation (in the plane) of the square through 360°.

becomes

R_{360} is the identity element, since the positions of the vertices are unchanged.

H: A flip (in 3-space) about a horizontal line through the center of the square.

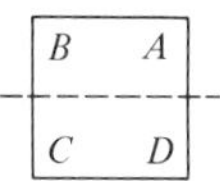
becomes
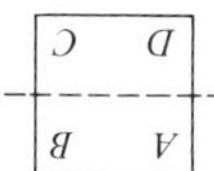

V: A flip (in 3-space) about a vertical line through the center of the square.

Then $(H \odot V)$ is that element which results when H (horizontal flip) is followed by V (vertical flip). Actual experiment will show that $(H \odot V) = R_{180}$, that $(R_{180} \odot V) = H$, and that $(H \odot H) = R_{360}$.

The reader should construct a model and carry out the operations just suggested, as well as supplying the remainder of the entries in the multiplication table of this group.

	R_{360}	R_{180}	H	V
R_{360}	R_{360}	R_{180}	H	V
R_{180}				
H			R_{360}	R_{180}
V				

After the table is completed, it will be possible to see that these four elements do form a group.

1. *Closure*: Only the elements R_{180}, R_{360}, H, V are needed to complete the table.
2. *Associative Law*: The four given elements are substitutions of the corner letters. Substitutions are associative, by the proof of Example 4.

$$R_{180} = \begin{pmatrix} A & B & C & D \\ C & D & A & B \end{pmatrix} \qquad R_{360} = \begin{pmatrix} A & B & C & D \\ A & B & C & D \end{pmatrix}$$

$$H = \begin{pmatrix} A & B & C & D \\ D & C & B & A \end{pmatrix} \qquad V = \begin{pmatrix} A & B & C & D \\ B & A & D & C \end{pmatrix}$$

3. *Existence of an Identity (Unity)*: R_{360}.
4. *Existence of an Inverse*: Since $u = R_{360}$ appears once, and only once, in each row and column of the "multiplication" table, it follows that

$x \odot a = R_{360}$ and $b \odot x = R_{360}$ are solvable for all a, b in

$$G = (R_{360}, R_{180}, V, H).$$

Examination shows that, if $x \odot y = R_{360}$, then $y \odot x = R_{360}$.

Problem Set 4.3

1. Complete the multiplication table for Example 4.

2. Complete the multiplication table for Example 5.

3. Make a multiplication table for Example 6.

   In Problems 4–16, determine whether or not each of the given systems is a group. If the set is *not* a group, specify which postulates are violated.

4. The even integers (zero, positive, negative), the operation being addition.

5. The odd integers, under multiplication.

6. The integers (mod 13), the operation being multiplication mod 13 and equality being congruence (mod 13).

7. Same as Problem 6, but with 0 (mod 13) omitted.

8. The integers mod 8, under addition (mod 8).

9. The numbers 1, -1, i, $-i$, under multiplication.

10. The numbers $1, \dfrac{1 + i\sqrt{3}}{2}, \dfrac{1 - i\sqrt{3}}{2}$ under multiplication.

11. To the elements of Example 6, add the two new elements:

    R_{90} = a rotation (in the plane) of the square through 90° clockwise,
    R_{270} = a rotation (in the plane) of the square through 270° clockwise.
    Does this enlarged system form a group?

12. To the elements of Example 6, add the two elements of Problem 11 and also the two elements

    D_1 = a flip about the diagonal from upper right to lower left.

D_2 = a flip about the other diagonal.

becomes

This system of eight elements forms a group known as the *octic group*.
(a) Find its multiplication table and show that Postulates 1, 3, 4 are satisfied.
(b) Represent each of the eight elements as a substitution (permutation) on the set of four vertex letters A, B, C, D, and hence prove that the set is associative.

13. The six expressions

$$t, \quad \frac{1}{t}, \quad 1 - t, \quad \frac{1}{1 - t}, \quad \frac{t - 1}{t}, \quad \frac{t}{t - 1}$$

where the operation is the substitution of the second factor for t in the first factor, thus

$$\frac{1}{1 - t} \odot \frac{1}{t} = \frac{1}{1 - \dfrac{1}{t}} = \frac{t}{t - 1}, \qquad \text{an element.}$$

(Readers familiar with the concept "cross ratio" will recognize this as a group of transformations under which the cross ratio remains invariant. Geometry is the study of properties left invariant by certain transformations.)

14. The integers mod 17, under

(a) multiplication mod 17,
(b) addition mod 17.

15. The integers $\not\equiv 0$ mod 23, under

(a) multiplication mod 23,
(b) addition mod 23.

16. The set of all polynomials in x, under the operation of addition. [$p(x) = c$ is a valid polynomial.]

17. Prove or disprove: The elements of an integral domain form a group under addition.

18. Do the elements of an integral domain form a group under multiplication?

19. Let the elements be points in a plane with real coordinates (x, y). Let the operation be defined as

$$(x_1, y_1) \odot (x_2, y_2) = (x_1 + x_2 e^{-y_1}, y_1 + y_2)$$

Equality is identity of points. Show that this set is a group.

20. Give an example of a mathematical system having one operation such that the system is *not* a group.

21. Explain why each of the following systems is *not* a group:

(a)

	a	b	c
a	a	b	c
b	b	a	c
c	c	c	c

(d)

	x	y	0
x	x	y	0
y	y	0	0
0	0	0	0

(b)

	a	b	c	d	e
a	a	b	c	d	e
b	b	e	e	b	e
c	c	e	e	c	e
d	d	b	c	a	e
e	e	e	e	e	e

(e)

	a_1	a_2	a_3	a_4	a_5
a_1	a_1	a_2	a_3	a_4	a_5
a_2	a_2	a_1	a_4	a_5	a_3
a_3	a_3	a_5	a_1	a_2	a_4
a_4	a_4	a_3	a_5	a_1	a_2
a_5	a_5	a_4	a_2	a_3	a_1

[HINT to the instructor: This is a loop, but not a group.]

(c)

	a	b
a	a	b
b	b	b

(f)

	b_1	b_2	b_3	b_4	b_5
b_1	b_4	b_1	b_5	b_3	b_2
b_2	b_3	b_5	b_2	b_1	b_4
b_3	b_1	b_2	b_3	b_4	b_5
b_4	b_2	b_4	b_1	b_5	b_3
b_5	b_5	b_3	b_4	b_2	b_1

Note that $b_3 \odot x = x$ for all x, but that $x \odot b_3 \neq x$ for certain x.

(g)

$\cdot$	d_1	d_2	d_3	d_4	d_5	d_6	d_7	d_8
d_1	d_1	d_2	d_3	d_4	d_5	d_6	d_7	d_8
d_2	d_2	d_1	d_4	d_5	d_6	d_7	d_8	d_3
d_3	d_3	d_4	d_5	d_6	d_6	d_8	d_2	d_1
d_4	d_4	d_5	d_6	d_7	d_8	d_3	d_1	d_2
d_5	d_5	d_6	d_7	d_8	d_2	d_1	d_3	d_4
d_6	d_6	d_7	d_8	d_3	d_1	d_2	d_4	d_5
d_7	d_7	d_8	d_1	d_2	d_3	d_4	d_5	d_6
d_8	d_8	d_3	d_2	d_1	d_4	d_5	d_6	d_7

[HINT: Does $d_2 \cdot (d_8 \cdot d_3) = (d_2 \cdot d_8) \cdot d_3$?]

22. Let (x, y) be a point in a plane. Consider the translation $(x, y) \to (x', y')$ where $x' = x + a$, $y' = y + b$, for a, b real numbers.

    (a) The three points $A(2, 1)$, $B(5, 7)$, $C(4, -2)$ are translated into three points A', B', C' by the translation $x' = x + 2$, $y' = y - 5$. Find the coordinates of A', B', C'.
    (b) Is the resulting triangle $A'B'C'$ congruent to the triangle ABC?
    (c) Show, by example, that there exist triangles which are congruent to ABC but which *cannot* be obtained from ABC by a translation of the form $x' = x + a$, $y' = y + b$ for any a, b.
    (d) Consider the set of *all* translations $x' = x + a$, $y' = y + b$, where a, b are real numbers. Does this set form a group if the operation is following one translation by a second, and equality is defined as identity of final position of points?

23. Continuing with the notation of Problem 22:

    (a) Define two triangles ABC and PQR to be "translationally equivalent" if there exist real numbers a and b such that $x' = x + a$, $y' = y + b$ takes ABC into PQR. Show that "translational equivalence" satisfies the three postulates of an equivalence relation given in Section 2.1.
    (b) Into what equivalence classes (Section 2.2) does "translational equivalence" separate the set of all triangles in the plane?
    ★(c) Generalize (a) and (b) to figures other than triangles.

*24. Analytic geometry provides a well-known connection between geometry and (elementary) algebra. There is also a fundamental connection between geometry and modern abstract algebra. Given a group G, of transformations (for example, the translations of Problems 22 and 23), two "objects" are said to be *equivalent* if one may be taken into the other by an element of G. Give an example of such a transformation and discuss the resulting geometry.

25. Discuss the logical difference in the phrase "$\cdots$ there exists an element $\cdots$" in Group Postulates 3 and 4.

4.4 ELEMENTARY PROPERTIES OF GROUPS

Since the associative law, $a \odot (b \odot c) = (a \odot b) \odot c$, is postulated for groups, it follows that $x^2 \odot x = x \odot x^2$, giving meaning to the expression x^3. Similarly, x^4 is given meaning as the common value of $x^3 \odot x = x \odot x^3 = x^2 \odot x^2$. It can be shown that, in general, $x^r \odot x^s = x^s \odot x^r$. Further discussion of this topic appears in Section 5.9. The general associative law, for k elements, may be proved using mathematical induction. (See Sections 1.6 and 1.7.) As in Boolean algebra, we define $x \odot y \odot z$ to be the common value of $(x \odot y) \odot z$ and $x \odot (y \odot z)$.

THEOREM 4.1   *The symbol x^n is meaningful in a group.*

To check his understanding of Theorem 4.1, the reader should show that

$$\left(\frac{1}{1-t}\right)^4 = \frac{1}{1-t}$$

where the group operation is the substitution of the second factor for t in the first factor, as in Problem 13 Set 4.3.

THEOREM 4.2   *The identity element u of a group is unique.*

Let us assume that both u and e are identity elements. Since u is an identity, $\qquad u \odot e = e.$
Since e is an identity, $u \odot e = u.$
Hence, $\qquad\qquad e = u.$ (Why? What postulate is used?)
Thus, the identity is unique.

THEOREM 4.3   *Each group element has a unique inverse.*

By Postulate 4, each element has an inverse. Assume that both b^* and b^+ are inverses of the same element b. Then

$$b \odot b^* = u$$

Multiply on the left by b^+, reassociate, and note that $b^+ \odot b = u$, where u is the identity element.

$$b^+ \odot (b \odot b^*) = b^+ \odot u$$

$$(b^+ \odot b) \odot b^* = b^+$$

$$u \odot b^* = b^+$$

$$b^* = b^+$$

Thus, the inverse of each element is unique. It is important that the reader realize that two quite different types of uniqueness are involved in Theorems 4.2 and 4.3. In the former, the identity u is an identity for the entire group. In the latter, *each element b has a unique inverse b^**, but no "universal inverse" is implied.

THEOREM 4.4   *The inverse of a product of group elements is the product of the inverses taken in reverse order.*

To show that $(a \odot b \odot c \cdots \odot p)^{-1} = p^{-1} \odot \cdots \odot c^{-1} \odot b^{-1} \odot a^{-1}$, we form the product

$$(p^{-1} \odot \cdots \odot c^{-1} \odot b^{-1} \odot a^{-1}) \odot (a \odot b \odot c \odot \cdots \odot p)$$

Reassociating,

$$(p^{-1} \odot \cdots \odot c^{-1} \odot b^{-1}) \odot (a^{-1} \odot a) \odot (b \odot c \odot \cdots \odot p)$$

$$= p^{-1} \odot \cdots \odot c^{-1} \odot b^{-1} \odot (u) \odot b \odot c \odot \cdots \odot p$$

$$= p^{-1} \odot \cdots \odot c^{-1} \odot (b^{-1} \odot b) \odot c \odot \cdots \odot p$$

$$= p^{-1} \odot \cdots \odot c^{-1} \odot (u) \odot c \cdots \odot p$$

and continuing

$$= p^{-1} \odot (u) \odot p = p^{-1} \odot p = u$$

As a special case, we have $(a \odot b)^{-1} = b^{-1} \odot a^{-1}$.

We note that $(b^m)^{-1} = (b^{-1})^m$ and *define* this common value to be b^{-m}. If b^0 is defined as u, the identity of the group, then b^k is defined in a consistent manner for all integers, positive, zero, and negative. It may be appropriate to point out this is truly a matter of definition, just as it was in elementary algebra, when x^{-m} and x^0 were defined for $x \neq 0$. *It is not a matter which is subject to proof.* All that can be proved is that, *if* the ordinary meaning of "x taken k times as a factor" is used for x^k, k a positive integer, and *if* the usual laws of arithmetic are to apply to negative and zero exponents, then it is appropriate to define $b^{-k} = (b^{-1})^k = (b^k)^{-1}$ and $b^0 = u$, where u is the identity element (unity) of the group to which b belongs.

THEOREM 4.5   *If G is a group which has only a finite number, n, of elements, then for each element $b \in G$, there exists a positive integer $r \leq n$ such that $b^r = u$, the group identity, and hence, $b^{-1} = b^{r-1}$.*

The method used to prove this theorem is of particular interest. The elements

$$u, b, b^2, b^3, \cdots, b^n$$

cannot all be distinct if the group contains only n elements. If u and one of the powers of b are equal, the theorem is proved. On the other hand, suppose that

$$b^s = b^t, \qquad 1 \leq t < s \leq n$$

Then, $b^s \odot (b^t)^{-1} = u$

$b^{(s-t)} = u$, where $1 \leq (s - t) < n$, and the theorem is proved.

Definitions: A group having only a finite number of elements is called a *finite group*. The number of elements in a finite group is the *order of the group*. The least positive integer r such that $b^r = u$ is the *period*[4] of b, if such an r exists. (Such an r exists if the group is finite; it may or may not exist if the group is infinite.) A *subgroup* of a group G is a set of elements of G which, by themselves, form a group. Every group G contains the subgroup, which consists of the identity element u alone, and also the subgroup which consists of the entire group G. The group G may or may not also contain other subgroups called *proper subgroups*.

---

[4] Some authors use the word "order" in place of "period" here.

Problem Set 4.4

1. Which of the systems given in Problems 4–16 of Set 4.3 are *finite* groups?

2. (a) Show that the residue classes 1, 5, 7, 11 modulo 12 form a group under multiplication (mod 12).
   (b) Show that 1, 5 form a proper subgroup.
   (c) Find another proper subgroup.

3. (a) Show that the nonzero residue classes modulo 17 form a group under multiplication mod 17.
   (b) Show that 1, 4, 13, 16 form a subgroup.
   (c) Find another proper subgroup.
   (d) Find a subgroup, S, of the subgroup given in (b). Is S also a subgroup of the original group?

4. Do the elements $z,\ -z,\ \dfrac{1}{z},\ -\dfrac{1}{z}$ form a group under the operation of substituting the second factor for z in the first factor?

5. (a) Do the elements $z,\ \dfrac{1}{z},\ \dfrac{z-1}{z+1},\ \dfrac{z+1}{z-1},\ \dfrac{1-z}{1+z},\ \dfrac{1+z}{1-z}$ form a group under the operation of substituting the second factor for z in the first factor?

   (b) If $-z,\ -\dfrac{1}{z}$ are added to the above six elements, is a group obtained under the given operation?

6. Do the integers modulo 19 form a group
   (a) under addition?
   (b) under multiplication?

7. Same as Problem 6, but modulo 14.

8. Let the symbol $\begin{pmatrix} 1 & 2 & 3 \\ 3 & 1 & 2 \end{pmatrix}$ be used to represent the permutation $\begin{cases} 1 \to 3 \\ 2 \to 1 \\ 3 \to 2 \end{cases}$. Write out the six possible permutations of the three symbols

1, 2, 3 and show that they form a group where the operation is following one permutation with another one.

9. Write out the $4! = 24$ possible permutations of 4 symbols and show that they form a group under the operation of following one permutation with another. Three or four students may wish to cooperate on this, since the $(24)^2$ multiplications are time consuming.

10. If $b = \begin{pmatrix} 1 & 2 & 3 & 4 & 5 & 6 \\ 2 & 3 & 4 & 5 & 6 & 1 \end{pmatrix}$, form the six permutations $b, b^2, b^3, b^4, b^5, b^6$.

11. Show that the six permutations of Problem 10 form a group. This group is a subgroup of order six of the group of all permutations of six symbols, which is a group of order 720. (What does this last sentence mean?)

12. Let C be the set of all complex numbers $a + bi$, such that $a^2 + b^2 = 1$. Does this set form a group under the operation of multiplication?

*13. Two sets of postulates are said to be equivalent if each set implies the other set. Show that the group postulates given are equivalent to the set below:

For every a, b, c in G

(1) $a \odot b$ is in G.
(2) $(a \odot b) \odot c = a \odot (b \odot c)$.
(3) There exist solutions x and y in G of the equations $a \odot x = b$, $y \odot a = b$.

14. Does the binary Boolean algebra $(0, 1)$ form a group (a) under the operation of addition? (b) under the operation of multiplication?

15. Does point set Boolean algebra form a group (a) under the operation of union, $\cup$? (b) under the operation of intersection, $\cap$?

16. Consider the set of all points in a plane under the operation vector addition $(x_1, y_1) \oplus (x_2, y_2) = (x_1 + x_2, y_1 + y_2)$. Is this a group?

17. Consider the set of all points in the plane under the operation $(x_1, y_1) \odot (x_2, y_2) = (x_1 + y_2, x_2 + y_1)$. Is this a group?

18. Let ABC be an equilateral triangle. As group elements, we take the following:

R_{120} = a rotation (in the plane) through $120°$, about the center of the triangle.

R_{240} = a rotation (in the plane) through $240°$, about the center of the triangle.

R_{360} = a rotation (in the plane) through $360°$, about the center of the triangle.

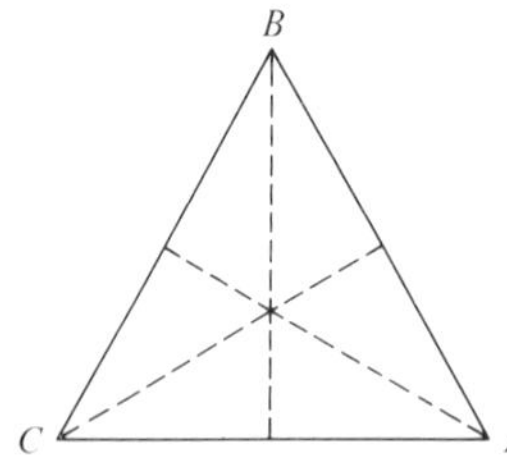

A = A flip (in space) about the altitude from A

B = A flip (in space) about the altitude from B

C = A flip (in space) about the altitude from C

   (a) Do these six elements form a group under the operation "following"?

$\star$(b) Compare the group so obtained with the group of substitutions (permutations) on three symbols given in Example 4, Section 4.3.

$\star$(c) Compare the group of part (a) with the group of Problem 13 Set 4.3.

19. Show that, for each integer n, it is possible to construct a group having exactly n elements (a group of order n) by considering

$$1, a, a^2, a^3, \cdots, a^{n-1}$$

where $a^n = 1$ and the operation is ordinary algebraic multiplication. Such a group is called *cyclic*, and the symbol a is called a *generator* of the group, since every group element is a power of a.

20. Show that there exist groups which are not generated by a single element, by considering the group consisting of the elements $1, 3, 5, 7$, where the operation is multiplication and reduction mod 8. [HINT: $1 \equiv (3)^2 \equiv (5)^2 \equiv (7)^2 \pmod{8}$. Hence, no element can generate all group elements.]

21. Show that $R_{360}, R_{270}, R_{180}, R_{90}$ form a subgroup of the octic group of Problem 12, Set 4.3.

4.5 ISOMORPHISM

Each of the following sets is a group having four elements (order 4).

(1) Elements: $1, 5, 7, 11$.   (Set 4.4, Problem 2)
Operation: multiplication and reduction mod 12.

(2) Elements: R_{360}, R_{180}, V, H.   (Section 4.3, Example 6)
Operation: following.

(3) Elements: $z, -z, \dfrac{1}{z}, -\dfrac{1}{z}$.   (Set 4.4, Problem 4)

Operation: substitution of the second element for z in the first. Abstractly, each of these groups has the same structure, which is displayed in the table below:

$\odot$	I	A	B	C
I	I	A	B	C
A	A	I	C	B
B	B	C	I	A
C	C	B	A	I

Noncyclic Four Group

The equation $B \odot A = C$ is interpreted as

$$\text{In (1):} \quad 7 \cdot 5 \equiv 11 \ (\text{mod } 12)$$
$$\text{In (2):} \quad V \odot R_{180} = H$$
$$\text{In (3):} \quad \frac{1}{z} \odot (-z) = -\frac{1}{z}$$

However, the group $(1, i, -1, -i)$, under ordinary multiplication, can *not* be made to fit this abstract pattern, since in the pattern $X^2 = I$, the identity, for each X in the group, but $(i)^2 = -1 \neq 1$. Thus, we have shown that groups of the same order do *not* necessarily have the same abstract structure. The abstraction of the group $(1, i, -1, -i)$ is as follows:

	I	A	B	C
I	I	A	B	C
A	A	B	C	I
B	B	C	I	A
C	C	I	A	B

Cyclic Four Group

where $B \odot A = C$ becomes $(-1) \cdot (i) = -i$.

The group 1, 4, 16, 13, under multiplication and reduction mod 17, is the same as the second (cyclic) four group. So is $R_{360}, R_{90}, R_{180}, R_{270}$, a subgroup of the group of Set 4.4, Problem 21.

It is true (you can prove it, if you wish) that every group of four elements is abstractly the same as either the noncyclic or the cyclic four group given above.

If two mathematical systems are abstractly identical (that is, identical except possibly for notation), they are said to be *isomorphic*. In particular, let G_a be a group having operation $\boxed{a}$ and elements $a_1, a_2, a_3, \cdots$, and G_b be a group having operation $\boxed{b}$ and elements $b_1, b_2, b_3, \cdots$. Let there exist a correspondence (mapping or function) $a_i \leftrightarrow b_i$ such that to each element $a_i \in G_a$ there corresponds one, and only one, element $b_i \in G_b$ and such that every $b_i \in G_b$ is the correspondent of exactly one element $a_i \in G_a$ (that is, the mapping is one-to-one reciprocal, or bi-unique). This correspondence is an *isomorphism*, $G_a \simeq G_b$, if "products" are preserved under the correspondence. That is, *if, for each a_i, a_j*

$$a_i \leftrightarrow b_i$$

$$a_j \leftrightarrow b_j$$

implies

$$(a_i \,\boxed{a}\, a_j) \leftrightarrow (b_i \,\boxed{b}\, b_j)$$

In the next problem set, the reader is asked to show that (1) isomorphism is an equivalence relation (reflexive, symmetric, and transitive), and (2) under any isomorphism between two groups, the group identity elements correspond, and (3) if $a_i \leftrightarrow b_i$, then $a_i^{-1} \leftrightarrow b_i^{-1}$.

Substitution (permutation) groups on n symbols having $n!$ elements (order $n!$) always exist. Example 4, Section 4.3, shows that the three symbols $A\ B\ C$ generate a group of $3! = 6$ elements:

$$\begin{pmatrix} A & B & C \\ A & B & C \end{pmatrix}, \quad \begin{pmatrix} A & B & C \\ B & C & A \end{pmatrix}, \quad \begin{pmatrix} A & B & C \\ C & A & B \end{pmatrix}$$

$$\begin{pmatrix} A & B & C \\ A & C & B \end{pmatrix}, \quad \begin{pmatrix} A & B & C \\ B & A & C \end{pmatrix}, \quad \begin{pmatrix} A & B & C \\ C & B & A \end{pmatrix}$$

However, the first *three* of these elements generate a subgroup having only 3 elements. An order n subgroup of the group of all permutations on n letters (order $n!$) is called a *regular permutation group of order n on n letters*.

We conclude this section with a truly remarkable theorem due to A. Cayley.

THEOREM 4.6 (CAYLEY)   *Every finite group G of order n is isomorphic to some regular substitution (permutation) group of order n on n letters.*

Our proof gives a method which could be used to actually construct the desired permutations.

Let a_1, a_2, $\cdots a_n$ be the distinct elements of the group G.

Let a_i be one of these elements.

Then

$$(a_1 \odot a_i), \quad (a_2 \odot a_i), \quad (a_3 \odot a_i), \quad \cdots, \quad (a_n \odot a_i)$$

are all distinct, since, if $a_r \odot a_i = a_s \odot a_i$, then upon multiplying on the right by a_i^{-1} we would have $a_r = a_s$, contrary to assumption. Since

$$(a_1 \odot a_i), \quad (a_2 \odot a_i), \quad (a_3 \odot a_i), \quad \cdots, \quad (a_n \odot a_i)$$

are n distinct elements of the group, every element of the group is represented among them. In short, they are the elements of the group in some order.

Let this order be:

$$a_{i_1}, \quad a_{i_2}, \quad a_{i_3}, \quad \cdots, \quad a_{i_n},$$

where $(a_1 \odot a_i) = a_{i_1}, (a_2 \odot a_i) = a_{i_2}$, etc.

Let a_i correspond to the permutation

$$a_i \leftrightarrow \begin{pmatrix} 1 & 2 & 3 & \cdots & n \\ i_1 & i_2 & i_3 & \cdots & i_n \end{pmatrix}$$

To complete the proof it is necessary to verify that the correspondence given above is actually an isomorphism; that is, that

$$(a_i \odot a_j) \leftrightarrow \begin{pmatrix} 1 & 2 & 3 & \cdots & n \\ i_1 & i_2 & i_3 & \cdots & i_n \end{pmatrix} \cdot \begin{pmatrix} 1 & 2 & 3 & \cdots & n \\ j_1 & j_2 & j_3 & \cdots & j_n \end{pmatrix}$$

Since $(a_i \odot a_j) = a_{ji}$, we have, from the construction of the permutation

$$(a_i \odot a_j) = a_{ji} \leftrightarrow \begin{pmatrix} 1 & 2 & 3 & \cdots & n \\ j_{i_1} & j_{i_2} & j_{i_3} & \cdots & j_{i_n} \end{pmatrix}$$

The isomorphism is proved if it can be shown that

$$\begin{pmatrix} 1 & 2 & 3 & \cdots & n \\ i_1 & i_2 & i_3 & \cdots & i_n \end{pmatrix} \cdot \begin{pmatrix} 1 & 2 & 3 & \cdots & n \\ j_1 & j_2 & j_3 & \cdots & j_n \end{pmatrix} = \begin{pmatrix} 1 & 2 & 3 & \cdots & n \\ j_{i_1} & j_{i_2} & j_{i_3} & \cdots & j_{i_n} \end{pmatrix}$$

This follows, since:

$$1 \rightarrow i_1 \rightarrow (\text{the } a_j \text{ transform of the element } i_1) = j_{i_1}$$
$$2 \rightarrow i_2 \rightarrow (\text{the } a_j \text{ transform of the element } i_2) = j_{i_2}$$
$$\vdots \quad \vdots \qquad \qquad \vdots \qquad \qquad \qquad \vdots$$
$$n \rightarrow i_n \rightarrow (\text{the } a_j \text{ transform of the element } i_n) = j_{i_n}$$

Why is the resulting group a regular permutation group of order n on n letters? This theorem is of great importance both in pure mathematics and in the applications of group theory, since permutation groups have been carefully investigated.

Problem Set 4.5

1. Show that group isomorphism is an equivalence relation as defined in Section 2.1.

2. Let G_1 be a group having identity u_1. Let G_2 be a group having identity u_2.

   (a) Show that, if $G_1 \simeq G_2$, then $u_1 \leftrightarrow u_2$.
   (b) Show that the converse of this theorem is false.

3. If G_1 and G_2 are isomorphic groups having corresponding elements $a_i \leftrightarrow b_i$, show that $a_i^{-1} \leftrightarrow b_i^{-1}$.

*4. Carry through the proof of Cayley's Theorem, 4.6, for the specific group given below:

	a_1	a_2	a_3
a_1	a_1	a_2	a_3
a_2	a_2	a_3	a_1
a_3	a_3	a_1	a_2

*5. Show that, in the proof of Cayley's theorem, the permutation which corresponds to a_i carries the first (unity) column of the multiplication table into the ith column.

6. The elements 1, 3, 5, 7 form a group under multiplication and reduction mod 8. Establish an isomorphism and show the correspondence between this group and one of the two abstract four groups (cyclic and noncyclic) given in this section.

7. The elements $1, \dfrac{-1 + i\sqrt{3}}{2}, \dfrac{-1 - i\sqrt{3}}{2}$ form a group under multiplication. The elements 0, 1, 2 form a group under addition and reduction mod 3. Are these two groups isomorphic?

8. The elements 1, 2, 3, 4 form a group under multiplication and reduction mod 5. To which abstract four group is it isomorphic?

9. The octic group (Set 4.3, Problem 12) has three subgroups of order 4

$$[R_{360}, H, V, R_{180}], \quad [R_{90}, R_{180}, R_{270}, R_{360}], \quad [R_{360}, D_1, D_2, R_{180}]$$

   Show that two of these subgroups are abstractly identical (isomorphic) to each other, but that the third one is not.

*10. The set of all even integers forms a group under addition. So does the set of all integers. Are these groups isomorphic?

11. Let G be the group consisting of the three elements 1, $-\frac{1}{2} + i\frac{\sqrt{3}}{2}$, $-\frac{1}{2} - i\frac{\sqrt{3}}{2}$, where the operation is ordinary multiplication and equality is equality in the usual (complex number) sense. Let H be the group having as elements the three equivalence classes into which congruence modulo 3 separates the integers and the operation addition with modular equality. Let K be the set of integers 1, 2, 3 with the operation of multiplication and equality as congruence mod 3.

    (a) Is K a group?
    (b) Are G and H isomorphic?
    (c) Is K isomorphic to either G or H?
    (d) Find a permutation group isomorphic to each of the groups mentioned in this problem.

12. Let M be the set of all *positive* real numbers under multiplication. Let L be the set of *all* real numbers under addition.

    (a) Show that M and L each are groups.
    (b) Show that M is isomorphic to L. [HINT: If you found it necessary to multiply several five-digit numbers together, what tables would you consult?]

13. Determine *all* possible (nonisomorphic) groups of order:

    (a) 2.   (b) 3.   (c) 4.   (d) 5.

*14. Show that there are exactly two distinct (nonisomorphic) groups of order 6.

15. Read "Theory of Braids" by E. Artin in the *American Scientist*, **38**, p. 112, and prepare a report thereon.

16. Every group is isomorphic to itself by the identity isomorphism. However, if G is the group of nonzero complex numbers under multiplication, $a + bi \leftrightarrow a - bi$ is also an isomorphism of G onto itself. Prove the latter statement. An isomorphism of a group onto itself is called an *automorphism*.

17. Show that every group with more than two elements has an automorphism (see Problem 16) other than the identity. [HINT: See Problem 4295, *American Mathematical Monthly*—Consult index.]

18. Would the correspondence of Problem 16 provide an automorphism of the additive group of complex numbers onto itself?

19. Read the article "On the Postulates Defining a Group" by J. V. Whittaker, in *American Mathematical Monthly* (November 1955), **62,** p. 636.

20. If G is the cyclic group containing 12 elements, how many of these 12 elements generate all of G? (See Problem 19, p. 142 for meaning of "generate.")

4.6 COSETS AND LAGRANGE THEOREM

Section 2.2 discussed the equivalence classes (baskets) into which an equivalence relation separates the objects of a set. These equivalence classes are used as elements of a mathematical system. For example, the set of all integers is separated into seven classes modulo 7, and these seven classes (baskets) are then treated as elements in the modulo 7 system. The integers are a group G under addition. The set, S, of all multiples of seven is a subgroup of G.

$$G(\cdots, -3, -2, -1, 0, 1, 2, 3, 4, \cdots) \supset S(\cdots, -14, -7, 0, 7, 14, 21, 28, \cdots).$$

We have defined

$$a \equiv b \pmod 7$$

to mean that there exists an element s of S such that $a = b + s$.

The concept of congruence and equivalence classes may profitably be generalized to abstract groups. Indeed one of the most important ideas in all of mathematics is that of clustering related items into the same equivalence class (basket) and then manipulating the baskets as if each member of the equivalence class behaved in the same way as every other member of the class.

The elementary school student who adds $\frac{1}{6}$ and $\frac{2}{15}$ by selecting $\frac{5}{30}$ as a representative of equivalence class $\frac{k}{6k}$ to which $\frac{1}{6}$ belongs, and $\frac{4}{30}$ as a representative of the equivalence class $\frac{2m}{15m}$ to which $\frac{2}{15}$ belongs and then uses the distributive property to obtain:

$$\frac{1}{6} + \frac{2}{15} = \frac{5}{30} + \frac{4}{30} = \frac{1}{30}(5 + 4) = \frac{9}{30}$$

is making use of this concept even though he would probably not verbalize it as we have.

The high-school student who, during one part of his geometry course, places all triangles similar to a given triangle in the same equivalence class is again using this principle. At a different point in his geometry course the same student will insist that, before two triangles can be placed in the same equivalence class, they must not only be similar in shape, but also equal in area (that is, be congruent). This variation in definition of equivalent sets seems to cause no serious difficulty in the student's thinking, and the concept proves useful in studying the structure of plane triangles.

Let us examine how one may go about separating the elements of a group into equivalence classes that will assist us in the study of group structure.

Let G be a group with elements $a, b, \cdots$ and an operation $\odot$ and let S be a subgroup of G. The relation

$$a \equiv b \pmod{S}$$

means that there exists an element $s \in S$ such that $a = b \odot s$.

Since G need not be commutative, we have defined *right congruence*. A similar definition involving $a = s \odot b$ may be made for *left congruence*. Congruence, under these definitions, is an equivalence relation. (This proof is requested in Problem 9, Set 4.6.) Numerical congruence discussed in Chapter 2 provides an important special case. Every cyclic group[5] is isomorphic either to the group of integers, I, under addition, or, if finite, is isomorphic to a modular (congruence) group, $I \pmod{S}$, where S is the subgroup of all multiples of some integer m; that is, to the modular systems studied in Chapter 2. This will not be proved here, but the proof is not beyond the reader's understanding. However, we prefer to devote the time to other material.

The effect of the relationship "congruence mod S," where S is a subgroup of G, is to separate the elements of G into equivalence classes (baskets). In group theory, it is usual to call such equivalence classes *cosets*. It is of

---

[5] A group is *cyclic* if there is an element a in the group such that each element of the group is equal to some power of a, that is, $\exists a \in G$, such that for each $b \in G$, $\exists k_b$ such that $b = a^{k_b}$.

extreme importance that these cosets are *distinct*. Essentially, this means that the equivalence relation is well defined. (See Section 4.2.)

Let us examine the cosets in detail for the octic group.

The octic group $G_8 = \{R_{360}, R_{270}, R_{180}, R_{90}, V, H, D_1, D_2\}$ was discussed in detail in Problem 12, Set 4.3 and elsewhere in this text. The set S consisting of the two elements $\{R_{360}, V\}$ forms a subgroup of G_8. According to the definition given above two elements a and b of G_8 belong to the same coset of G_8 mod S if and only if there exists an element $s \in S$ such that $a = b \odot s$.

Clearly one coset consists of the set $S\{R_{360}, V\}$ itself. (Why?)

$$S$$

$$\begin{Bmatrix} R_{360} \\ V \end{Bmatrix}$$

Let us consider an element b which belongs to G but which does *not* belong to S and form the product of this element b with each element s of S. If we select $b = D_1$, then from the multiplication table given below

G_8	R_{360}	R_{270}	R_{180}	R_{90}	H	V	D_1	D_2
R_{360}	R_{360}	R_{270}	R_{180}	R_{90}	H	V	D_1	D_2
R_{270}	R_{270}	R_{180}	R_{90}	R_{360}	D_2	D_1	H	V
R_{180}	R_{180}	R_{90}	R_{360}	R_{270}	V	H	D_2	D_1
R_{90}	R_{90}	R_{360}	R_{270}	R_{180}	D_1	D_2	V	H
H	H	D_1	V	D_2	R_{360}	R_{180}	R_{270}	R_{90}
V	V	D_2	H	D_1	R_{180}	R_{360}	R_{90}	R_{270}
D_1	D_1	V	D_2	H	R_{90}	R_{270}	R_{360}	R_{180}
D_2	D_2	H	D_1	V	R_{270}	R_{90}	R_{180}	R_{360}

one finds

$$S \qquad\qquad D_1 \odot S$$

$$\begin{Bmatrix} R_{360} \\ V \end{Bmatrix} \qquad D_1 \odot \begin{Bmatrix} R_{360} \\ V \end{Bmatrix} = \begin{Bmatrix} D_1 \\ R_{270} \end{Bmatrix}$$

so that $S = \{R_{360}, V\}$ and $D_1 \odot S = \{D_1, R_{90}\}$ are two of the cosets of G_8 mod S.

We next select another value of $b \in G$, but one which does not belong to either of the cosets already discussed. If we select $b = R_{180}$ and form $R_{180} \odot S$ we have

$$R_{180} \odot S = R_{180} \odot \begin{Bmatrix} R_{360} \\ V \end{Bmatrix} = \begin{Bmatrix} R_{180} \\ H \end{Bmatrix}$$

On the other hand if we had selected $b = D_2$ we would have obtained

$$D_2 \odot S = D_2 \odot \begin{Bmatrix} R_{360} \\ V \end{Bmatrix} = \begin{Bmatrix} D_2 \\ R_{90} \end{Bmatrix}$$

We have now exhausted the elements of G_8 and have separated the group

$$G_8 = \{R_{360}, R_{270}, R_{180}, R_{90}, V, H, D_1, D_2\}$$

into *disjoint cosets modulo a subgroup* $S = \{R_{360}, V\}$ as follows:

$$
\begin{array}{cccc}
S & D_1 \odot S & R_{180} \odot S & D_2 \odot S \\[4pt]
\begin{Bmatrix} R_{360} \\ V \end{Bmatrix} & \begin{Bmatrix} D_1 \\ R_{270} \end{Bmatrix} & \begin{Bmatrix} R_{180} \\ H \end{Bmatrix} & \begin{Bmatrix} D_2 \\ R_{90} \end{Bmatrix}
\end{array}
$$

We note that if we had selected R_{90} and formed $R_{90} \odot S$ the result

$$R_{90} \odot \begin{Bmatrix} R_{360} \\ V \end{Bmatrix} = \begin{Bmatrix} R_{90} \\ D_2 \end{Bmatrix}$$

is the same coset as was generated by D_2. We shall observe that, in general, since S is a sub*group* of G that if we generate two cosets $a \odot S$ and $b \odot S$, the cosets will either be identical or will be disjoint (that is, have no elements in common). This is a very important result.

THEOREM 4.7   If a group G is separated into cosets modulo a subgroup S of G, then any two cosets $a \odot S$ and $b \odot S$ are either disjoint or identical.

Let us represent the subgroup as

$$\left\{ \begin{array}{c} s_1 \\ s_2 \\ s_3 \\ s_4 \\ \vdots \\ s_k \\ \vdots \end{array} \right\}$$

$$\text{If} \quad a \odot \left\{ \begin{array}{c} s_1 \\ s_2 \\ s_3 \\ s_4 \\ \vdots \\ s_k \\ \vdots \end{array} \right\} = \left\{ \begin{array}{c} a \odot s_1 \\ a \odot s_2 \\ a \odot s_3 \\ a \odot s_4 \\ \vdots \\ a \odot s_k \\ \vdots \end{array} \right\} \quad \text{and} \quad b \odot \left\{ \begin{array}{c} s_1 \\ s_2 \\ s_3 \\ s_4 \\ \vdots \\ s_k \\ \vdots \end{array} \right\} = \left\{ \begin{array}{c} b \odot s_1 \\ b \odot s_2 \\ b \odot s_3 \\ b \odot s_4 \\ \vdots \\ b \odot s_k \\ \vdots \end{array} \right\}$$

and let us assume that some element, say $a \odot s_i \in \{a \odot S\}$ is equal to some element, $b \odot s_j \in \{b \odot S\}$. It then follows that

$$a \odot s_i = b \odot s_j$$

Since S is a subgroup, it follows that $s_i^{-1} \in S$ also and that

$$a = b \odot s_j \odot s_i^{-1}$$

Since S is closed under the operation $\odot$ it follows that $s_j \odot s_i^{-1} = s_k \in S$ for some k and hence that

$$a = b \odot s_j \odot s_i^{-1} = b \odot s_k \in \{b \odot S\}$$

which is exactly the definition of $a \equiv b \bmod S$ so that a and b are actually members of the same (not different) cosets. Hence any two cosets of $G \bmod S$ are either disjoint or identical. (Why?)

We recapitulate: Using the union, ∪, introduced in Section 3.11, we see that G is separated into distinct cosets having union G

$$G_8 = (R_{360}, V) \cup (R_{90}, D_2) \cup (R_{180}, H) \cup (R_{270}, D_1)$$

The only coset which is a group is (R_{360}, V), since only (R_{360}, V) contains the identity element of G_8.

If G is any group having a subgroup S, then (right) congruence modulo S separates G into cosets. No element is contained in two cosets, and each coset contains the same number of elements as S. The only coset which is a subgroup of G is S itself. (Prove this.)

The proofs of these statements are analogous to those just given for the group G_8. For example, to prove that no element is contained in two different cosets, assume element x is in the coset $\{a, as_2, \cdots\}$ and that x also appears in coset $\{b, bs_2, \cdots\}$. Then $\exists$ s_i, s_j in S such that $b \odot s_j = x$ and $x = a \odot s_i$. Hence, by the transitive law, $b \odot s_j = a \odot s_i$. Since S is a subgroup, s_j^{-1} is also in S. Thus, $b = a \odot s_i \odot s_j^{-1}$, which implies $\{bS\}$ and $\{aS\}$ are the same coset. If two cosets of the subgroup S of a group G have a common element, x, it follows that the two cosets are identical. Since congruence modulo a subgroup S divides G into cosets having no common elements and such that each coset contains the same number of elements, the Theorem of LaGrange follows at once. (Why?)

THEOREM 4.8 (LAGRANGE)[6]:   *The order (number of elements in) of a subgroup, S, of a finite group, G, is a divisor of the order of G.*

Problem Set 4.6

1. Determine the cosets into which the octic group G_8 is separated by the subgroup $S_4 = (R_{360}, R_{270}, R_{180}, R_{90})$.

2. Same as Problem 1 for the subgroup $S_2 = (R_{360}, H)$.

3. Let G be the group generated by the element b and its powers, under the restriction that $b^{12} = 1$, the group identity, where the operation

---

[6] Joseph Louis Lagrange (1736–1813) was a noted eighteenth century French mathematician. Interestingly enough Lagrange apparently showed little interest in or aptitude for mathematics until he was 17 years old—but by the time he was 23 his outstanding ability was recognized by the mathematical and scientific world of his day. Frederick the Great wrote to Lagrange in 1766 stating that "the greatest king in Europe (Frederick) wanted the greatest mathematician in Europe (Lagrange) at his court."

is multiplication. Let S be the subgroup consisting of the elements $1, b^4, b^8$. Find the cosets into which G is separated mod S.

*4. Show that, if a is an element of a finite group G, then there exists an integer k such that $a^k = 1$. [HINT: If the group G has g elements in it, then $1, a, a^2, a^3, a^4, \cdots, a^g$ cannot all be distinct. If $a^i = 1$, the proof is complete. If $a^i = a^m$, then either $a^{i-m} = 1$, or $a^{m-i} = 1$.]

5. Show that the cosets of integers modulo m form a commutative group of order m.

6. (a) Show that the cosets of integers which are relatively prime to 30 form a multiplicative group mod 30.
   (b) Show that the cosets of integers which are relatively prime to m form a multiplicative group modulo m.

7. Show that if G is a group and if G is separated into cosets by congruence modulo a subgroup S of G then one and only one of the cosets is itself a group.

8. If m is 24, is the group of Problem 7 isomorphic to a subgroup of that of Problem 5?

9. Prove that right congruence, as described above, is an equivalence relation.

10. Let C be a finite *cyclic* group containing g distinct elements (that is, of order g). Then from LaGrange's theorem it follows that the order of any subgroup of C is a divisor of g. Show that furthermore if d is a positive integer that divides g, then the cyclic group C has *one and only one* subgroup of *order d*.

11. Show, by giving counterexamples, that neither of these conclusions (a) "one" and (b) "only one" of Problem 10 can be guaranteed if C is merely a *finite group* rather than a finite *cyclic* group.

12. There are two theorems which you have encountered thus far in this chapter that are so important that your author has associated the name of a mathematician with each. State each of these theorems and then without consulting the proof given in the text, prove each.

13. An interesting exercise is to see how much *you* can determine about a particular group given just a very few facts about it. If, for example, I tell you that a given group G about which I am thinking contains an element b which is of period 15 you can deduce a number of other

properties of G. For example, either the order (number of elements) of G is ≥ 15 or G contains an infinitude of elements. You can also deduce the periods of each of 15 distinct elements of G. If I add that G also contains an element whose period is 14 you can now prove that G contains at least 210 elements and furthermore you can determine the periods of each of the 210 elements which you guarantee that G contains. You can also identify at least ten subgroups of G. Prove each of the above statements. Don't just state things that seem likely, either quote a theorem in this book to justify each assertion, or create a proof of your own. Are any of the subgroups mentioned above isomorphic?

14. What can *you* deduce if the two elements of Problem 13 have periods 17 and 5 respectively?

15. Make up some statements of your own concerning one or two elements of a group and swap problems with a classmate. Determine how much you can correctly deduce about each other's group and compare results.

16. (a) Prove that a subgroup of a cyclic group is always cyclic.
    (b) Show that a noncyclic group may have both cyclic and noncyclic subgroups, or may have only cyclic subgroups, but can*not* have only noncyclic subgroups.

17. Find a subgroup of the octic group G_8 other than the subgroup (R_{360}, V) used in this section and determine the disjoint cosets into which G_8 is separated modulo your subgroup.

*18. The "defining relations"

$$P^3 = Q^2 = R^2 = I \text{ (the identity)}$$

$$QP = PQR, \qquad RP = PQ, \qquad RQ = QR$$

are sufficient to enable you to create a 12-element group known as A_{12} the alternating group on four symbols. The twelve elements of this group are I, P, P^2, Q, PQ, P^2Q, R, PR, P^2R, QR, PQR, and P^2QR. Determine its multiplication table. [HINT: $P^2Q \odot PQR = P^2(QP)QR = P^2(PQR)QR = P^3QRQR = IQ(RQ)R = Q(QR)R = Q^2 \cdot R^2 = I \cdot I = I$].

*19. Find several nonisomorphic subgroups of group A_{12} of Problem 18.

*20. The group A_{12} of Problem 18 has C_3 the cyclic group of order 3 as one of its subgroups. Determine the disjoint cosets into which A_{12} is separated modulo C_3.

21. Let G be a group containing 26 elements, one of which is the identity element u. Either show that $b^{26} = u$ for every $b \in G$, or produce a counterexample.

4.7 FREE GROUPS

Among the various applications of group theory, a special type of group known as a *free group* has recently become important for its applications.[7] The elements of a *free group* consist of strings of symbols with integral exponents. If, for example, we consider all possible strings of the three symbols x, y, z with integral exponents some typical elements will be

$$x^4 y^{-2} x \qquad z^3 x^2 y z x z^{-2} y^3 \qquad z x y x y x$$

The usual compactifications permitted with integer exponents in a *non-commutative* system are permitted, thus:

$$x^2 x y x^{-3} y^2 y^4 x \text{ is the same as}$$

$$x^3 y x^{-3} y^6 x \text{ but is } not \text{ the same as}$$

$$y^7 x \text{ nor as } x y^7 \text{ which are also distinct.}$$

The element $x y^3 y^{-4} y x$ reduces to x^2. The null element x^0 consisting of no symbols is usually represented by the digit 1, or by the Greek letter ϵ or λ. We now have defined a set of elements consisting of strings of letters with integral exponents. Two elements are said to be equal if they can be reduced to the same form. If we introduce the operation $\odot$ of juxtaposition such that

$$z x y^2 z^{-1} \odot z y = z x y^2 z^{-1} z y = z x y^3$$

the resulting system can be shown to be a group.

The sophisticated reader will note that we have actually "slid over" some rather sticky questions. For example, it may be "obvious" that the same reduced form is obtained no matter in what order one attacks the problem of reducing a complicated word, but that does not obviate the

---

[7] See, for example, R. Crowell and R. H. Fox, *Introduction to Knot Theory* (Boston: Ginn, 1963), Chapters 3 and 4, and S. Ginsburg, *Mathematical Theory of Context-Free Languages* (New York: McGraw-Hill, 1966), pp. 2ff.

necessity of proving the validity of this observation, if a proof can be found. Similarly one really should prove that the operation $\odot$ of free group multiplication on strings of symbols is well defined (see page 123) and associative. We leave such proofs along with the proof that the equivalence classes set up by "reduces to the same reduced form" are actually disjoint and exhaustive to more sophisticated texts. Our purpose is to introduce groups with another type of element and to point out that such groups have important applications in other areas including topology and mathematical linguistics.

Problem Set 4.7

1. Show that, if associativity is assumed, the elements obtained as strings of the two symbols x and t with integral exponents do form a group under the operation of juxtaposition in which the element $a^0 = 1$ is the identity. Be sure to explain how to obtain the inverse of a given element. (This is an example of a "sticky question" on which your author is willing to accept a general "this is how to do it" proof rather than the tedious general induction necessary to meet today's standards of rigor, *providing that you, the student, realize that you are only presenting a "justification" and that there are mathematicians who, quite rightly, demand greater rigor.*) The purpose of this exercise is pedagogical, not to establish the validity of some new mathematics.

2. Prove that a free group has infinitely many elements.

3. The set of integers forms a group under addition. Create a "free group" that is isomorphic to the group of integers under addition.

4. (a) Create a free group that is *not* isomorphic to the set of integers under addition.
   (b) Does the free group which you created in 4(a) have a proper subgroup which is isomorphic to the integers under addition?

★4.8 QUOTIENT GROUPS. JORDAN-HÖLDER THEOREM

It has been pointed out that, since a group need not be commutative, different cosets may be obtained from left congruence $(a = s \odot b)$ and right congruence $(a = b \odot s)$. Examples show that for a given (non-commutative) group G, a certain subgroup, S, may yield the same cosets from right congruence as from left congruence, while a different subgroup, S^*, may not have this property. A subgroup S which separates G into the same cosets (equivalence classes) by right congruence as by left congruence, is called an *invariant* subgroup *of G*. It should be noted that invariance refers to a "set operation" not an "element operation." If the subgroup of elements $S = \{s_1, s_2, \cdots, s_k\}$ is denoted by $\{S\}$ then invariance means that $a \odot \{S\} = \{S\} \odot a$ for all $a \in G$. This does *not* imply that for an individual element s_i, $a \odot s_i = s_i \odot a$, but merely that for each $s_i \in S$, there exists an $s_j \in S$ such that $a \odot s_i = s_j \odot a$. Note that $\{S\} \odot \{S\} = \{S\}$. The importance of invariant subgroups is emphasized in the following theorem.

THEOREM 4.9   *The cosets (equivalence classes) of a group G, modulo a subgroup S form a group if, and only if, S is an invariant subgroup.*

Closure and associativity of cosets follow from the respective properties of the elements and the nature of an equivalence relation. The coset consisting of the subgroup S is the identity. A coset $\{a \cdot S\}$ has a well-defined coset inverse *if, and only if,* right and left congruence give the same cosets. The coset inverse is established by noting that if S is invariant, then

$$a \odot \{S\} = \{S\} \odot a$$

and

$$[a \odot \{S\}] \odot [a^{-1} \odot \{S\}] = [\{S\} \odot a] \odot [a^{-1} \odot \{S\}] = \{S\} \odot 1 \odot \{S\} = \{S\}$$

the coset identity.
Conversely, if

$$[a \odot \{S\}] \odot [a^{-1}\{S\}] = 1 \odot \{S\}$$

then,

$$a \odot \{S\} \odot a^{-1} = \{S\} \odot \{S\} = \{S\}$$

or

$$a \odot \{S\} = \{S\} \odot a$$

as desired.

The coset group formed by the equivalence classes modulo an invariant subgroup, H, of a group, G, is called a *factor group* or *quotient group* of G modulo H and is denoted by the symbol G/H or $G \pmod{H}$. The quotient group G/H is *not* necessarily isomorphic to a subgroup of G. It is, however, uniquely determined by G and the invariant subgroup H. The order of the quotient group is the order of G divided by the order of H, providing G is finite. (Why?)

An invariant subgroup H is called *maximal* in G if H is not properly contained in any other proper invariant subgroup of G. (This does *not* mean that a *maximal* invariant subgroup must be the largest in the sense of having the most elements or being the invariant subgroup of the largest order.) The cyclic group of order 6, $G(a,\ a^2, a^3,\ a^4,\ a^5,\ a^6 = 1)$ has subgroups of order 2, $S_2(a^3, 1)$, and of order 3, $S_3(a^2, a^4, 1)$, each of which is a maximal invariant subgroup of G. An important theorem, which sheds considerable insight into the structure of groups, deals with a chain of quotient groups and enables one to prove, for example, that there exist algebraic equations of degree five which have no solution in terms of radicals.

THEOREM 4.10 (JORDAN-HÖLDER)

Let G be of finite group having maximal invariant subgroups H_1 and K_1. Let H_1 have maximal invariant subgroup H_2 which has a maximal invariant subgroup H_3, and so on, obtaining two chains of subgroups of G

$$G \supset H_1 \supset H_2 \supset H_3 \supset \cdots \supset I$$

$$G \supset K_1 \supset K_2 \supset \cdots \supset I$$

where each subgroup is a maximal invariant subgroup of the subgroup just preceding it (but not necessarily even invariant in G nor in any other subgroups except its direct predecessor). Then:

(1) *The number of subgroups in each chain is the same.*

(2) *The quotient groups $\dfrac{G}{H_1}, \dfrac{H_1}{H_2}, \dfrac{H_2}{H_3},$*

$\cdots, \dfrac{H_m}{I}$ are isomorphic (in some

order) to the quotient groups

$\dfrac{G}{K_1}, \dfrac{K_1}{K_2}, \dfrac{K_2}{K_3}, \cdots \dfrac{K_m}{I}$

In the case of the cyclic group of order 6, $G(a, a^2, a^3, a^4, a^5, a^6 = 1)$, having chains

$$G \supset H_1(a^3, 1) \supset I(1)$$

and

$$G \supset K_1(a^2, a^4, 1) \supset I(1)$$

the quotient groups are

$$\frac{G}{H_1} = \text{(cyclic 3 group)}, \quad \frac{H_1}{I} = \text{(cyclic 2 group)}$$

$$\frac{G}{K_1} = \text{(cyclic 2 group)}, \quad \frac{K_1}{I} = \text{(cyclic 3 group)}$$

$$\frac{G}{H_1} \simeq \frac{K_1}{I}, \qquad \text{while} \qquad \frac{H_1}{I} \simeq \frac{G}{K_1}$$

A proof of the Jordan–Hölder Theorem will be found in *Introduction to Abstract Algebra* by C. MacDuffee (New York: Dover, 1956) on page 65 and elsewhere. R. Bruck and others have generalized this theorem to nonassociative systems called *loops*. These ideas are studied further in courses of modern abstract algebra.

Problem Set 4.8

1. Determine the invariant subgroups of the octic group G_8. Find the corresponding quotient groups.

2. Determine the various quotient group chains of the octic group G_8. Display the isomorphism described in the Jordan-Hölder Theorem.

*4.9 SUGGESTIONS FOR INDEPENDENT INVESTIGATION

Galois Theory

Evariste Galois (1811–1832) was a hot-blooded young man who twice failed his college entrance examination (L'Ecole Polytechnique in Paris). He had been imprisoned as a revolutionist and political extremist and had led an extremely frustrating existence—yet on the night before *he was killed in a framed duel when he was only* 20 *years old*, he wrote out some of his mathematical ideas—ideas in which, among other things, he applied group theory to show that there exist polynomial equations which can*not* be solved in terms of the coefficients using only addition, subtraction, multiplication, division, and the extraction of roots as can quadratic, cubic, and quartic equations. If you would be interested in reading more about the life of Galois, read the book *Whom the Gods Love* by Leopold Infeld[8] or consult some of the standard mathematical histories such as *Men of Mathematics* by E. T. Bell.[9] If you wish a simplified and very readable account of part of the mathematics developed by Galois during his brief lifetime, your author recommends the charming little book *Galois and the Theory of Groups* by H. G. & L. R. Lieber or *Abstract Algebra and Solutions by Radicals* by J. E. & M. W. Maxfield.

Further Readings

We have only begun the study of groups in this chapter. Entire books have been written on the Theory of Groups. Below you will find a list of a few of these. Consult your school library under the call numbers:

Dewey: 512.86
Library of Congress: QA 171

to see what it contains.

Problem Set 4.9

1. Consult another text on algebra or group theory. Find one theorem that has *not* been mentioned in this text and explain the meaning (*not* the proof) of the selected theorem to a mathematics student or teacher other than your instructor.

2. What is a quasigroup?

3. What is a semigroup? Is the concept of a semigroup more or less general than the concept of a group?

4. What is a loop? Is every loop a group or is every group a loop or neither (or, heaven forbid, both)?

---

[8] New York: McGraw-Hill, 1948.
[9] New York: Simon & Schuster, 1937.

Selected Reading List, Chapter 4
GROUPS

The following books and journals will provide additional reading for students interested in term projects or independent study.

Books

Dixon, J. D., *Problems in Group Theory*. Waltham, Mass.: Blaisdell, 1967.

Kurosh, A., *Theory of Groups*. New York: Chelsea, Vol. 1, 1956, Vol. 2, 1960.

Lieber, H. G. and L. R. *Galois and the Theory of Groups*. New York: Science Press, 1956.

Maxfield, J. and M. *Abstract Algebra and Solutions by Radicals*. Philadelphia: Saunders, 1971.

Articles

Curtis, C. W., "The Classical Groups as a Source of Algebraic Problems," *The American Mathematical Monthly*, **74,** No. 1, Part 2 (Jan. 1967), pp. 80–94.

Deskins, W. E., "A Characterization of Finite Supersolvable Groups," *The American Mathematical Monthly*, **75,** No. 2 (Feb. 1968), p. 180.

DiPaola, J. W., "When is a Totally Symmetric Loop a Group?," *The American Mathematical Monthly*, **76,** No. 3 (Mar. 1969), p. 249.

Feit, W., "A Group-Theoretic Proof of Wilson's Theorem," *The American Mathematical Monthly*, **65,** No. 2 (Feb. 1958), p. 120.

Frame, J. S., "Symmetry Groups and Molecular Structure," *Pi Mu Epsilon Journal*, Spring 1959, p. 463.

Frank, T. S., "On Groups, Quasi and Otherwise," *The American Mathematical Monthly*, **74,** No. 8 (Oct. 1967), pp. 938–942.

Mech, W. P., "Graphs of Groups," *Journal of Undergraduate Mathematics*, **1,** No. 2 and **2,** No. 1 (1969, 1970). An interesting two part paper written by an undergraduate student.

Oakley, C. O., and R. J. Wisner, "Flexagons," *The American Mathematical Monthly*, **64,** No. 3 (March, 1957), p. 143.

Spitznagel, E. L., Jr., "A New Look at the Fifteen Puzzle," *Mathematics Magazine*, **40,** No. 4 (Sept. 1967), pp. 171–174.

Zassenhaus, H., "What Makes a Loop a Group," *The American Mathematical Monthly*, **75,** No. 2 (Feb. 1968), p. 139.

"An Abelian Group," Problem E 1996, *The American Mathematical Monthly*, **75,** No. 8 (Oct. 1968), p. 904.

"An Algebraic Theory of English Pronominal Reference," *Semiotica* (1969), pp. 397–421.

"An Impossible Group Table," Problem E 1957, *The American Mathematical Monthly*, **75,** No. 5 (May 1968), p. 548.

"Group Under Multiplication," Problem E 1914, *The American Mathematical Monthly*, **75,** No. 1 (Jan. 1968), p. 83.

"Postulates for a Group," Problem E 2026, *The American Mathematical Monthly*, **75,** No. 10 (Dec. 1968), p. 1121.

Notes on Chapter 4

MATRICES

5

5.1 INTRODUCTION

There are many algebras in which $A \cdot B$ and $B \cdot A$ are not the same thing. The groups studied in Chapter 4 provide numerous examples. Portions of modern physics, chemistry, psychology, and statistics are based on such "noncommutative" systems, as they are called. The following simple experiment provides a physical example in which $A \cdot B$ and $B \cdot A$ are not identical.

Experiment: Place two closed books flat on the table in front of you with their faces upward and their spines (bound edges) on the left. This is the normal position in which a book might lie before it is opened. The book will remain closed throughout the experiment.

Rotate the first book through 90° about its bottom edge. (It will now be standing upright on the table.) Now rotate the same book through 90° about its spine. Leave the book in this position.

Rotate the second book through 90° about its spine. (If the book were released at this point, it would fall open in reading position.) Now rotate it through 90° about its bottom edge.

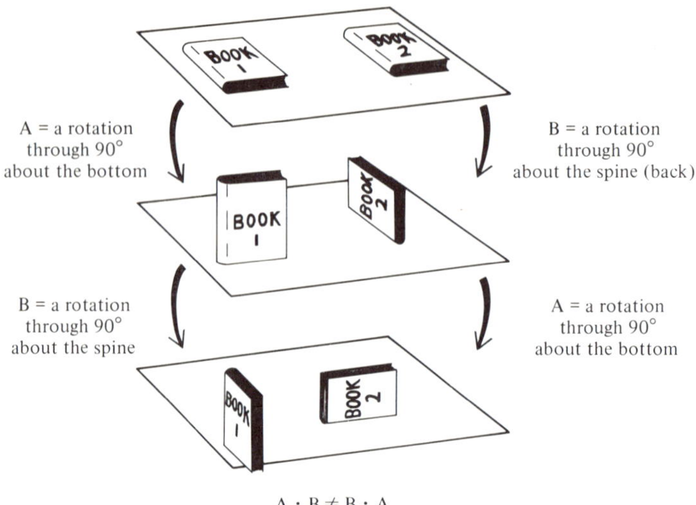

$$A \cdot B \neq B \cdot A$$

Note that the two books are not in the same final position. Each book has been rotated through 90° about its bottom edge and 90° about its spine, but the order was not the same and the results are different. It is possible to use matric theory to forecast the result of these operations, and of much more complicated rotations in three-dimensional, four-dimensional, or higher-dimensional space.

A matrix is a rectangular array of numbers (*elements*) for which multiplication is defined in a special way. A matrix should not be confused with a determinant, which is a single number or value associated with a square array. (See Chapter 7.) A matrix is the array itself. Two matrices are said to be equal if, and only if, the elements in corresponding positions are equal. For example, if the elements are ordinary integers,

$$\begin{bmatrix} 7 & 2 \\ 3 & 5 \end{bmatrix} = \begin{bmatrix} 5+2 & 2 \\ 3 & 5 \end{bmatrix}, \quad \text{but} \quad \begin{bmatrix} 7 & 2 \\ 3 & 5 \end{bmatrix} \neq \begin{bmatrix} 7 & 3 \\ 2 & 5 \end{bmatrix}$$

The matrices $\begin{bmatrix} 3 & 9 \\ -2 & 8 \end{bmatrix}$ and $\begin{bmatrix} -3 & 3 \\ 4 & 2 \end{bmatrix}$ are unequal if the elements are integers, but they are equal if the elements are the integers mod 6. (Why? Would they be equal if the elements were integers mod 5?, mod 3?) Thus, to discuss matrices, it is necessary first to consider the set from which the elements of the matrix are to be selected. Equality of matrices depends upon the equivalence relation used in the set from which the elements of the matrix are selected. If nothing is said to indicate otherwise, we assume that the elements are selected from the complex numbers or from some subset of the complex numbers such as the rational numbers or the integers.

5.2 MATRIX PRODUCT

$$\text{If } M = \begin{bmatrix} a & b \\ c & d \end{bmatrix} \quad \text{and} \quad N = \begin{bmatrix} w & x \\ y & z \end{bmatrix}$$

are two matrices, then their product $M \cdot N$ is defined to be

$$M \cdot N = \begin{bmatrix} a & b \\ c & d \end{bmatrix} \cdot \begin{bmatrix} w & x \\ y & z \end{bmatrix} = \begin{bmatrix} aw + by & ax + bz \\ cw + dy & cx + dz \end{bmatrix}$$

The element in the first (horizontal) row and second (vertical) column of the product $M \cdot N$ is a sum of elements, each of which is the product of an element from the first row of M multiplied by a corresponding element from the second column of N.
Thus,

$$\begin{bmatrix} a & b \\ * & * \end{bmatrix} \cdot \begin{bmatrix} * & x \\ * & z \end{bmatrix} = \begin{bmatrix} * & ax + bz \\ * & * \end{bmatrix}$$

In a similar fashion, the element in row R and column C of the product $M \cdot N$ is the sum of the products of the elements of the Rth row of M multiplied by the corresponding elements of the Cth column of N.

$$\text{If } A = \begin{bmatrix} 1 & -1 \\ 3 & 2 \end{bmatrix} \quad \text{and} \quad B = \begin{bmatrix} 7 & 4 \\ 5 & 8 \end{bmatrix}$$

$$\text{then} \quad A \cdot B = \begin{bmatrix} 1 & -1 \\ 3 & 2 \end{bmatrix} \cdot \begin{bmatrix} 7 & 4 \\ 5 & 8 \end{bmatrix}$$

$$= \begin{bmatrix} (1)(7) + (-1)(5) & (1)(4) + (-1)(8) \\ (3)(7) + (2)(5) & (3)(4) + (2)(8) \end{bmatrix} = \begin{bmatrix} 2 & -4 \\ 31 & 28 \end{bmatrix}$$

However,

$$B \cdot A = \begin{bmatrix} 7 & 4 \\ 5 & 8 \end{bmatrix} \cdot \begin{bmatrix} 1 & -1 \\ 3 & 2 \end{bmatrix}$$

$$= \begin{bmatrix} (7)(1) + (4)(3) & (7)(-1) + (4)(2) \\ (5)(1) + (8)(3) & (5)(-1) + (8)(2) \end{bmatrix} = \begin{bmatrix} 19 & 1 \\ 29 & 11 \end{bmatrix}$$

Thus, in the system of matrices, $A \cdot B$ and $B \cdot A$ are not necessarily the same. Although the matrices discussed above have 2 rows and 2 columns, later we shall find important applications for $m \times n$ matrices having m rows and n columns.

The reader may check his or her understanding of matrix multiplication by showing that

$$\begin{bmatrix} 3 & -5 \\ 1 & 2 \end{bmatrix} \cdot \begin{bmatrix} 7 & 4 \\ 6 & 10 \end{bmatrix} = \begin{bmatrix} -9 & -38 \\ 19 & 24 \end{bmatrix}$$

Matrices larger than the 2×2 matrices just described, which contain hundreds of elements, are used in many practical applications. At the Naval Ordnance Testing Station, matrices are used in computations involving rocket and projectile flight. Matrices are used in modern economic theory. The branch of psychology known as factor analysis applies matrix methods. Systems of 35 (or more) equations in 35 (or more) unknowns which arise in industrial research may be neatly solved using matrix methods. Competent biologists and geneticists find matrix methods helpful in the study of the complex interrelations of heredity and genetics. Large laboratories and oil refineries often ask universities to recommend graduates who are facile in the use of matrices.

The widespread utility of matrices stems from their unusual method of multiplication in which each element of the product matrix is obtained through the interaction of several elements of the original matrices. This will be examined in greater detail after elementary properties have been studied.

5.3 PAULI MATRICES

The Pauli matrices, used in the study of electron spin in quantum mechanics, have an interesting arithmetic. If $i^2 = -1$, the Pauli matrices are:

$$I = \begin{bmatrix} 1 & 0 \\ 0 & 1 \end{bmatrix}, \qquad A = \begin{bmatrix} -i & 0 \\ 0 & i \end{bmatrix}, \qquad B = \begin{bmatrix} 0 & 1 \\ -1 & 0 \end{bmatrix},$$

$$C = \begin{bmatrix} 0 & -i \\ -i & 0 \end{bmatrix}, \qquad D = \begin{bmatrix} -1 & 0 \\ 0 & -1 \end{bmatrix}, \qquad E = \begin{bmatrix} i & 0 \\ 0 & -i \end{bmatrix},$$

$$F = \begin{bmatrix} 0 & -1 \\ 1 & 0 \end{bmatrix}, \qquad G = \begin{bmatrix} 0 & i \\ i & 0 \end{bmatrix}$$

The Pauli matrices form a closed set under matrix multiplication; that is, the product of two or more Pauli matrices is again a Pauli matrix. (Try it and see.)

Problem Set 5.3

In Problems 1–10, form the products $A \cdot B$ and $B \cdot A$ for the given matrices:

1. $A = \begin{bmatrix} 14 & 1 \\ -2 & 7 \end{bmatrix}$ $\qquad\qquad B = \begin{bmatrix} 3 & 20 \\ 5 & -6 \end{bmatrix}$

2. $A = \begin{bmatrix} 3 & -5 \\ 1 & 2 \end{bmatrix}$ $\qquad\qquad B = \begin{bmatrix} 7 & 4 \\ 6 & 10 \end{bmatrix}$

3. $A = \begin{bmatrix} 1 & 2 \\ 9 & 8 \end{bmatrix}$ $\qquad\qquad B = \begin{bmatrix} 1 & 3 \\ 4 & 9 \end{bmatrix}$

4. $A = \begin{bmatrix} 4 & 1 \\ 7 & 5 \end{bmatrix}$ $\qquad\qquad B = \begin{bmatrix} 2 & 7 \\ 9 & 2 \end{bmatrix}$

5. $A = \begin{bmatrix} 1 & 3 \\ 2 & 9 \end{bmatrix}$ $\qquad\qquad B = \begin{bmatrix} -5 & 8 \\ 4 & 17 \end{bmatrix}$

6. $A = \begin{bmatrix} 1 & 4 \\ 17 & 9 \end{bmatrix}$ $\qquad\qquad B = \begin{bmatrix} 20 & 30 \\ 40 & 50 \end{bmatrix}$

7. $A = \begin{bmatrix} 81 & 76 \\ 45 & 22 \end{bmatrix}$ $\qquad$ $B = \begin{bmatrix} 1 & 2 \\ -1 & 3 \end{bmatrix}$

8. $A = \begin{bmatrix} 1 & 0 \\ 0 & 1 \end{bmatrix}$ $\qquad$ $B = \begin{bmatrix} 9 & 13 \\ 27 & 65 \end{bmatrix}$

9. $A = \begin{bmatrix} 1 & 0 \\ 1 & 0 \end{bmatrix}$ $\qquad$ $B = \begin{bmatrix} 2 & 9 \\ 11 & -15 \end{bmatrix}$

10. $A = \begin{bmatrix} 15 & 7 \\ 3 & 14 \end{bmatrix}$ $\qquad$ $B = \begin{bmatrix} 0 & 0 \\ 1 & 1 \end{bmatrix}$

Problems 11–20 consist of reworking Problems 1–10 when it is understood that: (a) the mod 7 system is used; (b) the mod 4 system is used. Reduce the matrices first, then perform the multiplications in the modular system. Check your result by reducing the integral answers to Problems 1–10 mod 7 and mod 4.

21. Construct a multiplication table for the Pauli matrices defined in Section 5.3.

22. (a) Thus far, only one operation (multiplication) has been defined for the Pauli matrices. Section 1.4 lists a set of eight postulates which are satisfied by integral domains. Some of these postulates deal with addition, and hence cannot yet have meaning for the set of Pauli matrices. Are any of the postulates which deal with multiplication satisfied by the Pauli matrices? Do not attempt to prove that multiplication is associative. It is, but the proof is not easy. [HINT: The Pauli matrix I is the multiplicative identity.]

    (b) Do the Pauli matrices form a group?

23. (a) Make a multiplication table for the subsystem of the Pauli matrices consisting of the four matrices I, B, D, F.

    (b) Show that this subsystem is closed and commutative, under multiplication, and has a unity (multiplicative identity). Show that that the cancellation law holds. Is it a group?

    (c) Is it isomorphic to the cyclic four group given in Section 4.5?

24. Show that a matrix of the form $\begin{bmatrix} b & 0 \\ 0 & b \end{bmatrix}$ commutes with all 2×2 matrices even though matrices in general are not necessarily commutative.

25. Find the most general matrix which will commute with $\begin{bmatrix} 4 & 1 \\ 3 & 0 \end{bmatrix}$.

$$\left[\text{HINT:}\quad \begin{bmatrix} 4 & 1 \\ 3 & 0 \end{bmatrix} \cdot \begin{bmatrix} a & b \\ c & d \end{bmatrix} = \,?;\qquad \begin{bmatrix} a & b \\ c & d \end{bmatrix} \cdot \begin{bmatrix} 4 & 1 \\ 3 & 0 \end{bmatrix} = \,?\right]$$

26. Do the matrices $\begin{bmatrix} 1 & 0 \\ 0 & 1 \end{bmatrix}$ and $\begin{bmatrix} 1 & 0 \\ 1 & 0 \end{bmatrix}$ form a group under matrix multiplication?

27. Do the matrices $\begin{bmatrix} -1 & 0 \\ 0 & 1 \end{bmatrix}$ and $\begin{bmatrix} 1 & 0 \\ 0 & 1 \end{bmatrix}$ form a group?

28. Do the Pauli matrices I, A, C, D, E, G form a group?

5.4 SQUARE MATRICES

Generalizing from 2×2 square matrices to $n \times n$ square matrices, in which multiplication is defined in a similar row by column fashion, presents no new difficulty.

EXAMPLE 1

$$A \cdot B = \begin{bmatrix} 1 & 2 & 4 \\ 3 & -2 & 5 \\ 4 & 2 & -1 \end{bmatrix} \cdot \begin{bmatrix} 0 & 1 & -2 \\ 3 & 0 & 1 \\ 4 & 0 & 2 \end{bmatrix}$$

$$= \begin{bmatrix} 1\cdot 0 + 2\cdot 3 + 4\cdot 4 & 1\cdot 1 + 2\cdot 0 + 4\cdot 0 & 1\cdot(-2) + 2\cdot 1 + 4\cdot 2 \\ 3\cdot 0 - 2\cdot 3 + 5\cdot 4 & 3\cdot 1 - 2\cdot 0 + 5\cdot 0 & 3\cdot(-2) - 2\cdot 1 + 5\cdot 2 \\ 4\cdot 0 + 2\cdot 3 - 1\cdot 4 & 4\cdot 1 + 2\cdot 0 - 1\cdot 0 & 4\cdot(-2) + 2\cdot 1 - 1\cdot 2 \end{bmatrix} \cdot$$

$$\begin{bmatrix} 22 & 1 & 8 \\ 14 & 3 & 2 \\ 2 & 4 & -8 \end{bmatrix} = P$$

The reader is expected to compute the product of these matrices for himself, on scratch paper, to be certain that he follows the procedure. It is essential to do one's own pencil work rather than to nod "yes" to the steps supplied in the text. The reader gains useful insight into the process. For example, it is *not* just coincidence that the first column of the matrix A is identical with the second column of the product matrix P. Do not continue until you are certain that for any 3×3 matrix M, the product $M \cdot B$ will have this property if the second column of B is

$$\begin{bmatrix} * & 1 & * \\ * & 0 & * \\ * & 0 & * \end{bmatrix}$$

The $n \times n$ matrix I_n, which has 1's down the main (upper left to lower right) diagonal and zeros elsewhere, has the property demanded of the unity (multiplicative identity) in Section 1.4, namely: $M \cdot I_n = I_n \cdot M = M$, where M is an $n \times n$ matrix. The reader is asked to prove this in Problem 1 of the next set. It may help to consider:

$$I_n \cdot M = \begin{bmatrix} 1 & 0 & 0 & 0 & 0 & 0 & \cdots & 0 \\ 0 & 1 & 0 & 0 & 0 & 0 & \cdots & 0 \\ 0 & 0 & 1 & 0 & 0 & 0 & \cdots & 0 \\ 0 & 0 & 0 & 1 & 0 & 0 & \cdots & 0 \\ 0 & 0 & 0 & 0 & 1 & 0 & \cdots & 0 \\ 0 & 0 & 0 & 0 & 0 & 1 & \cdots & 0 \\ 0 & 0 & 0 & 0 & \cdot & \cdot & \cdots & 0 \\ \vdots & & & & & & & \vdots \\ 0 & 0 & 0 & 0 & 0 & 0 & \cdots & 1 \end{bmatrix} \begin{bmatrix} A_{11} & A_{12} & \cdots & A_{1n} \\ A_{21} & A_{22} & \cdots & A_{2n} \\ A_{31} & A_{32} & \cdots & A_{3n} \\ \cdot & & & \cdot \\ \cdot & & & \cdot \\ \cdot & & & \cdot \\ \cdot & & & \cdot \\ A_{n1} & A_{n2} & \cdots & A_{nn} \end{bmatrix} = \;?$$

It is necessary to consider this product in the commuted order $M \cdot I_n$ to complete the proof. The use of the "Kronecker delta" δ_{ij} is helpful. By definition $\delta_{ij} = 0$ if $i \neq j$ and $\delta_{ii} = 1$, thus the element in row i, column j of the identity matrix I is δ_{ij}.

THEOREM 5.1 *The matrix I_n, containing ones down the main diagonal and zeros elsewhere, has the property that*

$$I_n \cdot M = M \cdot I_n = M$$

for every $n \times n$ matrix M.

The reader should gain considerable understanding of the effect of certain so-called elementary matrices as he works the next set of exercises. Finally, Problems 21–34 sum up much of the theory of row and column transformations. You are encouraged to conjecture what these effects are as soon as you are ready to do so. Then, be sure to *prove* or disprove your conjecture. This may easily be your first opportunity to discover significant mathematical theorems for yourself. Often, texts state theorems for the student to prove. In this problem set you have an opportunity to *invent* the theorems before attempting to prove them.

Problem Set 5.4

1. Prove Theorem 5.1.

Use the matrices given below to form the products requested in Problems 2–20.

$$A = \begin{bmatrix} 0 & 1 & 0 \\ 1 & 0 & 0 \\ 0 & 0 & 1 \end{bmatrix}, \qquad B = \begin{bmatrix} 1 & 0 & 0 \\ 0 & 1 & 0 \\ 1 & 0 & 1 \end{bmatrix}, \qquad C = \begin{bmatrix} 4 & 0 & 0 \\ 0 & -1 & 0 \\ 0 & 0 & 1 \end{bmatrix},$$

$$D = \begin{bmatrix} 0 & 0 & 1 \\ 1 & 0 & 0 \\ 0 & 0 & 0 \end{bmatrix}, \qquad E = \begin{bmatrix} 1 & 5 & -1 \\ 2 & 4 & -2 \\ 3 & 2 & 6 \end{bmatrix}, \qquad F = \begin{bmatrix} 2 & 1 & 4 \\ 3 & 6 & 0 \\ 7 & 2 & 1 \end{bmatrix},$$

$$G = \begin{bmatrix} 1 & 1 & 0 \\ 4 & 0 & 2 \\ 2 & 6 & 1 \end{bmatrix}, \qquad X = \begin{bmatrix} a & b & c \\ d & e & f \\ g & h & i \end{bmatrix}.$$

2. (a) $A \cdot E$ (b) $A \cdot F$ 12. (a) $E \cdot B$ (b) $F \cdot B$

3. (a) $A \cdot G$ (b) $A \cdot X$ 13. (a) $G \cdot B$ (b) $X \cdot B$

4. (a) $B \cdot E$ (b) $B \cdot F$ 14. (a) $E \cdot C$ (b) $F \cdot C$

5. (a) $B \cdot G$ (b) $B \cdot X$ 15. (a) $G \cdot C$ (b) $X \cdot C$

6. (a) $C \cdot E$ (b) $C \cdot F$ 16. (a) $E \cdot D$ (b) $F \cdot D$

7. (a) $C \cdot G$ (b) $C \cdot X$ 17. (a) $G \cdot D$ (b) $X \cdot D$

8. (a) $D \cdot E$ (b) $D \cdot F$ 18. (a) $E \cdot F$ (b) $F \cdot E$

9. (a) $D \cdot G$ (b) $D \cdot X$ 19. (a) $G \cdot E$ (b) $E \cdot G$

10. (a) $E \cdot A$ (b) $F \cdot A$ 20. (a) $F \cdot G$ (b) $G \cdot F$

11. (a) $G \cdot A$ (b) $X \cdot A$

21. On the basis of Problems 2(a), 2(b), and 3(a), make a conjecture on the effect of left multiplication by the matrix A. Does Problem 3(b) prove your conjecture?

22. Using Problems 4 and 5, make and verify a conjecture about left multiplication by the matrix B.

23. Using Problems 6 and 7, make and verify a conjecture about left multiplication by the matrix C.

24. Using Problems 8 and 9, make and verify a conjecture about left multiplication by the matrix D.

25.–28. Using the results of Problems 10–17, make and verify conjectures concerning the effect of right multiplication by matrices A, B, C, D, respectively.

29. Consider the results of Problems 21–28. Verify the conjecture that the four matrices A, B, C, D have a certain effect on the rows of X, when X is multiplied on the left by one of them, while if X is multiplied on the right by that matrix, it has a related effect upon the columns of X.

30. Try to generalize the results of Problem 29 to more general matrices than A, B, C, and D.

31. Show that, if the identity matrix has two of its rows interchanged (as in A, for example), then the product (changed identity) $\cdot (X)$ is the matrix X with the corresponding rows interchanged.

32. Show that, if the identity matrix has one of its rows multiplied by k, then the product (changed identity) $\cdot (X)$ is the matrix X with the corresponding row multiplied by k.

33. Show that, if the identity matrix has one of its rows added to another row, then the product (changed identity) $\cdot (X)$ is the matrix X with the corresponding row additions.

34. Make conjectures similar to those made in Problems 31–33, but involving changes on the columns of the identity matrix and the commuted product $(X) \cdot$ (changed identity).

Find the product of the 6×6 matrices formed in the manner indicated from the matrices given before Problem 2, where θ is the 3×3 matrix having zero for each element, and I the 3×3 identity matrix.

35. $\begin{bmatrix} A & \theta \\ \theta & B \end{bmatrix} \cdot \begin{bmatrix} E & F \\ G & C \end{bmatrix}$ 37. $\begin{bmatrix} A & \theta \\ \theta & I \end{bmatrix} \cdot \begin{bmatrix} E & G \\ G & C \end{bmatrix}$

36. $\begin{bmatrix} C & \theta \\ \theta & D \end{bmatrix} \cdot \begin{bmatrix} E & F \\ G & C \end{bmatrix}$ 38. $\begin{bmatrix} C & \theta \\ \theta & D \end{bmatrix} \cdot \begin{bmatrix} \theta & E \\ F & \theta \end{bmatrix}$

★39. Prove that the matrix $\begin{bmatrix} 0 & 0 \\ -1 & 1 \end{bmatrix}$ cannot be an element of a group whose operation is matrix multiplication and having the matrix I as the identity element of the group. Does the matrix given above form a one-element group?

★40. Can the matrix $\begin{bmatrix} 1 & 0 & 0 \\ 0 & 0 & 0 \\ 0 & 0 & -1 \end{bmatrix}$ appear as an element of a group whose operation is matrix multiplication?

★5.5 A PROOF OF THE ASSOCIATIVITY OF MATRICES, USING Σ NOTATION

In the last Problem Set, the reader was given an opportunity to make certain generalizations concerning matrices. Before summarizing and further generalizing these observations, in Sections 5.6–5.8 we prove the associative and distributive laws for matrices having elements taken from an integral domain. If the reader and his instructor are *both* willing to accept the fact that the multiplication of matrices is an associative operation, $(A \cdot B) \cdot C = A \cdot (B \cdot C)$, this section may be omitted. The author's personal feeling is that, for many students, the introductory purpose of this course is better served by assuming the associative law at this point and continuing with Section 5.6. Good students will wish to study this section whether or not it is assigned in general.

Mathematicians use the Greek letter Σ to represent a sum; thus, $\sum_{i=1}^{n} x_i$ represents the sum $x_1 + x_2 + x_3 + \cdots + x_n$. It is also common practice to let $A = (a_{rc})$ represent a matrix having a_{13} as the element in the first row and third column. In general, a_{rc} is the element in the rth row and cth column. If $A = (a_{rc})$ and $B = (b_{rc})$ are two matrices whose product is defined, then the element in the jth row and kth column of the product matrix $(A \cdot B)$ is c_{jk}.

$$
\begin{bmatrix}
* & * & \cdot & \cdot & * \\
\cdot & & & & \cdot \\
\cdot & & & & \cdot \\
\cdot & & & & \cdot \\
a_{j1} & a_{j2} & \cdot & \cdot & a_{jn} \\
\cdot & & & & \cdot \\
\cdot & & & & \cdot \\
\cdot & & & & \cdot \\
* & & & & *
\end{bmatrix}
\begin{bmatrix}
* & \cdots & b_{1k} & \cdots & * \\
& & & & \\
* & \cdots & b_{2k} & & * \\
& & b_{3k} & & \\
\cdot & & \cdot & & \cdot \\
\cdot & & \cdot & & \cdot \\
\cdot & & \cdot & & \cdot \\
* & & b_{nk} & & *
\end{bmatrix}
=
\begin{bmatrix}
* & \cdots & \cdots & * \\
& & & \\
\cdot & & c_{jk} & \cdot \\
\cdot & & & \cdot \\
\cdot & & & \cdot \\
* & \cdots & \cdots & *
\end{bmatrix}
$$

where
$$
c_{jk} = a_{j1}b_{1k} + a_{j2}b_{2k} + a_{j3}b_{3k} + \cdots + a_{jn}b_{nk}
$$

$$
= \sum_{i=1}^{n} a_{ji}b_{ik}
$$

Hence,

$$
(a_{rc}) \cdot (b_{rc}) = \left[\left(\sum_{i=1}^{n} a_{ri}b_{ic} \right)_{rc} \right]
$$

This notation is often used in advanced work and is well adapted to giving neat appearing proofs. A proof that matrix multiplication is associative, and which makes use of this notation, follows.

We wish to show that $(A \cdot B) \cdot D = A \cdot (B \cdot D)$.

Let $\qquad (A \cdot B) \cdot D = H \qquad$ and $\qquad A \cdot (B \cdot D) = K$

Also, let $\qquad A \cdot B = S \qquad$ and $\qquad B \cdot D = T$

so that $\qquad S \cdot D = H \qquad$ and $\qquad A \cdot T = K$

Then
$$h_{rc} = \sum_i s_{ri} d_{ic}$$

where
$$s_{ri} = \sum_i a_{rj} b_{ji}, \qquad h_{rc} = \sum_i \left(\sum_j a_{rj} b_{ji} \right) d_{ic}$$

and
$$k_{rc} = \sum_j a_{rj} t_{jc}$$

where
$$t_{jc} = \sum_i b_{ji} d_{ic}, \qquad k_{rc} = \sum_j a_{rj} \left(\sum_i b_{ji} d_{ic} \right)$$

Since i and j are independent, and since a_{rj} is not affected by i, we have, for each r, c,
$$h_{rc} = \sum_{i,j} a_{rj} b_{ji} d_{ic} = k_{rc}$$

or $H = K$.

Thus $(A \cdot B) \cdot D = A \cdot (B \cdot D)$.

Problem 16 of Problem Set 5.8 asks for a similar proof of the distributive law.

Problem Set 5.5

1. Work through the above proof using 3×3 matrices.

2. You already know that in general $A \cdot B$ and $B \cdot A$ are *not* equal for arbitrary $n \times n$ matrices. Set up the first few steps using the notation of this section for an "attempted proof" that $A \cdot B = B \cdot A$. Explain the denouement.

3. A matrix formed by replacing all the ones of the identity matrix by k's is called a scalar matrix. Use the notation of this section to prove that multiplying a matrix M by a scalar matrix K (on either side—the proofs differ slightly) has the effect of multiplying each element of the matrix M by the common diagonal element, k, of K. It is usual to denote the scalar product $K \cdot M$ as $k \cdot M$.

*4. Define $\delta_{ij} = \begin{cases} 1 & \text{if} \quad i = j \\ 0 & \text{if} \quad i \neq j \end{cases}$. This is known as Kronecker's delta,

after the famous German mathematician L. Kronecker (1823–1891).

(a) Show that $I = (\delta_{rc})$ and that $\sum_{i=1}^{n} (\delta_{ri} \cdot \delta_{ic}) = \delta_{rc}$.

5.6 ELEMENTARY ROW OPERATIONS

In speaking of a row or a column of a matrix, we refer to all of the elements in that particular row or column. We now define *elementary row operations on a matrix*.

Definition. The elementary row operations on a matrix M are:

(1) The interchange of two rows.
(2) The multiplication of a row by a nonzero constant. (Really, by a constant having an inverse in the system from which the elements are chosen. In the case of the real, rational, or complex fields, this means a nonzero constant.)
(3) The addition of k times one row to another. (For example, replacing the fourth row by the elementwise sum of the fourth plus eight times the second row.)

An elementary (row) matrix is a matrix which produces one of the elementary (row) transformations when used as a (left) multiplier.

The symbol E_1 is used to represent the matrix resulting from a single elementary row transformation of type one on the $n \times n$ identity matrix.

Examples of Possible E_1 Matrices

$$\begin{bmatrix} 1 & 0 & 0 \\ 0 & 0 & 1 \\ 0 & 1 & 0 \end{bmatrix}, \qquad \begin{bmatrix} 0 & 1 \\ 1 & 0 \end{bmatrix}, \qquad \begin{bmatrix} 0 & 0 & 1 & 0 \\ 0 & 1 & 0 & 0 \\ 1 & 0 & 0 & 0 \\ 0 & 0 & 0 & 1 \end{bmatrix}$$

The product matrix $E_1 \cdot M$ equals the matrix obtained by performing the corresponding row operation on M. (See Problem Set 5.4.)

The symbol E_2 is used to represent the matrix resulting from a single elementary row transformation of type two on the $n \times n$ identity matrix.

Examples of Possible E_2 Matrices

$$\begin{bmatrix} 1 & 0 & 0 \\ 0 & -3 & 0 \\ 0 & 0 & 1 \end{bmatrix}, \quad \begin{bmatrix} 1 & 0 \\ 0 & 3/17 \end{bmatrix}, \quad \begin{bmatrix} 1 & 0 & 0 & 0 \\ 0 & 1 & 0 & 0 \\ 0 & 0 & 1 & 0 \\ 0 & 0 & 0 & 74/91 \end{bmatrix}$$

As before, matrix $E_2 \cdot M$ equals the matrix obtained by performing the corresponding row operations on M.

The symbol E_3 is used to represent the matrix resulting from a single elementary row transformation of type three on the $n \times n$ identity matrix.

Examples of Possible E_3 Matrices

$$\begin{bmatrix} 1 & 0 & 0 & 0 \\ 0 & 1 & 0 & 0 \\ 0 & 0 & 1 & 0 \\ 0 & 8 & 0 & 1 \end{bmatrix}, \quad \begin{bmatrix} 1 & -5 \\ 0 & 1 \end{bmatrix}, \quad \begin{bmatrix} 1 & 0 & 0 \\ 0 & 1 & -\frac{1}{2} \\ 0 & 0 & 1 \end{bmatrix}$$

The matrix $E_3 \cdot M$ equals the matrix obtained by performing the corresponding row operations on M.

Similar operations are defined for columns.

Since each of the elementary row operations may be "undone" by another elementary row transformation of the same type (but not necessarily the same transformation), it follows that each elementary matrix has an inverse matrix which is also an elementary matrix. This fact leads to the following theorem.

THEOREM 5.2 $\qquad$ *If M is an elementary matrix, or a product of elementary matrices, then there exists a matrix M^{-1} such that $M \cdot M^{-1} = M^{-1} \cdot M = I$, the identity matrix.*

The proof that there exists a matrix M^{-1} such that $M \cdot M^{-1} = I$ follows at once, since, if

$$M = E_i \cdot E_j \cdot \ \cdots \ \cdot E_r \cdot E_s$$

then

$$M^{-1} = E_s^{-1} \cdot E_r^{-1} \cdot \ \cdots \ \cdot E_j^{-1} \cdot E_i^{-1}$$

has the property that

$$M \cdot M^{-1} = M^{-1} \cdot M = I$$

since matrix multiplication is associative. (See Section 5.6.)

Since, in general, elementary row operations change the matrix, it is generally *not* correct to say that $A = B$, where B is obtained by applying elementary row and/or column transformations to A. Instead we shall write $A \cong B$. The symbol $\cong$ is read "is equivalent to." If only row operations are used, the matrices are said to be row equivalent. The reader will be asked in Problem 21, Set 5.6 to show that "equivalence" of square matrices is an equivalence relation according to the definition given in Section 2.1.

In Section 5.4 and Problem Set 5.4, the reader was given an opportunity to discover and prove certain important theorems for 3×3 matrices. These theorems are now stated for matrices in general.

THEOREM 5.3   *Let a series of elementary row operations be performed on an identity matrix I to obtain $E \cong I$. If the same series of elementary row operations is performed (in the same order) on a matrix M, the resulting matrix is the same as the product $E \cdot M$.*

THEOREM 5.4   *Let a series of elementary column operations be performed on an identity matrix I to obtain $E \cong I$. If the same series of elementary column operations is performed (in the same order) on a matrix M, the resulting matrix is the same as the product $M \cdot E$.*

These theorems indicate the difference in effect between right and left multiplication by a matrix, and may even suggest a reason why matrix multiplication is not commutative.

The proof of Theorems 5.3 and 5.4 for a general $n \times n$ matrix is accomplished by multiplying M on the left by the sequence of elementary matrices needed to produce the change in M, and then using the associative law several times.

For example:

$$E'_1(E_3[E_1 M]) = (E'_1 E_3 E_1) M = EM$$

Similar arguments can be made for column operations and right multipliers.

Problem Set 5.6

1. How many different 3×3 elementary matrices of type one (row interchange) are there?

2. Do elementary row matrices of type one on 3×3 matrices commute with one another?

3. How many different 3×3 elementary matrices are there of type two (multiplication of each element in a given row by a constant which has an inverse in the system):

   (a) If the elements of the matrix are rational numbers?
   (b) If the elements of the matrix are integers modulo 7?
   (c) If the elements of the matrix are integers modulo 12?

4. Do elementary matrices of type two commute with one another under conditions (a), (b), (c) of Problem 3?

5. How many elementary matrices of type three are there under conditions (a), (b), (c) of Problem 3?

6. Do elementary matrices of type three commute with one another under conditions (a), (b), (c) of Problem 3?

7. Find 4×4 matrices L_1, L_2, L_3 which will accomplish, respectively, elementary row transformations of type one, two, and three when they are used as left multipliers.

8. Find 4×4 matrices R_1, R_2, R_3 which will accomplish, respectively, elementary column transformations of types one, two, and three when they are used as right multipliers.

9. Find a matrix N such that

$$N \cdot \begin{bmatrix} 1 & 2 & 3 \\ 7 & 9 & 5 \\ 4 & 1 & 2 \end{bmatrix} = \begin{bmatrix} 1 & 2 & 3 \\ 8 & 11 & 8 \\ 3 & -1 & -1 \end{bmatrix}$$

   [HINT: The rows of the product matrix may be obtained from rows of the right factor by a series of elementary row operations.]

10. Prove that, if a matrix M can be transformed into the identity matrix I by a series of elementary row transformations, and if the same

series of row transformations is applied in the same order to a matrix I to obtain a matrix Q, then $Q = M^{-1}$. This gives a reasonable method of computing M^{-1} if the latter exists. If M^{-1} does not exist (consider $\begin{bmatrix} 8 & 6 \\ 12 & 9 \end{bmatrix}$, for example), then it is impossible to transform M into an identity matrix by elementary row transformations.

11. Show that the process of finding M^{-1}, by the operation suggested in Problem 10, can be simplified. One need not keep track of the row operations performed, if an $n \times 2n$ matrix (M, I), where I is the $n \times n$ identity matrix, is formed and then reduced to the equivalent matrix (I, Q) by elementary *row* operations. Why does $Q = M^{-1}$?

In Problems 12–18, use the method of Problem 11 to compute the inverse of the given matrices, or to prove that no inverse exists. If an inverse is found, check by showing that $M \cdot M^{-1} = I$.

12. $\begin{bmatrix} 1 & 3 & 2 \\ 4 & 9 & 6 \\ 1 & 8 & 1 \end{bmatrix}$

13. $\begin{bmatrix} 1 & 2 & 3 \\ 3 & 21 & 1 \\ 1 & 1 & 1 \end{bmatrix}$

14. $\begin{bmatrix} 10 & 5 & 6 \\ 1 & 9 & 2 \\ 8 & 3 & 1 \end{bmatrix}$

15. $\begin{bmatrix} 1 & 2 & 3 & 4 \\ 5 & 4 & 9 & 1 \\ 3 & 1 & 2 & 7 \\ 4 & 6 & 1 & 8 \end{bmatrix}$

16. $\begin{bmatrix} 6 & 2 & 3 & 2 \\ 2 & 1 & 0 & 1 \\ 2 & 1 & 1 & 1 \\ 6 & 0 & -1 & 1 \end{bmatrix}$

17. $\begin{bmatrix} 1 & 0 & 0 & 0 & 0 \\ 0 & 2 & 0 & 0 & 0 \\ 0 & 0 & 3 & 0 & 0 \\ 0 & 0 & 0 & 4 & 0 \\ 0 & 0 & 0 & 0 & 5 \end{bmatrix}$

18. $$\begin{bmatrix} 1 & 0 & 0 & 1 & 2 & 3 \\ 0 & 1 & 0 & 3 & 2 & 1 \\ 0 & 0 & 1 & 1 & 1 & 1 \\ 1 & 3 & 2 & 10 & 5 & 9 \\ 4 & 9 & 6 & 1 & 9 & 2 \\ 1 & 8 & 1 & 8 & 3 & 1 \end{bmatrix}$$

19. Prove that, if a matrix A has a two-sided inverse, then this inverse is unique. [HINT: Let X and Y be two matrices such that $XA = I = AX$, $YA = I = AY$. Since $XA = YA$, $(XA)X = (YA)X$. By using the associative law and the fact that $AX = I$, it is now possible to conclude $X = Y$.] It may be of more than casual interest to note that there are matrices (having infinitely many elements) in which there exist several one-sided inverses, but no two-sided inverse; i.e., there exist several matrices X such that $XA = I$, but none such that $AX = I$. This does not occur in the case of square, finite-dimensional matrices having exact elements—the type discussed here. Students interested in learning more about infinite matrices may consult the booklet *Infinite Matrices* published by the Galois Institute. Non-square matrices are discussed in Section 5.10.

20. Explain why it would be improper to use the cancellation law on $X \cdot A = Y \cdot A$ in Problem 19 to conclude $X = Y$.

21. Does the operation $\cong$ obey the three postulates of an equivalence relation given in Section 2.1?

22. Show that, if $A \cong B$, then there exist matrices P and Q, each of which has an inverse, such that $P \cdot A \cdot Q = B$.

23. Let A and B each be $n \times n$ matrices such that $A \cdot B = 0$. Does it then follow that $B \cdot A = 0$? Prove that it does follow, or give a counter-example.

24. Let T be an $n \times n$ matrix having each element below its main diagonal (upper left to lower right) zero, and nonzero elements on its main diagonal. Show that T is row equivalent to the identity matrix.

25. State and prove a converse of the theorem of Problem 24. Could there be more than one possible converse?

26. Read the account of various methods of matrix inversion given by D. Greenspan, in *American Mathematical Monthly*, **62** (1955), pp. 303–319.

5.7 ADDITION OF MATRICES

Matrices of the same size are added element-wise. Matrices of different sizes cannot be added.

EXAMPLE 2

$$\begin{bmatrix} 2 & 7 & 3 \\ 5 & -3 & 2 \\ 4 & 6 & -1 \end{bmatrix} + \begin{bmatrix} 1 & 2 & 5 \\ -3 & 4 & 6 \\ 4 & 3 & 3 \end{bmatrix} = \begin{bmatrix} 3 & 9 & 8 \\ 2 & 1 & 8 \\ 8 & 9 & 2 \end{bmatrix}$$

Since addition is element-wise, matrices have the same additive properties as does the system from which the elements are chosen. (Prove this.) A similar statement concerning multiplicative properties would be false. (Why?)

5.8 DOMAIN PROPERTIES OF SQUARE MATRICES

Section 1.4 lists postulates for an integral domain. These postulates are not all satisfied by the set of all $n \times n$ (say 5×5) matrices with elements taken from an integral domain—not even if the elements are taken from the rational, real, or complex numbers, for example. Our next inquiry is, "which of the domain postulates are necessarily satisfied by the set of all $n \times n$ matrices?"

(1) Since the sum and the product of two $n \times n$ matrices is again an $n \times n$ matrix, the closure postulate is satisfied.
(2) Addition is commutative, providing the elements from which they are formed are commutative under addition. It was shown in Section 5.2 that matrix multiplication is *not* commutative.
(3) Matrices are associative under both addition and multiplication, providing the elements from which they are formed have appropriate properties. (See Section 5.5.) The integers, rational numbers, real numbers, and the complex numbers are all associative. Can you think of any systems which are not associative? They do exist.

(4) The matrix having zeros in every position is the additive identity or zero matrix. (Prove this.)

(5) The matrix I_n, having ones down the main diagonal and zeros elsewhere, is the multiplicative identity or unity. (Theorem 5.1.)

(6) If M is a matrix, then the matrix $(-1) \cdot M$, (that is, the matrix M with the sign of each element changed) is the additive inverse of M. (Prove this.)

(7) The cancellation law does *not* hold. This may be shown by examining one counterexample. (Why?) Consider the matrices:

$$C = \begin{bmatrix} 1 & 2 \\ 3 & 6 \end{bmatrix} \neq z, \qquad A = \begin{bmatrix} 0 & 4 \\ 4 & 6 \end{bmatrix}, \qquad B = \begin{bmatrix} 2 & 6 \\ 3 & 5 \end{bmatrix}$$

Then,

$$C \cdot A = \begin{bmatrix} 1 & 2 \\ 3 & 6 \end{bmatrix} \cdot \begin{bmatrix} 0 & 4 \\ 4 & 6 \end{bmatrix} = \begin{bmatrix} 8 & 16 \\ 24 & 48 \end{bmatrix}$$

and

$$C \cdot B = \begin{bmatrix} 1 & 2 \\ 3 & 6 \end{bmatrix} \cdot \begin{bmatrix} 2 & 6 \\ 3 & 5 \end{bmatrix} = \begin{bmatrix} 8 & 16 \\ 24 & 48 \end{bmatrix}$$

but

$$A = \begin{bmatrix} 0 & 4 \\ 4 & 6 \end{bmatrix} \neq \begin{bmatrix} 2 & 6 \\ 3 & 5 \end{bmatrix} = B$$

(8) The distributive law, like the associative law, is proved using the $\sum$ notation presented in Section 5.5.

Thus the system of all $n \times n$ square matrices with elements in an integral domain satisfies the postulates for an integral domain *with the exceptions of the commutative law of multiplication and the cancellation law.* A system satisfying these postulates is called a *ring with unity.* Such systems are of considerable importance.

In the next problem set, the reader will be asked to discover whether or not the matrix product $a \cdot b = 0$ implies that either $a = 0$ or $b = 0$, as in the case in an integral domain. (Theorem 1.1.)

Matrices, in general, need not be commutative under multiplication, and hence one may rightfully wonder what is meant by M^3. It could mean $M \cdot M^2$, or it could mean $M^2 \cdot M$. The fact that matrix multiplication is associative obviates this decision.

Since

$$a \cdot (b \cdot c) = (a \cdot b) \cdot c$$

it follows that

$$M \cdot (M \cdot M) = (M \cdot M) \cdot M$$

or

$$M \cdot M^2 = M^2 \cdot M$$

Hence, while not all matrices have commutative multiplication, nevertheless *there are some matrices which do commute*. It will be shown that the only matrices which commute with *all* other matrices are of the form $(k\delta_{ij})$; that is, matrices having k, where the identity matrix has ones, and zeros elsewhere. All powers of a given matrix (or even polynomials in a given matrix), however, commute with one another.

Problem Set 5.8

1. (a) Provide counterexamples for each integral domain postulate which does not hold for the set of all 3×3 matrices having rational elements.
   (b) Give an illustration using 3×3 matrices for each integral domain postulate which does hold for the set of all 3×3 matrices having rational elements.
   $\star$(c) Try to find an integral domain postulate, other than the cancellation law and the commutative law of multiplication, which is violated in the set of matrices with elements taken from the integers mod 6.

2. Let $D = \begin{bmatrix} 1 & 0 & -35 \\ 0 & 1 & 18 \\ 0 & 0 & 0 \end{bmatrix}$. Find D^2, D^3, and D^{317}.

3. Let $F = \begin{bmatrix} -1 & 1 & -1 & 1 \\ -3 & 2 & -1 & 0 \\ -3 & 1 & 0 & 0 \\ -1 & 0 & 0 & 0 \end{bmatrix}$. Show that $F^3 = I$.

4. Find F^{125}.

5. Show that the matrices $\begin{bmatrix} 4 & 0 \\ 0 & 4 \end{bmatrix}$, $\begin{bmatrix} 1 & 0 \\ 0 & 1 \end{bmatrix}$, and $\begin{bmatrix} 3 & -2 \\ -1 & 2 \end{bmatrix}$ all satisfy the matrix equation $X^2 - 5X + 4 = 0$; that is,

$$X^2 - \begin{pmatrix} 5 & 0 \\ 0 & 5 \end{pmatrix} \cdot X + \begin{pmatrix} 4 & 0 \\ 0 & 4 \end{pmatrix} = \begin{pmatrix} 0 & 0 \\ 0 & 0 \end{pmatrix}.$$

This is quite different from equations in the real, or even the complex, numbers, where an equation may *not* have more solutions than its degree.

6. Let $G = \begin{bmatrix} 0 & 1 & 0 & 0 & 0 \\ 0 & 0 & 1 & 0 & 0 \\ 0 & 0 & 0 & 1 & 0 \\ 0 & 0 & 0 & 0 & 1 \\ 0 & 0 & 0 & 0 & 0 \end{bmatrix}$. Find G^2, G^3, G^4, G^5, G^6, and G^{273}.

7. Let M be a square matrix. Show that $M^s \cdot M^t = M^t \cdot M^s$ for integral exponents s and t.

8. An integral domain has the property that a product $F \cdot G$ of two elements from the domain is zero *if, and only if,* at least one of these elements is zero. This rule is *not* valid in the mod 12 system. For example, $10 \neq 0 \pmod{12}$ and $6 \neq 0 \pmod{12}$, but $10 \cdot 6 = 60 \equiv 0 \pmod{12}$. You have shown in this section that a product of two matrices is the zero matrix if either factor is the zero matrix. You are now asked to discover for yourself whether or not the fact that the product of two square matrices is the zero matrix necessarily means that at least one of the matrices is the zero matrix, if the elements of the matrix are taken from an integral domain. Either prove that this is so, or produce a counterexample using 3×3 matrices.

9. Find a matrix T such that

$$T \cdot M = \begin{bmatrix} 1 & 2 & 3 \\ 5 & 7 & 9 \\ 6 & 6 & 6 \end{bmatrix} \quad \text{where} \quad M = \begin{bmatrix} 1 & 2 & 3 \\ 4 & 5 & 6 \\ 3 & 3 & 3 \end{bmatrix}$$

Also, find the matrix $M \cdot T$. Can you forecast what $M \cdot T$ will be after seeing T, but before forming the product $M \cdot T$?

10. Same as Problem 9, when

$$M = \begin{bmatrix} 1 & 2 & 3 \\ -2 & -3 & -4 \\ 1 & 2 & 1 \end{bmatrix} \quad \text{and} \quad T \cdot M = \begin{bmatrix} 0 & 0 & 0 \\ 2 & 3 & 4 \\ 1 & 2 & 3 \end{bmatrix}$$

11. Same as Problem 9, when $M = \begin{bmatrix} 4 & 3 \\ 1 & 2 \end{bmatrix}$ and $T \cdot M = \begin{bmatrix} 1 & 0 \\ 0 & 1 \end{bmatrix}$. [HINT:

Let $T = \begin{bmatrix} a & b \\ c & d \end{bmatrix}$ and determine a set of four equations in four un-

knowns which must be satisfied.]

12. Same as Problem 9, with $M = \begin{bmatrix} 1 & 1 \\ 0 & 1 \end{bmatrix}$ and $T \cdot M = \begin{bmatrix} 5 & 7 \\ 0 & 2 \end{bmatrix}$.

13. Same as Problem 9, with $M = \begin{bmatrix} 2 & 1 & 0 \\ 2 & 2 & 2 \\ 1 & 0 & 1 \end{bmatrix}$ and $T \cdot M = \begin{bmatrix} 0 & 0 & 0 \\ 2 & 1 & 0 \\ 3 & 1 & 1 \end{bmatrix}$.

14. Same as Problem 9, with $M = \begin{bmatrix} 1 & 2 & 1 \\ 2 & 3 & 2 \\ 3 & 4 & 0 \end{bmatrix}$ and $T \cdot M = \begin{bmatrix} 0 & 1 & 1 \\ 3 & 4 & 0 \\ \frac{1}{2} & \frac{2}{3} & 0 \end{bmatrix}$.

*15. Find a 2×2 matrix X having integral elements such that X satisfies the equation $X \cdot \begin{bmatrix} 1 & 2 \\ 0 & 1 \end{bmatrix} = \begin{bmatrix} 1 & 2 \\ 0 & 1 \end{bmatrix} \cdot X$. Is X unique?

16. (a) Use the notation of Section 5.5 to prove that 3×3 matrices obey the distributive law

$$A \cdot (B + C) = A \cdot B + A \cdot C$$

*(b) Try to produce a similar proof for $n \times n$ matrices.

5.9 PROOFS AND THEIR STRUCTURE

Mathematics studies the basic *structure* of various intellectual systems and attempts to prove theorems based on these systems. Let us see what is implied by the last section. We know that the set of all n by n square matrices whose elements are taken from an integral domain do *not* themselves form an integral domain since the operation of multiplication is not commutative (violates Postulate 2) and the cancellation law (Postulate 7) is not satisfied. In Chapter 1 (Theorem 1.1) we showed that the integral domain postulates are sufficient to guarantee that if $A \cdot B = 0$, then either $A = 0$ or $B = 0$ or possibly both. Problem 8, Set 5.8 noted that the conclusion of the theorem does not hold in the modulo 12 system, which is not an integral domain, and you were asked to determine whether or not it held in the set of 3 by 3 matrices over an integral domain. Perhaps you were lucky, and found a counterexample by trial and error. Finding counterexamples is a fortuitous way of solving certain problems when it works, but if luck is not with you it may not prove fruitful—especially if the conjecture happens to be valid.

In Chapter 1 we proved Theorem 1.1 as follows:

Given: The postulates of an integral domain, and $A \cdot B = 0$.
To prove: Either $A = 0$, or $B = 0$, or both.

Proof:
Either $A = 0$ or $A \neq 0$
If $A = 0$ the theorem is satisfied.
If $A \neq 0$, then
$A \cdot B = 0 = A \cdot 0$ by hypothesis and Postulate 4
$A \cdot B = A \cdot 0$ transitive property
$\therefore B = 0$ by Postulate 7, since $A \neq 0$ by hypothesis.

Hence if A, B are elements of an integral domain such that $A \cdot B = 0$ it follows that either $A = 0$ or $B = 0$, or possibly both. Note that Postulate 7 was specifically used in the last step of the proof of Theorem 1.1. Since Postulate 7 does *not* hold for 3×3 matrices, we may say that the *given proof* does not hold for 3×3 matrices *but this does not indicate that the theorem itself fails* for 3 by 3 matrices. (Why not?)

It is still a reasonable conjecture (but not true) that "If the product $A \cdot B$ of two 3×3 matrices is the zero matrix, then at least one of the matrices A or B must also be the zero matrix." (You can easily show that the *converse* of this conjecture is valid, but that is of little help in proving or disproving the conjecture. See Section 1.5 if in doubt.) If we seek a counterexample it might be well to seek it among matrices which violate integral domain Postulate 7, even though we have no guarantee that we shall find a counterexample there. Mathematicians often find it helpful to examine proofs of related theorems to suggest either methods of proof or possible hiding places for counterexamples for a conjecture. Once the structure of a mathematical system is *fully* understood, proofs or counterexamples are much easier to find, but it is in the investigation of only partially understood structures that the real fun of mathematics is found. In such cases it often helps to ask, "What postulates that were used to prove a related theorem in another algebraic system are violated here?" It is not difficult to create numerous examples of matrices A and B such that $A \cdot B = [0]$ but $A \neq [0]$ and $B \neq [0]$. Two such examples are:

$$\begin{bmatrix} 1 & 2 & 4 \\ 2 & 3 & 9 \\ 1 & 1 & 5 \end{bmatrix} \cdot \begin{bmatrix} -12 & 0 & 6 \\ 2 & 0 & -1 \\ 2 & 0 & -1 \end{bmatrix} = \begin{bmatrix} 0 & 0 & 0 \\ 0 & 0 & 0 \\ 0 & 0 & 0 \end{bmatrix}$$

and

$$\begin{bmatrix} 4 & 0 & 8 \\ 7 & 0 & 7 \\ 3 & 0 & 10 \end{bmatrix} \cdot \begin{bmatrix} 0 & 0 & 0 \\ 5 & 2 & 9 \\ 0 & 0 & 0 \end{bmatrix} = \begin{bmatrix} 0 & 0 & 0 \\ 0 & 0 & 0 \\ 0 & 0 & 0 \end{bmatrix}$$

Problem Set 5.9

1. Find additional examples of nonzero matrices whose product is zero.

2. A *real number* may or may not have a square root which is also a real number. (Give an example of each.) If a real number does have a real square root, then it may have exactly two distinct square roots (as do positive real numbers) or, in the case of the real number zero, it may have a unique square root. In no case does a real number have more than two distinct real square roots. The reader is asked to show that there are 2 by 2 matrices B with real elements such that the matrix equation

$$[x] \cdot [x] = [B]$$

has *infinitely many* real matrix solutions.

$$\left[\text{HINT:} \quad \begin{bmatrix} 0 & 6 \\ \frac{1}{6} & 0 \end{bmatrix}^2 = \begin{bmatrix} 1 & 0 \\ 0 & 1 \end{bmatrix} \right]$$

3. How *many* matrices with real elements are there which are solutions of the matrix equation

$$[x] \cdot [x] = \begin{bmatrix} 4 & 0 \\ 0 & 4 \end{bmatrix}$$

Are you *sure* of your answer? (Why?)

4. Either *prove* that $\begin{bmatrix} 0 & 0 \\ 0 & 0 \end{bmatrix}$ is the *only* real matrix solution of the matrix equation

$$[x] \cdot [x] = \begin{bmatrix} 0 & 0 \\ 0 & 0 \end{bmatrix}$$

or find another solution.

5. In real numbers if $Z_1 = (A + B)^2 = A^2 + 2AB + B^2$

$$\text{and } Z_2 = (B + A)^2 = B^2 + 2BA + A^2$$

$$\text{then } Z_1 = Z_2.$$

It is not obvious whether or not this is also true if A and B are matrices, (where $2AB$ means $A \cdot B + A \cdot B$). One argument could be based on the fact that matrix addition is commutative and that hence $A + B = B + A$ and thus $(A + B)^2 = (B + A)^2$ and therefore $Z_1 = Z_2$. On the other hand it is well known that there exist matrices A and B such that $A \cdot B$ and $B \cdot A$ are different and hence it is easy to produce matrices A and B such that $A^2 + 2AB + B^2 \neq B^2 + 2BA + A^2$. Which argument is correct? Where does the fallacy lie? Prove your assertion.

6. Show that in the system of 3 by 3 matrices it is *not* in general true that

$$(A + B) \cdot (A - B) = A^2 - B^2$$

7. In the real number system the equation $AX^2 + BX + C = 0$ sometimes has no real solution, but if it has a real solution we can obtain this solution either by use of the quadratic formula or by an algorithm which involves factoring. The process is not as simple when A, B, and C are 2 by 2 matrices and we are seeking a matrix solution of the equation. None the less, in some cases such solutions exist.

(a) Show that $X = \begin{bmatrix} 3 & 1 \\ -1 & 2 \end{bmatrix}$ is a solution of the matrix equation

$$X^2 + \begin{bmatrix} -5 & 0 \\ 0 & -5 \end{bmatrix} \cdot X + \begin{bmatrix} 7 & 0 \\ 0 & 7 \end{bmatrix} = \begin{bmatrix} 0 & 0 \\ 0 & 0 \end{bmatrix}$$

(b) Attempt to solve the matrix equation

$$[A] \cdot [X]^2 + [B] \cdot [X] + [C] = [0]$$

for some 2 by 2 matrices $[A]$, $[B]$, and $[C]$ of your own choice.
(c) Explain exactly where the difficulty arises in part b.
(d) Consider the technique you used to *prove* the quadratic formula for real (or even complex) coefficients. Can you find a place in which you used either the cancellation law or the fact that in real (and complex) numbers a product is zero if *and only if* at least one of the factors is zero.
(e) If appropriate, revise and/or extend your answer to part (c) in view of your answer to part (d).

8. Instead of considering the set of all 2 by 2 matrices with real elements, let us consider the subset S consisting of those matrices in M which are of the form $\begin{bmatrix} \cos\theta & \sin\theta \\ -\sin\theta & \cos\theta \end{bmatrix}$ for some real θ. Show that the matrices in this set form a *group* under matrix multiplication. Is the group a commutative group?

9. Consider the set C of all matrices of the form $\begin{bmatrix} a & b \\ -b & a \end{bmatrix}$ where a and b are real numbers. What theorems can *you* prove about the set C?

10. Consider the set W of all matrices of the form $\begin{bmatrix} a & 0 \\ 0 & \dfrac{1}{a} \end{bmatrix}$ where a is a real number other than zero. Prove that W is a group under the operation of matrix multiplication. Make some conjectures about the set W and either prove or disprove your conjectures.

11. Problems 8–10 examined three subsets S, C, and W of the set M of all 2 by 2 matrices with real elements. Which, if any, of these sets are

subsets of one another? Use the set inclusion notation discussed in Section 3.11. Can you find a matrix that belongs to all four sets? Can you find a matrix in each superset that does not belong to any of the above sets which are subsets of the given superset?

12. Prove or disprove:

    The set C of Problem 9 is isomorphic (see Section 4.5) to the set of complex numbers under multiplication.

13. Make some conjectures about the set of Pauli matrices (Section 5.3) and prove or disprove your conjectures. Is there a subset of the Pauli matrices which is isomorphic to a subset of the set C of Problem 9?

14. Let M be the set of 2 by 2 matrices with real elements. We define a new operation $\odot$ on matrices in M as follows:

    $A \odot B = A \cdot B - B \cdot A$ where $\cdot$ is regular matrix multiplication.

    Prove the following theorems:

    (1) M is closed under the new operation $\odot$
    (2) $A \odot B = -B \odot A$
    (3) $A \odot A = [0]$
    (4) $A \odot (B \odot C) + B \odot (C \odot A) + C \odot (A \odot B) = [0]$
    (5) $A \odot I = I \odot A = [0]$
    (6) Show that there is no "unit" matrix U such that

    $$U \odot A = A \odot U = A \qquad \text{for all } A \in M$$

    (7) Show that there exist matrices in $\dot{M}$ such that

    $$A \odot (B \odot C) \neq (A \odot B) \odot C$$

    (8) Show that the distributive laws

    $$A \odot (B + C) = (A \odot B) + (A \odot C)$$

    and

    $$(A + B) \odot C = (A \odot C) + (B \odot C)$$

    are valid for the new operation $\odot$

Students interested in further study of the operation $\odot$ are encouraged to do so. Actually this is an example of a nonassociative algebra known as *Lie Algebra* (named after the geometer Sophus Lie, whose surname is pronounced "Lee"). It has important applications in mathematical genetics and in quantum mechanics. The *Mathematical Reviews* will show that Lie Algebras are the subject of current research interest in mathematics.

15. The *Jordan Product* of two matrices of M is defined as

$$A \textcircled{j} B = \tfrac{1}{2}(A \cdot B + B \cdot A)$$

(a) Show that Jordan multiplication is commutative but *not* associative.

(b) What other properties can *you* discover about Jordan multiplication?

(c) See if you can find out anything about the mathematician Jordan for whom the Jordan Product is named.

5.10 MORE GENERAL MATRICES AND VECTORS

It was an easy generalization from 2×2 square matrices to $n \times n$ square matrices in which multiplication is defined in a similar row by column fashion. In many applications, it is desirable to consider matrices which are *not square*. In the multiplication of nonsquare matrices, the same row by column multiplication is used, but some restriction on the relative sizes of the two matrices is necessary before the row by column product can be found. Before reading on, see if *you* can decide what this restriction should be. The matrices need not be the same size. Indeed, unless the matrices are square they *cannot* be the same size and have either product meaningful. (Why?) If the matrix product $A \cdot B$ is to have meaning, the number of *columns in matrix A* must be the same as the number of *rows in matrix B*. In this case, the row-column rule discussed in Section 5.2 is still valid. The product $(A \cdot B)$ may be different in size from either A or B. A matrix having only one row (or having only one column) is often called a *vector*. Familiar examples are (x, y) in the plane or (x, y, z) in three space.

EXAMPLE 3

$$\begin{bmatrix} 1 & 2 \\ 4 & 9 \\ 3 & 7 \\ 2 & 6 \end{bmatrix} \cdot \begin{bmatrix} 2 \\ 3 \end{bmatrix} = \begin{bmatrix} 1(2) + 2(3) \\ 4(2) + 9(3) \\ 3(2) + 7(3) \\ 2(2) + 6(3) \end{bmatrix} = \begin{bmatrix} 8 \\ 35 \\ 27 \\ 22 \end{bmatrix}$$

EXAMPLE 4

$$\begin{bmatrix} 1 & -2 & 3 & 4 & 1 \\ 4 & -1 & 0 & 0 & 1 \\ 7 & 1 & 6 & 2 & 0 \end{bmatrix} \cdot \begin{bmatrix} 2 & 1 \\ 3 & 2 \\ 0 & 5 \\ 1 & 2 \\ 2 & 0 \end{bmatrix} = \begin{bmatrix} 2 & 20 \\ 7 & 2 \\ 19 & 43 \end{bmatrix}$$

EXAMPLE 5

$$\begin{bmatrix} 1 & 2 & 1 & 3 & 4 \\ 2 & 1 & 9 & 0 & 7 \end{bmatrix} \cdot \begin{bmatrix} 1 \\ 2 \\ 3 \\ 4 \\ 5 \end{bmatrix} = \begin{bmatrix} 40 \\ 66 \end{bmatrix}$$

EXAMPLE 6

$$\begin{bmatrix} 4 \\ 3 \end{bmatrix} \cdot \begin{bmatrix} 1 & 6 & 2 & 9 & 3 \end{bmatrix} = \begin{bmatrix} 4 & 24 & 8 & 36 & 12 \\ 3 & 18 & 6 & 27 & 9 \end{bmatrix}$$

It is convenient to write k in place of the matrix $k = (k\delta_{ij})$ where δ_{ij} is the Kronecker delta defined in Problem 4, Set 5.6.

$$K = \begin{bmatrix} k & 0 & 0 & 0 & 0 & \cdots & 0 \\ 0 & k & 0 & & & & 0 \\ 0 & 0 & k & 0 & & & 0 \\ 0 & & 0 & k & 0 & & 0 \\ 0 & & & 0 & k & \ddots & \vdots \\ \vdots & & & & & \ddots & 0 \\ 0 & 0 & 0 & 0 & \cdots & 0 & k \end{bmatrix}$$

Either k or the matrix K is called a *scalar*. Thus, $k \cdot (M)$ is the matrix M with each element multiplied by k. With this interpretation, the polynomial $f(x) = 3x^2 - 6x$ has meaning for x, a matrix. (See Problem 5, Set 5.8.)

EXAMPLE 7

$$2 \cdot \begin{bmatrix} -1 & 3 \\ 4 & 9 \\ 2 & 7 \end{bmatrix} = \begin{bmatrix} -2 & 6 \\ 8 & 18 \\ 4 & 14 \end{bmatrix}, \qquad 3 \cdot \begin{bmatrix} -3 & 2 & 1 \\ 4 & 9 & 6 \\ 8 & 1 & 1 \end{bmatrix} = \begin{bmatrix} -9 & 6 & 3 \\ 12 & 27 & 18 \\ 24 & 3 & 3 \end{bmatrix},$$

$$5 \cdot \begin{bmatrix} 7 \\ 1 \\ 2 \\ 3 \end{bmatrix} = \begin{bmatrix} 35 \\ 5 \\ 10 \\ 15 \end{bmatrix}$$

EXAMPLE 8

If $T = \begin{bmatrix} 2 & 7 \\ 6 & 9 \end{bmatrix}$, find $f(T)$, where $f(x) = 3x^2 - 6x$.

$$f(T) = 3 \cdot \begin{bmatrix} 2 & 7 \\ 6 & 9 \end{bmatrix}^2 - 6 \cdot \begin{bmatrix} 2 & 7 \\ 6 & 9 \end{bmatrix} = 3 \begin{bmatrix} 46 & 77 \\ 66 & 123 \end{bmatrix} - 6 \begin{bmatrix} 2 & 7 \\ 6 & 9 \end{bmatrix}$$

$$= \begin{bmatrix} 138 & 231 \\ 198 & 369 \end{bmatrix} + \begin{bmatrix} -12 & -42 \\ -36 & -54 \end{bmatrix} = \begin{bmatrix} 126 & 189 \\ 162 & 315 \end{bmatrix}$$

If A is a *square* matrix, then either A has no inverse, or A has a unique, two-sided inverse A^{-1} such that $A^{-1}A = AA^{-1} = I$. (Problem 19, Set 5.6.) If A is not square, we can no longer guarantee that the left inverse is also a right inverse. It is also true that if the *elements* of A are taken from a system (such as the floating-point number system used on most modern computers) which does *not* satisfy the postulates for a field (such as the field of rational numbers or the field of real numbers) then we may *not* be able to guarantee that the inverse is unique or that it is two-sided. This is important since much of today's applied work using matrices uses computers.

EXAMPLE 9

$$A = \begin{bmatrix} 2 & 1 \\ 1 & 1 \\ 0 & 0 \end{bmatrix}$$

then

$$\begin{bmatrix} 1 & -1 & 4 \\ -1 & 2 & 7 \end{bmatrix} \cdot \begin{bmatrix} 2 & 1 \\ 1 & 1 \\ 0 & 0 \end{bmatrix} = \begin{bmatrix} 1 & 0 \\ 0 & 1 \end{bmatrix} = I$$

However,

$$\begin{bmatrix} 2 & 1 \\ 1 & 1 \\ 0 & 0 \end{bmatrix} \cdot \begin{bmatrix} 1 & -1 & 4 \\ -1 & 2 & 7 \end{bmatrix} = \begin{bmatrix} 1 & 0 & 15 \\ 0 & 1 & 11 \\ 0 & 0 & 0 \end{bmatrix} \neq I$$

Thus, we have a matrix L such that $L \cdot A = I$, but $A \cdot L \neq I$. Furthermore, $(L \cdot A)$ and $(A \cdot L)$ are not even the same size! Although we shall not prove it here, there is actually no matrix R such that $A \cdot R = I$ for the given matrix A. Another curious thing about the matrix L such that $L \cdot A = I$ is that L is *not* unique. Actually,

$$L = \begin{bmatrix} 1 & -1 & x \\ -1 & 2 & y \end{bmatrix}$$

satisfies $L \cdot A = I$ for all values of x and y.

A matrix L such that $L \cdot A = I$ is called a *left inverse* of A. The square matrix I produced will have the same number of *columns* as A. (Why?)

A matrix R such that $A \cdot R = I$ is called a *right inverse* of A. The square matrix I produced will have the same number of *rows* as A. (Why?)

Let A be a matrix having r rows and c columns, with real or rational elements (or elements from any field).

(1) If $r < c$, A has no left inverse. It may or may not have a right inverse. If a right inverse exists, it may or may not be unique.

(2) If $r = c$, either A has no inverse, or A has a unique two-sided inverse.

(3) If $r > c$, A may or may not have a left inverse. If a left inverse exists, it may or may not be unique. It has no right inverse.

Problem Set 5.10

1. Let $A = \begin{bmatrix} 2 & 3 & 7 \\ 1 & 4 & 9 \end{bmatrix}$. Determine matrices C_1, C_2, C_3 such that

    (a) $A \cdot C_1$ has fewer elements in it than does A. (Note: not *smaller* elements, *fewer* elements, that is, the matrix $M = A \cdot C_1$ may have say four elements rather than six elements.)

    (b) $A \cdot C_2$ has the same number of elements as A. (Must $A \cdot C_2$ then be the same size as A, or could $A \cdot C_2$ contain 3 rows and 2 columns?)

    (c) $A \cdot C_3$ has more elements than A.

2. Let $X = \begin{bmatrix} 1 & 2 & -1 \\ 3 & 4 & -2 \\ 5 & 6 & -3 \end{bmatrix}$. Determine matrices B_1, B_2, B_3 such that

    (a) $B_1 \cdot X$ contains fewer elements than X.

    (b) $B_2 \cdot X$ contains the same number of elements as X.

    (c) $B_3 \cdot X$ contains more elements than X.

In Problems 3–16, perform the indicated multiplications where possible. If the multiplications cannot be performed, indicate this.

$$A = \begin{bmatrix} 1 & 2 \\ 7 & 9 \\ 3 & 4 \\ 1 & 6 \end{bmatrix}, \quad B = \begin{bmatrix} 2 & 1 \\ 3 & 5 \end{bmatrix}, \quad C = [-4, \ -2]$$

$$D = \begin{bmatrix} 4 \\ -2 \end{bmatrix}, \quad E = \begin{bmatrix} 1 & 7 & 9 \\ 3 & 2 & 18 \end{bmatrix}, \quad F = \begin{bmatrix} 1 \\ 2 \\ 3 \\ 4 \end{bmatrix}$$

3. $C \cdot E$ 7. $B \cdot C$ 11. $F \cdot C$ 15. $D \cdot C$

4. $C \cdot A$ 8. $B \cdot D$ 12. $D \cdot F$ 16. $A \cdot D$

5. $A \cdot E$ 9. $E \cdot F$ 13. $E \cdot B$

6. $E \cdot A$ 10. $F \cdot A$ 14. $B \cdot E$

17. Form the products $(A \cdot D) \cdot C$ and $A \cdot (D \cdot C)$. Is the same result obtained in each case?

18. If $g(x) = \begin{bmatrix} 3 & 0 \\ 0 & 3 \end{bmatrix} x^2 + \begin{bmatrix} -7 & 0 \\ 0 & -7 \end{bmatrix} x$, find $g(T)$, where $T = \begin{bmatrix} 2 & 7 \\ 6 & 9 \end{bmatrix}$

19. David purchases the following items:

   2 books at 6¢ each; 7 pencils at 10¢ each; 3 erasers at 15¢ each; 5 reams of paper at $1.85 each.

   He writes the expression (2, 7, 3, 5) to help him recall the number of items, and (.06, .10, .15, 1.85) to help him recall the prices paid. Form the matrix product

$$\begin{bmatrix} 2 & 7 & 3 & 5 \end{bmatrix} \cdot \begin{bmatrix} 0.06 \\ 0.10 \\ 0.15 \\ 1.85 \end{bmatrix}$$

   What significance does the product matrix have for David?

20. Each of five scholarship candidates takes two tests. The results presented in the table below are the raw scores (not percentages) achieved on these tests.

	Test 1	Test 2
Adams	9	6
Boyle	6	3
Chase	8	1
Dunn	2	9
Zilch	9	4

   It is decided to weight the second test only $\frac{2}{3}$ as much as the first test. Show that the multiplication of the test result matrix by the matrix $\begin{bmatrix} 3 \\ 2 \end{bmatrix}$,

$$\begin{bmatrix} 9 & 6 \\ 6 & 3 \\ 8 & 1 \\ 2 & 9 \\ 9 & 4 \end{bmatrix} \cdot \begin{bmatrix} 3 \\ 2 \end{bmatrix}$$

produces a ranking of the data in the form desired.

Use the matrix B given before Problem 3 in working Problems 21, 22, 23.

21. If $f(X) = X^2 - \begin{bmatrix} 1 & 0 \\ 2 & 5 \end{bmatrix} \cdot X + \begin{bmatrix} 2 & -5 \\ 1 & 16 \end{bmatrix}$, find $f(B)$.

22. Find $h(B)$ where $h(x) = 4x^3 - 3x$.

23. Find $P(B)$ where $P(x) = (2 - x)(5 - x)$, where the constants are interpreted as scalar matrices.

★24. (a) Let $T = \begin{bmatrix} 0 & 1 \\ -1 & 0 \end{bmatrix}$ and let $P = [x, y]$. Show that $P \cdot T = [-y, x]$.

(b) Consider the effect upon the point $P(x, y)$ in the xy plane of rotating the *axes* through $-90°$ (that is, of rotating the point $+90°$ about $(0, 0)$). Can you see any relationship between this and part (a) of this problem?

★(c) Show that, if a point (a, b) is rotated through an angle θ about the origin (or, equivalently, the axes rotated through an angle $-\theta$ about the origin), then the new coordinates of the point will be the same as that obtained in the matrix product

$$[a, b] \cdot \begin{bmatrix} \cos\theta & \sin\theta \\ -\sin\theta & \cos\theta \end{bmatrix}$$

★25. Find a matrix M such that the point obtained in the matrix product $[x, y] \cdot M$ is the point obtained by rotating $P(x, y)$ through $+30°$

about the origin. Consider three or four actual points and show that your matrix M has the desired effect in these cases.

26. Find a matrix B^{-1} such that $B \cdot B^{-1} = I_2$ for the B given before Problem 3.

27. Find the multiplicative inverse of the matrix $\begin{bmatrix} a & b \\ c & d \end{bmatrix}$.

28. A is a matrix having r_1 rows and c_1 columns. B is a matrix having r_2 rows and c_2 columns. The product $A \cdot B$ has one row and one column.

    (a) What values are possible for r_1, r_2, c_1, c_2?
    (b) Answer part (a) under the additional restriction that $B \cdot A$ must be meaningful.

29. How many of the Pauli matrices satisfy the equation $X^2 + I = 0$?

30. Since scalars commute with matrices, a more general distributive law can be proved, where A, B, and C are matrices and k_1 and k_2 are scalars. Namely

$$A \cdot (k_1 B + k_2 C) = k_1 A \cdot B + k_2 A \cdot C$$

Let P be an $n \times n$ matrix with rational elements, and I be the $n \times n$ identity matrix. Let a and b be rational numbers. Show, using the extended distributive law, that, if $Q = aP + bI$, $P \cdot Q = Q \cdot P$.

31. If A and B are square matrices, then if they are the same dimension, the matrix product may be formed, and is also of the same dimension. However, if A and B are not square, it may be possible to form the product in such a manner that the product matrix is smaller than either original matrix. Can the product matrix possibly be of larger dimension than either original matrix? Give examples.

32. Show that, for nonsquare matrices, if $(A \cdot B) \cdot C$ is defined, then $A \cdot (B \cdot C)$ is also defined. State which of the laws of matrices could not hold if this were not true.

33. Can the product of two nonsquare matrices be a square matrix? Give examples.

34. If possible, determine a matrix L such that $L \cdot A = I$, where I is an identity matrix and $A = \begin{bmatrix} 3 & 4 & 7 & 2 \\ 1 & 9 & 3 & 0 \end{bmatrix}$. Also, form $A \cdot L$.

35. If possible, determine a matrix R such that $A \cdot R = I$, where I is an

identity matrix and A is the matrix given in Problem 34. Also, form $R \cdot A$.

36. Find either a left or a right inverse for the matrix below, if one exists, and show that the other does not exist.

$$\begin{bmatrix} 3 & 5 & 7 \\ 2 & 1 & 9 \end{bmatrix}$$

37. Same as Problem 36 for the matrix

$$\begin{bmatrix} 3 & 1 \\ 2 & 5 \\ 4 & 9 \end{bmatrix}$$

38. Does the matrix $\begin{bmatrix} 3 & 1 & 5 \\ 6 & 2 & 10 \end{bmatrix}$ have a one-sided inverse? Explain.

39. Discuss carefully and determine whether or not the matrices $Z_1 = A \cdot (B + C)$ and $Z_2 = (A \cdot B + A \cdot C)$ always exist when A, B, and C are not square. Is it possible that

   (a) Z_1 might exist and Z_2 not be defined?
   (b) Z_2 might be defined and Z_1 not be defined?
   (c) Can you show that if both Z_1 and Z_2 are defined, then they are equal (that is, that the distributive law holds)?
   (d) Can you show that if either Z_1 or Z_2 exists, they both exist and are equal?

5.11 APPLICATIONS OF MATRIX NOTATION

Matrices are important both in mathematics and in their many applications in other fields. Modern physics uses matrix theory in quantum mechanics and in the study of atomic and crystal structure. Most branches of engineering are following the lead of the aeronautical and electrical engineer by using matrix techniques to solve problems involving intricate interrelationships. (The interesting text *Elementary Matrices*, by Frazer, Duncan, and Collar, was written by three aeronautical engineers. It shows how the differential equations and other problems of aerodynamics, stress, and structure are solved using matrix methods.) *Multiple Factor Analysis* by Thurstone uses matrix methods in modern psychology. The study of electrical networks, oscillation theory, damped vibrations, circuit analysis, and many other branches of engineering and physical sciences are simplified by the use of matrix methods. The biological sciences (particularly in studies of growth problems and of heredity) as well as sociology, economics, and industrial management also use matrix methods in modern research. In short, whatever a student's major interest may be, the study of the fundamentals of matrix theory will enable him to read more of the current and future literature in his own and related fields than would be possible otherwise. This short text will not study the details of these many applications. Instead, we examine two examples of sufficient generality to be understandable to readers of varied backgrounds, but which may indicate the types of interrelations in which matrix theory plays so important a role.

EXAMPLE 10

Let us assume that a steel mill has orders for three types of steel and that the orders are for 7 units of No. 1 steel, 4 units of No. 2 steel, and 11 units of No. 3 steel. A unit may represent a pound or 1000 tons; it is unimportant for our example, although of considerable importance to the steelmaker. It is possible to represent the order by means of the row matrix (vector), $D = (7, 4, 11)$.[1]

A number of different raw materials, including pig iron, coke, limestone, manganese, furnace time, and labor are needed to make steel. The amounts of raw materials (in some suitable units) needed to make one unit of each of these different types of steel may be concisely represented in a matrix of the following form:

	Pig iron	Coke	Lime-stone	Manga-nese	Furnace time	Labor	
No. 1 steel	7	3	5	0	9	4	
No. 2 steel	8	3	1	1	8	7	$= M$
No. 3 steel	9	1	4	5	15	12	

In an actual commercial problem, these elements would possibly be decimal fractions rather than integers. Each row of the matrix M forms a 1×6 matrix (vector) giving the relative amounts of each raw material needed for a given type of steel. Similarly, each column of M is a 3×1 (column matrix or vector) giving the total amount of a given raw material needed to make one unit of each type of steel.

[1] The commas are unnecessary, but a convenience.

The matrix product

$$D \cdot M = [7, 4, 11] \cdot \begin{bmatrix} 7 & 3 & 5 & 0 & 9 & 4 \\ 8 & 3 & 1 & 1 & 8 & 7 \\ 9 & 1 & 4 & 5 & 15 & 12 \end{bmatrix}$$

	Pig iron	Coke	Lime-stone	Manga-nese	Furnace time	Labor
= [	180,	44,	83,	59,	260,	188]

gives the total amount of each raw material needed to complete the order $D = (7, 4, 11)$.

If the cost of one unit of each of the six raw materials is known, this may be expressed as a 6×1 matrix

$$C = \begin{bmatrix} 3 \\ 2 \\ 1 \\ 4 \\ 10 \\ 6 \end{bmatrix} \begin{matrix} \text{pig iron} \\ \text{coke} \\ \text{limestone} \\ \text{manganese} \\ \text{furnace time} \\ \text{labor} \end{matrix}$$

where pig iron costs \$3 per unit, coke \$2 per unit, etc.

The matrix product

$$M \cdot C = \begin{bmatrix} 7 & 3 & 5 & 0 & 9 & 4 \\ 8 & 3 & 1 & 1 & 8 & 7 \\ 9 & 1 & 4 & 5 & 15 & 12 \end{bmatrix} \cdot \begin{bmatrix} 3 \\ 2 \\ 1 \\ 4 \\ 10 \\ 6 \end{bmatrix} = \begin{bmatrix} 146 \\ 157 \\ 275 \end{bmatrix} \begin{matrix} \text{No. 1 steel} \\ \text{No. 2 steel} \\ \text{No. 3 steel} \end{matrix}$$

gives the total cost of making one unit of each type of steel.

The matrix product

$$D \cdot M \cdot C = [7, 4, 11] \cdot \begin{bmatrix} 7 & 3 & 5 & 0 & 9 & 4 \\ 8 & 3 & 1 & 1 & 8 & 7 \\ 9 & 1 & 4 & 5 & 15 & 12 \end{bmatrix} \cdot \begin{bmatrix} 3 \\ 2 \\ 1 \\ 4 \\ 10 \\ 6 \end{bmatrix} = [4675] = T$$

gives the total cost, T, of order $D = [7, 4, 11]$. The reader should note that the triple product $D \cdot M \cdot C$ may be evaluated either as $[D \cdot M] \cdot C$ or as $D \cdot [M \cdot C]$, and that although the intermediate results are dissimilar, the associative law states that the same final matrix is obtained in either case. Before continuing, complete the evaluation:

$$[D \cdot M] \cdot C = \left[[7, 4, 11] \cdot \begin{bmatrix} 7 & 3 & 5 & 0 & 9 & 4 \\ 8 & 3 & 1 & 1 & 8 & 7 \\ 9 & 1 & 4 & 5 & 15 & 12 \end{bmatrix} \right] \cdot \begin{bmatrix} 3 \\ 2 \\ 1 \\ 4 \\ 10 \\ 6 \end{bmatrix}$$

$$= [180, 44, 83, 59, 260, 188] \cdot \begin{bmatrix} 3 \\ 2 \\ 1 \\ 4 \\ 10 \\ 6 \end{bmatrix} = [4675]$$

$$D \cdot [M \cdot C] = [7, 4, 11] \cdot \begin{bmatrix} 7 & 3 & 5 & 0 & 9 & 4 \\ 8 & 3 & 1 & 1 & 8 & 7 \\ 9 & 1 & 4 & 5 & 15 & 12 \end{bmatrix} \cdot \begin{bmatrix} 3 \\ 2 \\ 1 \\ 4 \\ 10 \\ 6 \end{bmatrix}$$

$$= [7, 4, 11] \cdot \begin{bmatrix} 146 \\ 157 \\ 275 \end{bmatrix} = [4675]$$

This problem could have been worked without the use of matrix notation. However, it sets the stage for variations in which matrix notation is even more valuable, particularly in situations where machine computation is available. Reasonable questions such as the following arise:

(1) If the order matrix $D = (7, 4, 11)$ is replaced by an arbitrary order matrix $D = (x, y, z)$, what values (production percentages) for X, Y, Z, where $x + y + z = 100$ will produce the smallest total cost, T?

(2) Same as Problem 1, with the exception that the elements in the "furnace time" column of the matrix $(D \cdot M)$ must not exceed a fixed amount.

(3) If the matrix $P = \begin{bmatrix} 2.10 \\ 3.15 \\ 8.00 \end{bmatrix}$ $\begin{matrix} \text{No. 1 steel} \\ \text{No. 2 steel} \\ \text{No. 3 steel} \end{matrix}$ represents the selling prices of the three types of steel, then

$$D \cdot P = (x, y, z) \cdot \begin{bmatrix} 2.10 \\ 3.15 \\ 8.00 \end{bmatrix}$$

represents the total income in selling x units of No. 1 steel, y units of No. 2 steel, and z units of No. 3 steel. To maximize the *profit*, we wish to determine values of x, y, z which will make

$$D \cdot M \cdot C - D \cdot P = D \cdot (M \cdot C - P) = \text{profit}$$

as large as possible. Ways of doing this are studied in more advanced courses. It is usual to have additional restrictions, such as that the total labor and total furnace time may not exceed fixed top values (upper bounds).

It may be that both cost C and selling price P are dependent upon many other factors. In this case, each of these matrices is obtained as a product of other matrices, much as the matrix giving the total cost of one unit of each type of steel was obtained as the product $M \cdot C$.

EXAMPLE 11

When a small amount of liquid is introduced into a closed system, a fixed percentage of the liquid will change into a vapor state, and a given percentage of the vapor will change back into a liquid state. The process will be repeated indefinitely. A similar analysis applies in population study, where a given portion of the city population moves into the country and a portion of the country population moves into the city each year, and in other situations. To make the problem specific, let us assume that $\frac{1}{4}$ of the liquid present at the beginning of the day turns into vapor during the day, and that an amount equal to $\frac{1}{10}$ of vapor present at the beginning of the day turns into liquid during the day. This situation is indicated by the following matrix

$$
\begin{array}{ccc}
 & \begin{array}{c}\text{Portion}\\\text{into}\\\text{liquid}\end{array} & \begin{array}{c}\text{Portion}\\\text{into}\\\text{vapor}\end{array} \\
\text{Liquid} & \begin{bmatrix} \frac{3}{4} & \frac{1}{4} \\[6pt] \frac{1}{10} & \frac{9}{10} \end{bmatrix} \\
\text{Vapor} & &
\end{array}
$$

At first glance, you may feel that eventually all the substance will turn into vapor. This is *not true*.

Let L_0 be the proportion of the substance originally in liquid state and V_0 the proportion originally in vapor state. This is represented by the matrix (L_0, V_0). In a similar fashion, let (L_i, V_i) represent the proportion in liquid and vapor states at the end of the ith day.

At the end of day 1,

$$(L_1, V_1) = (L_0, V_0) \cdot \begin{bmatrix} \frac{3}{4} & \frac{1}{4} \\ \frac{1}{10} & \frac{9}{10} \end{bmatrix}$$

At the end of day 2,

$$(L_2, V_2) = (L_1, V_1) \cdot \begin{bmatrix} \frac{3}{4} & \frac{1}{4} \\ \frac{1}{10} & \frac{9}{10} \end{bmatrix} = (L_0, V_0) \cdot \begin{bmatrix} \frac{3}{4} & \frac{1}{4} \\ \frac{1}{10} & \frac{9}{10} \end{bmatrix} \cdot \begin{bmatrix} \frac{3}{4} & \frac{1}{4} \\ \frac{1}{10} & \frac{9}{10} \end{bmatrix}$$

$$= (L_0, V_0) \cdot \begin{bmatrix} \frac{3}{4} & \frac{1}{4} \\ \frac{1}{10} & \frac{9}{10} \end{bmatrix}^2$$

At the end of day 3,

$$(L_3, V_3) = (L_2, V_2) \cdot \begin{bmatrix} \frac{3}{4} & \frac{1}{4} \\ \frac{1}{10} & \frac{9}{10} \end{bmatrix} = (L_0, V_0) \cdot \begin{bmatrix} \frac{3}{4} & \frac{1}{4} \\ \frac{1}{10} & \frac{9}{10} \end{bmatrix}^3$$

At the end of day k,

$$(L_k, V_k) = (L_0, V_0) \cdot \begin{bmatrix} \frac{3}{4} & \frac{1}{4} \\ \frac{1}{10} & \frac{9}{10} \end{bmatrix}^k \longrightarrow (\tfrac{2}{7}, \tfrac{5}{7})$$

The truly amazing fact is that no matter what the original proportions (L_0, V_0) may be, for large k,

$$(L_k, V_k) = (L_0, V_0) \cdot \begin{bmatrix} \frac{3}{4} & \frac{1}{4} \\ \frac{1}{10} & \frac{9}{10} \end{bmatrix}^k$$

approaches the same value, namely $(\tfrac{2}{7}, \tfrac{5}{7})$. This means that no matter what liquid-vapor distribution is assumed in the beginning, after a long time approximately $\tfrac{2}{7}$ of the substance will be liquid and $\tfrac{5}{7}$ will be vapor. An equilibrium position is obtained since $(\tfrac{2}{7})\,(\tfrac{1}{4}) = \tfrac{1}{14}$ of the total original liquid changes from liquid state into vapor state, and $(\tfrac{5}{7}) \cdot (\tfrac{1}{10}) = \tfrac{1}{14}$ of the total original liquid changes from vapor into liquid state, when

$$(L, V) = (\tfrac{2}{7}, \tfrac{5}{7})$$

(See Problems 1 and 12, Set 9.6, for further proof.)

The above example is a much simplified example of a *Markov chain*. In *Markov chains*, the probability for a given instance is a function of the outcome of the immediately preceding experiment. Arguments of this type are currently coming into great importance in the social and biological sciences, as well as in diffusion studies in physics, chemistry, and geology.

There are assumptions in each of these examples which make them seem unreal—for example, the percentage of persons moving to and from the suburbs is not constant, but rather it is a function of several variables. The percentages of liquid-vapor change are also functions rather than true constants. Similarly, the price of steel is not constant, but is a function of both supply and demand. The next step, of course, is to make the idealized situation more realistic. This is often done by introducing more matrix functions in place of constant ones used. Since World War II, the use of large (miscalled "Giant Brain") computing machines has made such computations feasible in cases of immense complexity.

EXAMPLE 12

One of the most important electrical network configurations is the "four-terminal network," having two input terminals and two output terminals. Much of the apparatus with which an electrical engineer deals is of this general nature. In the equations

$$E_1 = aE_2 + bI_2$$

$$I_1 = cE_2 + dI_2$$

where E_1 and E_2 are the input and output voltages, I_1 and I_2 are the corresponding currents and a, b, c, d are the network parameters (a and d are dimensionless, while b and c have the impedance [ohm] and admittance [mho] dimensions).

In terms of matrices,

$$\begin{bmatrix} E_1 \\ I_1 \end{bmatrix} = \begin{bmatrix} a & b \\ c & d \end{bmatrix} \cdot \begin{bmatrix} E_2 \\ I_2 \end{bmatrix}$$

The matrix $A = \begin{bmatrix} a & b \\ c & d \end{bmatrix}$ is called the *transfer matrix* of the network. The solution of the simultaneous linear equations given by

$$\begin{bmatrix} E_1 \\ I_1 \end{bmatrix} = A \cdot \begin{bmatrix} E_2 \\ I_2 \end{bmatrix}$$

is

$$A^{-1} \cdot \begin{bmatrix} E_1 \\ I_1 \end{bmatrix} = \begin{bmatrix} E_2 \\ I_2 \end{bmatrix} \qquad \text{(Why?)}$$

Another important feature of transfer matrices is that, if A_1 and A_2 are the transfer matrices of two four-terminal networks which are connected in cascade, the transfer matrix of the combination is the product of the transfer matrices of the original networks, in the same order:

$$A_1 \cdot A_2 = A_{1,2}$$

A typical problem of this type would be

	Network 1	Network 2
a	1.00	2.50
b	10.00 ohms	5.00 ohms
c	0.25 mhos	1.00 mhos
d	3.00	1.50

Then,

$$\begin{bmatrix} 1.00 & 10.00 \\ 0.25 & 3.00 \end{bmatrix} \cdot \begin{bmatrix} 2.50 & 5.00 \\ 1.00 & 1.50 \end{bmatrix} = \begin{bmatrix} 12.5 & 20.0 \\ 3.62 & 5.75 \end{bmatrix}$$

EXAMPLE 13

The impedances in an open-circuit transistor network may be

$$Z = \begin{bmatrix} A_{ee} & Z_{ec} \\ Z_{ce} & A_{cc} \end{bmatrix} = \begin{bmatrix} 500 & 250 \\ 35000 & 18000 \end{bmatrix}$$

where e denotes the emitter terminal and c the collector terminal. The admittance matrix is then Z^{-1}.

$$Z^{-1} = \begin{bmatrix} 72 & -1 \\ -140 & 2 \end{bmatrix} \times 10^{-3}$$

The so-called *indefinite admittance matrix* of a network is obtained from Z^{-1} by adding a third row and column with elements so chosen that the sum of each row and of each column is zero. The student should show that this is always possible. The indefinite admittance matrix of the above transistor network is then

$$\begin{bmatrix} 72 & -1 & -71 \\ -140 & 2 & 138 \\ 68 & -1 & -67 \end{bmatrix}$$

This matrix provides a rapid method of computing operating performance of grounded-base, grounded-emitter, or grounded-collector and ungrounded amplifier stages. If a terminal is grounded, the corresponding row and column of the matrix is deleted. If a conductance is connected between two terminals, the new admittance matrix is obtained by adding and subtracting the conductance value to matrix elements corresponding to these terminals. This is not the proper place to go into the details of the electrical engineering involved. The purpose is to demonstrate a few of the many uses of matrix theory. Le Corbeiller's book *Matrix Analysis of Electric Networks* (Cambridge, Mass.: Harvard Press, 1950) provides many others, and R. A. Howard gives a nice matrix derivation of the relation $(Z^{-1} = C_t \cdot Z \cdot C)$ between branch impedance and mesh impedance in a network containing emf sources in Volume 22, page 93 (Feb. 1954) of *American Journal of Physics*.

Additional discussion of the applications of matrix theory, not only to electrical engineering, but also to vibration problems, elastic structure theory, differential equations, and similar engineering problems will be found in Chapter 13 of the first volume and in Chapter 12 of the second volume of *Modern Mathematics for the Engineer*, edited by E. F. Beckenbach (New York: McGraw-Hill, 1956, 1961).

5.12 MAPPINGS AND TRANSFORMATIONS

In two-dimensional (plane) space, a point is represented as a pair of coordinates (x_o, y_o). In three-dimensional (solid) space, a point is represented by a triple of coordinates (x_o, y_o, z_o). In four-dimensional (hyper) space, a point is a quadruple of coordinates (w_o, x_o, y_o, z_o) and in n-dimensional (hyper) space, a point is represented by an n-tuple of coordinates $(w_o, x_o, y_o, z_o, q_o, \cdots, t_o)$. Each n-tuple is a vector or row matrix.

In analytic geometry, one studies the effect of rotating the coordinate axes on the coordinates of points (or on the equations of loci). This results in the equations

$$x' = x \cos \theta + y \sin \theta$$

$$y' = -x \sin \theta + y \cos \theta$$

in which x' and y' represent the new coordinates of the point (x, y) after the x, y axes have been rotated through an angle θ.

These relations may be expressed neatly in matrix form as

$$\begin{bmatrix} x' \\ y' \end{bmatrix} = \begin{bmatrix} \cos\theta & \sin\theta \\ -\sin\theta & \cos\theta \end{bmatrix} \cdot \begin{bmatrix} x \\ y \end{bmatrix}$$

Much more general transformations than rotations of axes may be represented by matrices. For example, $\begin{bmatrix} x' \\ y' \end{bmatrix} = \begin{bmatrix} k & 0 \\ 0 & k \end{bmatrix} \cdot \begin{bmatrix} x \\ y \end{bmatrix}$ represents a uniform stretching away from the origin such that each point is moved to a new point, k times as far from the origin as before. If $0 < |k| < 1$, it is a uniform *compression* rather than a stretching. If $k < 0$, the stretching or compression is followed by a "reflection" through the origin.

The transformation

$$\begin{bmatrix} x' \\ y' \end{bmatrix} = \begin{bmatrix} 1 & 0 \\ a & 1 \end{bmatrix} \cdot \begin{bmatrix} x \\ y \end{bmatrix}$$

represents *shearing* motion parallel to the y axis. Each point is moved along a line parallel to the y axis a distance proportional to its abscissa. A rectangle with vertices $(0, 0)$, $(2, 0)$, $(2, 1)$, $(0, 1)$ is sheared into a parallelogram with vertices $(0, 0)$, $(2, 2a)$, $(2, 2a + 1)$, $(0, 1)$. Experiment with actual points until you understand the nature of the shear transformation. A shear parallel to the x-axis is obtained using the transformation

$$\begin{bmatrix} x' \\ y' \end{bmatrix} = \begin{bmatrix} 1 & a \\ 0 & 1 \end{bmatrix} \cdot \begin{bmatrix} x \\ y \end{bmatrix} = \begin{bmatrix} x + ay \\ y \end{bmatrix}$$

More complicated transformations can often be expressed as a series of simple transformations, and, importantly, a single matrix which does the entire complicated transformation in one multiplication may be obtained as the product of the simple transformation matrices in the reverse order from which they were applied. That is, if P is operated on by four transformations, T_1, T_2, T_3, T_4 in that order, the result is $R = T_4 \cdot \{T_3 \cdot [T_2 \cdot (T_1 \cdot P)]\}$. However, since matrix multiplication is associative, $R = (T_4 \cdot T_3 \cdot T_2 \cdot T_1) \cdot P = T \cdot P$ for some matrix T. For example, a rotation through $30°$, followed by a stretching of $k = 4$ units away from the origin, followed by a y-direction shearing motion of proportionality

3 may be obtained as

$$\begin{bmatrix} x' \\ y' \end{bmatrix} = \begin{bmatrix} 1 & 0 \\ 3 & 1 \end{bmatrix} \cdot \begin{bmatrix} 4 & 0 \\ 0 & 4 \end{bmatrix} \cdot \begin{bmatrix} \sqrt{3}/2 & \frac{1}{2} \\ -\frac{1}{2} & \sqrt{3}/2 \end{bmatrix} \cdot \begin{bmatrix} x \\ y \end{bmatrix}$$

or

$$\begin{bmatrix} x' \\ y' \end{bmatrix} = \begin{bmatrix} 2\sqrt{3} & 2 \\ 6\sqrt{3} - 2 & 6 + 2\sqrt{3} \end{bmatrix} \cdot \begin{bmatrix} x \\ y \end{bmatrix}$$

Show that this gives the desired results in the case of several specific points. Note that the transformation corresponding to the matrix nearest the point vector, $\begin{bmatrix} x \\ y \end{bmatrix}$, is applied first.

EXAMPLE 14

Find a shearing parallel to the y axis and a stretching transformation $\begin{bmatrix} k & 0 \\ ak, & k \end{bmatrix}$ which will transform the line $y = 6x + 5$ into the line $y = 10$. In this case, the point $(x, 6x + 5)$ must be carried into the point $(?, 10)$, where the ? represents some function of x, but not necessarily x itself.

$$\begin{bmatrix} k & 0 \\ ak & k \end{bmatrix} \cdot \begin{bmatrix} x \\ 6x + 5 \end{bmatrix} = \begin{bmatrix} kx \\ akx + 6kx + k5 \end{bmatrix} = \begin{bmatrix} kx \\ (ak + 6k)x + 5k \end{bmatrix} = \begin{bmatrix} x' \\ y' \end{bmatrix}$$

which should equal $\begin{bmatrix} ? \\ 0 \cdot x + 10 \end{bmatrix}$ to have the desired effect.

Equating the constant terms and coefficients of x, one obtains

$$ak + 6k = 0, \qquad 5k = 10$$

or

$$k = 2, \qquad ak = -12$$

and the desired transformation matrix is

$$\begin{bmatrix} 2 & 0 \\ -12 & 2 \end{bmatrix}$$

The stretching factor is $k = 2$ and the shear factor is $a = -6$.

EXAMPLE 15

Determine what happens to the locus $y = 2x + 3$ under stretching transformation $\begin{bmatrix} 5 & 0 \\ 0 & 5 \end{bmatrix}$.

$$\begin{bmatrix} 5 & 0 \\ 0 & 5 \end{bmatrix} \cdot \begin{bmatrix} x \\ 2x + 3 \end{bmatrix} = \begin{bmatrix} 5x \\ 10x + 15 \end{bmatrix} = \begin{bmatrix} x' \\ y' \end{bmatrix}$$

Hence, since $x' = 5x$ and $y' = 10x + 15$, the equation becomes $y' = 10\left(\dfrac{x'}{5}\right) + 15$ or $y' = 2x' + 15$. That is, the line $y = 2x + 3$ becomes the line $y = 2x + 15$.

EXAMPLE 16

Determine what happens to the locus $y = 2x + 3$ under the shear $\begin{bmatrix} 1 & 0 \\ 3 & 1 \end{bmatrix}$.

$$\begin{bmatrix} 1 & 0 \\ 3 & 1 \end{bmatrix} \cdot \begin{bmatrix} x \\ 2x + 3 \end{bmatrix} = \begin{bmatrix} x \\ 5x + 3 \end{bmatrix} = \begin{bmatrix} x' \\ y' \end{bmatrix}$$

Thus $x' = x$ and $y' = 5x + 3$. That is, the line $y = 2x + 3$, becomes the line $y = 5x + 3$.

One of the beauties of the matrix notation is that it generalizes so nicely. Transformations in a plane have been discussed. Similar discussions for three-dimensional space can be given. In fact, similar discussions are used in studies of four-dimensional, five-dimensional, six-dimensional, and higher-dimensional spaces. Six-dimensional space has practical applications in hydrodynamics, thermodynamics, and electronics of solid bodies. It is convenient to use three-dimensional vectors representing forces at each point of a three-dimensional surface or body. When these are summed, a sextuple integral results. Talk of fourth and higher dimensions is often shrouded in an umbral mystery. These powerful concepts are useful, and even essential, in modern physics, chemistry, engineering, and mathematics. Although they are more general than the usual high-school geometry, they are not mysterious. Rather, let us say their power is awe inspiring.

The xy plane of freshman analytic geometry is an example of a two-dimensional vector space. A plane curve is said to be symmetric with respect to the x axis if, for each vector, or point (a, b) on the curve, the vector $(a, -b)$ also lies on the curve. That is, the correspondence

$$(x, y) \rightarrow (x, -y)$$

carries each point of the curve into a point of the curve. However, the correspondence $(x, y) \rightarrow (x, -y)$ is applied to every vector (point) of the plane, carrying each vector (x, y) into its "x-axis mirror image," $(x, -y)$. This correspondence is called a "reflection of the plane through the x axis" for obvious reasons. The correspondence $(x, y) \rightarrow (x, -y)$ may be accomplished using the matrix $R = \begin{bmatrix} 1 & 0 \\ 0 & -1 \end{bmatrix}$, since

$\begin{bmatrix} 1 & 0 \\ 0 & -1 \end{bmatrix} \cdot \begin{bmatrix} x \\ y \end{bmatrix} = \begin{bmatrix} x \\ -y \end{bmatrix}$. This is but another example of the very general theory of matrix transformations of a vector space. The word *mapping* is often used to describe such transformations, which are of singular mathematical importance. If ξ represents an n coordinate vector (i.e., n row by one column matrix) and T is an n row square matrix, then $T \cdot \xi$ is the n coordinate vector into which ξ is transformed (mapped) by T. The reader may wish to investigate what happens when T is not square.

A polynomial with real coefficients and two variables x and y with each term of second degree is called *quadratic form* in x and y. It is often convenient to express such forms in matrix notation.

EXAMPLE 17

Express $3x^2 + 8xy - 5y^2$ in matrix form.

$$[x, y] \cdot \begin{bmatrix} 3 & 4 \\ 4 & -5 \end{bmatrix} \cdot \begin{bmatrix} x \\ y \end{bmatrix} = 3x^2 + 8xy - 5y^2$$

There are other possible matrix expressions of the given form, but the one given is perhaps the most useful. The matrix $\begin{bmatrix} 3 & 4 \\ 4 & -5 \end{bmatrix}$ has the property that, if its rows and columns are transposed, the same matrix results. In the notation of Section 5.5, this is stated as $a_{rc} = a_{cr}$. Such a matrix is called *symmetric*. In courses in matrix theory, considerable theory can be developed concerning such matrices. Not only is the analytic geometry of conic sections developed by considering certain types of matrices which will transform the quadratic form matrix into a diagonal matrix, but solid analytic geometry, statistics, dynamics, and certain problems in maxima and minima each yield to leverage applied through the use of quadratic forms.

A *translation* of axis often simplifies problems and results. It is quite feasible to accomplish both a translation and rotation in the same operation by using the transformation

$$x' = x \cos \theta + y \sin \theta - a$$

$$y' = -x \sin \theta + y \cos \theta - b$$

To express this in matrix form it is convenient to use

$$\begin{bmatrix} x' \\ y' \\ 1 \end{bmatrix} = \begin{bmatrix} \cos \theta & \sin \theta & -a \\ -\sin \theta & \cos \theta & -b \\ 0 & 0 & 1 \end{bmatrix} \cdot \begin{bmatrix} x \\ y \\ 1 \end{bmatrix}$$

Note that the 1 annexed to the bottom of the column matrix (vector) remains 1 after multiplication by a matrix having bottom row 0, 0, 1. (Why?)

A transformation of the type induced by the matrix

$$\begin{bmatrix} \cos \theta & \sin \theta & -a \\ -\sin \theta & \cos \theta & -b \\ 0 & 0 & 1 \end{bmatrix}$$

is known as a *Euclidean plane transformation*. Euclidean geometry is the study of those properties (distance, angle, area, etc.) of curves (often, but not necessarily, restricted to plane polygons and conic sections) which are invariant (unchanged) under Euclidean plane transformations. Differential geometry studies those properties (curvature, torsion, etc.) of space curves which remain invariant under Euclidean space transformations. Each of these studies may be generalized by considering properties (concurrence of lines and points, cross ratio, etc.) which remain invariant under more general (projective) transformations of the type

$$\begin{bmatrix} x' \\ y' \end{bmatrix} = \begin{bmatrix} a & b \\ c & d \end{bmatrix} \cdot \begin{bmatrix} x \\ y \end{bmatrix}, \quad \text{where} \quad ad - bc \neq 0$$

This is known as projective geometry.

The important branch of mathematics known as *topology* is a study of properties which remain invariant under still more general transformations. This powerful and fascinating branch of mathematics is currently producing remarkable results in many fields. (See *Mathematical Reviews.*)

The theory of substitution of variables may be approached using matrix theory. For example:

$$\begin{array}{rll} & \textit{Equations} & \begin{array}{c}\textit{Coefficient}\\\textit{Matrices}\end{array} \\[2ex] \text{If} \quad W = 3U - 5V & & \begin{bmatrix} 3 & -5 \\ 1 & 2 \end{bmatrix} \\[2ex] Z = 1U + 2V, & & \\[3ex] \text{where} \quad U = 7x + 4y & & \begin{bmatrix} 7 & 4 \\ 6 & 10 \end{bmatrix} \\[2ex] V = 6x + 10y & & \end{array}$$

Then,

$$W = 3(7x + 4y) - 5(6x + 10y) = -9x - 38y \qquad \begin{bmatrix} -9 & -38 \\ 19 & 24 \end{bmatrix}$$

$$Z = 1(7x + 4y) + 2(6x + 10y) = 19x + 24y$$

Note that the product of the coefficient matrices yields the same result with less effort:

$$\begin{bmatrix} 3 & -5 \\ 1 & 2 \end{bmatrix} \cdot \begin{bmatrix} 7 & 4 \\ 6 & 10 \end{bmatrix} = \begin{bmatrix} -9 & -38 \\ 19 & 24 \end{bmatrix}$$

This is an example of a substitution of one equation into another, performed using the coefficient matrices.

It would be unfair to leave our brief discussion of matrices without pointing out that it has been only an introduction. To give more than an introduction is impossible in a short book; indeed, it would defeat the purpose of the book. It is the author's hope that each of you will be able to take one of the regular courses in matrix theory offered at most universities. The final chapter on matrix theory has not yet been written. A glance into almost any issue of *Mathematical Reviews* (see your school library) will reveal current research papers dealing with matrices. For example, matrix methods are known which sum every convergent infinite series to its normal Cauchy sum, but which will also obtain consistent sums for certain series which are ordinarily divergent! This extension of the definition of convergence has been fruitful in modern mathematics and has greatly aided our understanding of its nature.

Chapter 9 introduces several more advanced ideas involving matrices, including the Hamilton–Cayley theorem, characteristic roots and vectors, derivatives and integrals of matrices as used in aircraft flutter analysis, and infinite series of powers of matrices. The reader who is already familiar with determinant theory may wish to study Chapter 9 next.

Problem Set 5.12

1. Use the rotation matrix with $\theta = 30°$ to determine the x', y' coordinates of points having x, y coordinates $A(4\sqrt{3}, 10\sqrt{3})$, $B(6, 2)$, $C(\sqrt{3}, 8)$, $D(2, -15)$, $E(4\sqrt{3}, 61)$, and $F(-4, 12)$. Draw diagrams and use plane geometry to demonstrate that the values obtained are correct in at least one case.

2. Use stretching transformation matrices with $k = 3$, -7, and $\frac{1}{2}$ to obtain the x', y' coordinates of the points having x, y coordinates $A(1, 1)$, $B(3, 4)$, $C(2, -7)$, after the corresponding stretching away from the origin has taken place. Make sketches to show what has happened for each value of k.

3. The same as Problem 2 for the shearing parallel to the y-axis transformation with $a = 3$, -7, and $\frac{1}{2}$.

4. Determine what happens to each of the following simple geometric loci under a stretching transformation with $k = 3$.

   (a) A circle with center $(0, 0)$ and radius 7.
   (b) A circle with center $(0, 5)$ and radius 5.
   (c) A circle with center $(3, 0)$ and radius 3.
   (d) A line parallel to the x axis and three units above it.
   (e) A line parallel to the y axis and two units to the right of it.

5. The same as Problem 4 for $k = \frac{1}{2}$.

6. The same as Problem 4 for $k = -2$.

7. Determine what happens to each of the following simple geometric loci under a shearing parallel to the x-axis transformation with $a = 3$.

   (a) A square with vertices $(0, 0)$, $(1, 0)$, $(1, 1)$, $(0, 1)$.
   (b) A rectangle with vertices $(0, 0)$, $(5, 0)$, $(5, 1)$, $(0, 1)$.
   (c) A line parallel to the x axis and three units above it.
   (d) A line parallel to the y axis and seven units to the right of it.
   (e) A circle with center $(0, 0)$ and radius 1.

8. The same as Problem 7, with $a = \frac{1}{2}$.

9. The same as Problem 7, with $a = -1$.

10. Demonstrate that the matrix

$$\begin{bmatrix} 2\sqrt{3} & 2 \\ 6\sqrt{3} - 2 & 6 + 2\sqrt{3} \end{bmatrix}$$

   given in the text does produce the same effect as the rotation, stretching, and shearing transformations mentioned, by applying it to the points given in Problem 2. Is the order in which these transformations are performed important? Use diagrams and plane geometry to demonstrate.

11. Find what the new coordinates of the points of Problem 2 will be, if the origin is translated to the point $(3, -5)$ and then the axes rotated through (a) 45° (b) 60°. Does it make a difference whether the translation is done before or after the rotation?

*12. See if you can describe the effect of the transformations given below on a point in the x, y plane.

(a) $\begin{bmatrix} 1 & 0 \\ 1 & 0 \end{bmatrix}$
(b) $\begin{bmatrix} 2 & 0 \\ 0 & 0 \end{bmatrix}$
(c) $\begin{bmatrix} 1 & 1 \\ 0 & 2 \end{bmatrix}$
(d) $\begin{bmatrix} -1 & 1 \\ 1 & -1 \end{bmatrix}$

(e) $\begin{bmatrix} 0 & 1 \\ 0 & 1 \end{bmatrix}$
(f) $\begin{bmatrix} 0 & 1 \\ 1 & 0 \end{bmatrix}$
(g) $\begin{bmatrix} 3 & 0 \\ 0 & -3 \end{bmatrix}$
(h) $\begin{bmatrix} -1 & 0 \\ 0 & -1 \end{bmatrix}$

(i) $\begin{bmatrix} \dfrac{\sqrt{3}}{2} & \dfrac{1}{2} \\ -\dfrac{1}{2} & \dfrac{\sqrt{3}}{2} \end{bmatrix}$
(j) $\begin{bmatrix} \dfrac{1}{\sqrt{2}} & \dfrac{1}{\sqrt{2}} \\ -\dfrac{1}{\sqrt{2}} & \dfrac{1}{\sqrt{2}} \end{bmatrix}$

13. Consult the journal *Mathematical Reviews* and count the number of articles concerning matrix theory which appear in any one month's issue.

14. Examine the recent books and the periodicals in your school library and find an application of matrix theory to *your own field of interest*. The following sources have proved particularly fruitful.

    (a) The various reviews: (*Mathematical Reviews, Physical Reviews, Psychological Reviews*, etc.)
    (b) Professional journals: (*Biometrica, Electrical Engineering, Social Science, Scientific American, Econometrica, Journal of Applied Physics*, etc.)
    (c) Texts, such as *Mathematical Thinking in the Social Sciences* by P. F. Lazarsfeld; *Theory of Games* by J. C. C. McKinsey or by Kuhn and Tucker; *Modern Mathematics for the Engineer* by Beckenbach.
    (d) Reports of large companies, such as Bell Laboratories, General Electric, Douglas Aircraft, Carter Oil, Lockheed, etc.

15. Read and report on the article "What is a Matrix" by C. C. MacDuffee in *American Mathematical Monthly* **50,** p. 360.

16. Read and report on the article by Murnaghan in *Scientific Monthly,* **68,** No. 4, p. 268.

*17. Read and report on "On the representation of a Lorentz Transformation by Means of Two-rowed Matrices" by Murnaghan in *American Mathematical Monthly,* **38,** p. 504.

18. Show that rotations and shears are not necessarily commutative; that is, a rotation followed by a shear may produce results different than the same shear followed by the rotation. [HINT: Find an example.]

19. Show that a stretching transformation commutes not only with all other stretching transformations, but also with rotations and shears. [HINT: The matrix of a stretching transformation is of a special type whose properties you have already studied.]

20. The matrix identity $B^{-1} - A^{-1} = A^{-1} \cdot (A - B) \cdot B^{-1}$ is very helpful in establishing error bounds in computer numerical analysis. Show that the given identity holds for all non-singular n by n matrices A and B with elements selected from the set of real numbers.

Selected Reading List, Chapter 5
MATRICES

The following books and journals will provide additional reading for students interested in term projects or independent study.

Books

MacDuffee, C. C., *Vectors and Matrices*, Carus Monograph No. 7 (Mathematical Association of America).

MacDuffee, C. C., *Theory of Matrices* (One of the Ergebnisse Series, reprinted by Chelsea.)

Perlis, S., *Theory of Matrices*, Reading, Mass.: Addison-Wesley, 1952.

Articles

Alspach, B., M. Goldberg and J. W. Moon, "The Group of the Composition of Two Tournaments," *Mathematics Magazine*, **41,** No. 2 (Mar. 1968), pp. 77–80.

Cox, R. H., "A Proof of the Schroeder-Bernstein Theorem," *American Mathematical Monthly*, **75,** No. 5 (May 1968), p. 508.

Eliezer, C. J., "A Note on Group Commutators of 2×2 Matrices," *American Mathematical Monthly*, **75,** No. 10 (Dec. 1968), p. 1090.

Moran, D. A., "Pascal Matrices," *Mathematics Magazine*, **40,** No. 1 (Jan. 1967), pp. 12–14.

Sinkhorn, R., "Concerning the Jordan Normal Form," *Mathematics Magazine*, **41,** No. 2 (Mar. 1968), pp. 91–94.

Swartz, W., "Integration by Matrix Inversion," *American Mathematical Monthly*, **65,** No. 4 (Apr. 1958), p. 282.

Wilansky, A., "Spectral Decomposition of Matrices for High School Students," *Mathematics Magazine*, **41,** No. 2 (Mar. 1968), pp. 51–59.

Wilf, H., "Curve Fitting Matrices," *American Mathematical Monthly*, **65,** No. 4 (Apr. 1958), p. 272.

Williams, V. C. and F. S. Cater, "On the Rank of a Matrix," *Mathematics Magazine*, **41,** No. 5 (Nov. 1968), pp. 249–250.

"A Matrix Property," Problem E 2052, *American Mathematical Monthly*, **76,** No. 2 (Feb. 1969), p. 191.

See Chapter 9 for additional work on matrices and page 309 for additional reading suggestions.

Notes on Chapter 5

Notes on Chapter 5

LINEAR SYSTEMS

6

6.1 SYSTEMS OF LINEAR EQUATIONS

The solution of a system of linear equations in several unknowns is often accomplished by "addition and subtraction" of equations. For example, the system

$$\begin{cases} 3x + 2y + 2z = 3 \\ 2x - 3y + 3z = 1 \\ 2x + 3y - 2z = -3 \end{cases}$$

may be reduced, after a suitable number of algebraic gyrations, to the system

$$\begin{cases} 1x = -1 \\ 1y = 1 \\ 1z = 2 \end{cases}$$

which is called a solution. However, the first system is equivalent to the matrix equation

$$\begin{bmatrix} 3 & 2 & 2 & 3 \\ 2 & -3 & 3 & 1 \\ 2 & 3 & -2 & -3 \end{bmatrix} \cdot \begin{bmatrix} x \\ y \\ z \\ -1 \end{bmatrix} = \begin{bmatrix} 0 \\ 0 \\ 0 \end{bmatrix}$$

while the second (solution) system is represented as

$$\begin{bmatrix} 1 & 0 & 0 & -1 \\ 0 & 1 & 0 & 1 \\ 0 & 0 & 1 & 2 \end{bmatrix} \cdot \begin{bmatrix} x \\ y \\ z \\ -1 \end{bmatrix} = \begin{bmatrix} 0 \\ 0 \\ 0 \end{bmatrix}$$

The reduction of the matrix

$$\begin{bmatrix} 3 & 2 & 2 & 3 \\ 2 & -3 & 3 & 1 \\ 2 & 3 & -2 & -3 \end{bmatrix} \qquad \text{to the matrix} \qquad \begin{bmatrix} 1 & 0 & 0 & -1 \\ 0 & 1 & 0 & 1 \\ 0 & 0 & 1 & 2 \end{bmatrix}$$

may be performed using elementary row operations. (See Section 5.6.) The steps are essentially those which would be used by any beginning algebra student, but the amount of writing is reduced since the variables and connecting signs are omitted. There are also other, more important, advantages. Example 1, below, is solved both by using equations and by using the detached coefficient (matrix) method, so that the reader may note the similarities.

Note that, if there exist values of x, y, z such that

$$\begin{bmatrix} 3 & 2 & 2 & 3 \\ 2 & -3 & 3 & 1 \\ 2 & 3 & -2 & 3 \end{bmatrix} \cdot \begin{bmatrix} x \\ y \\ z \\ -1 \end{bmatrix} = \begin{bmatrix} 0 \\ 0 \\ 0 \end{bmatrix}$$

then it is also true that, for an elementary matrix E_i

$$[E_i] \cdot \begin{bmatrix} 3 & 2 & 2 & 3 \\ 2 & -3 & 3 & 1 \\ 2 & 3 & -2 & 3 \end{bmatrix} \cdot \begin{bmatrix} x \\ y \\ z \\ -1 \end{bmatrix} = \begin{bmatrix} 0 \\ 0 \\ 0 \end{bmatrix}$$

for the same values of x, y, z, since, if $(M \cdot X) = (0)$, then

$$(E \cdot M) \cdot X = E \cdot (M \cdot X) = E \cdot (0) = (0)$$

where E, M, X, and (0) are matrices or vectors of the proper size. The converse is also true; that is, $E \cdot M \cdot X = (0)$ implies $M \cdot X = (0)$ if E is an elementary matrix. Let us write the equation just given as

$$[E_i] \cdot [A] \cdot \begin{bmatrix} x \\ y \\ z \\ -1 \end{bmatrix} = \begin{bmatrix} 0 \\ 0 \\ 0 \end{bmatrix}$$

If E_i can be determined such that $[E_i \cdot A]$ is the matrix, $\begin{bmatrix} 1 & 0 & 0 & -1 \\ 0 & 1 & 0 & 1 \\ 0 & 0 & 1 & 2 \end{bmatrix}$

then the given equation becomes

$$\begin{bmatrix} 1 & 0 & 0 & -1 \\ 0 & 1 & 0 & 1 \\ 0 & 0 & 1 & 2 \end{bmatrix} \cdot \begin{bmatrix} x \\ y \\ z \\ -1 \end{bmatrix} = \begin{bmatrix} 0 \\ 0 \\ 0 \end{bmatrix}$$

or

$$1 \cdot x \qquad\qquad = -1$$

$$1 \cdot y \qquad = \quad 1$$

$$1 \cdot z = \quad 2$$

that is,
$$\begin{bmatrix} x \\ y \\ z \\ -1 \end{bmatrix} = \begin{bmatrix} -1 \\ 1 \\ 2 \\ -1 \end{bmatrix} \quad \text{and} \quad \begin{bmatrix} 3 & 2 & 2 & 3 \\ 2 & -3 & 3 & 1 \\ 2 & 3 & -2 & 3 \end{bmatrix} \cdot \begin{bmatrix} -1 \\ 1 \\ 2 \\ -1 \end{bmatrix} = \begin{bmatrix} 0 \\ 0 \\ 0 \end{bmatrix}$$

as desired. Since it is simple to transform (A) to the desired form $(E \cdot A)$ by row transformations, this method is the one usually used. It is common practice to state this result as $(x, y, z) = (-1, 1, 2)$.

EXAMPLE 1

$$\text{Solve} \qquad \begin{cases} 3x - y - z = -\tfrac{1}{2} \\[2mm] 2x + y + 4z = \quad 0 \\[2mm] 4x + 3y + 5z = \quad 6 \end{cases}$$

Solutions Using Equations: *Solution Using Matrices:*

(1) $3x - y - z = -\tfrac{1}{2}$

(2) $2x + y + 4z = \quad 0$

(3) $4x + 3y + 5z = \quad 6$

$$\begin{bmatrix} 3 & -1 & -1 & -\tfrac{1}{2} \\ 2 & 1 & 4 & 0 \\ 4 & 3 & 5 & 6 \end{bmatrix} \cdot \begin{bmatrix} x \\ y \\ z \\ -1 \end{bmatrix} = \begin{bmatrix} 0 \\ 0 \\ 0 \end{bmatrix}$$

One of the many possible procedures for solving this system of linear equations is outlined below. Since only the coefficient matrix is used, only the coefficient matrix has been written.

Add -1 times the second row (or equation) to the first row (or equation).
Add -2 times row two to row three.

$$1x - 2y - 5z = -\tfrac{1}{2}$$
$$2x + 1y + 4z = 0$$
$$1y - 3z = 6$$

$$\begin{bmatrix} 1 & -2 & -5 & -\tfrac{1}{2} \\ 2 & 1 & 4 & 0 \\ 0 & 1 & -3 & 6 \end{bmatrix}$$

Add -2 times (new) row one to (new) row two.
Interchange the two bottom rows.

$$1x - 2y - 5z = -\tfrac{1}{2}$$
$$1y - 3z = 6$$
$$5y + 14z = 1$$

$$\begin{bmatrix} 1 & -2 & -5 & -\tfrac{1}{2} \\ 0 & 1 & -3 & 6 \\ 0 & 5 & 14 & 1 \end{bmatrix}$$

Add -5 times row two to row three.

$$1x - 2y - 5z = -\tfrac{1}{2}$$
$$1y - 3z = 6$$
$$29z = -29$$

$$\begin{bmatrix} 1 & -2 & -5 & -\tfrac{1}{2} \\ 0 & 1 & -3 & 6 \\ 0 & 0 & 29 & -29 \end{bmatrix}$$

Multiply row three by $\tfrac{1}{29}$.

$$1x - 2y - 5z = -\tfrac{1}{2}$$
$$1y - 3z = 6$$
$$1z = -1$$

$$\begin{bmatrix} 1 & -2 & -5 & -\tfrac{1}{2} \\ 0 & 1 & -3 & 6 \\ 0 & 0 & 1 & -1 \end{bmatrix}$$

Add 3 times row three to row two.
Add 5 times row three to row one.

$$1x - 2y = -\tfrac{11}{2}$$
$$1y = 3$$
$$1z = -1$$

$$\begin{bmatrix} 1 & -2 & 0 & -\tfrac{11}{2} \\ 0 & 1 & 0 & 3 \\ 0 & 0 & 1 & -1 \end{bmatrix}$$

Add 2 times row two to row one.

$$
\begin{array}{rcl}
1x & = & \frac{1}{2} \\[2mm]
1y & = & 3 \\[2mm]
1z & = & -1
\end{array}
\qquad
\begin{bmatrix} 1 & 0 & 0 & \frac{1}{2} \\ 0 & 1 & 0 & 3 \\ 0 & 0 & 1 & -1 \end{bmatrix}
\cdot
\begin{bmatrix} x \\ y \\ z \\ -1 \end{bmatrix}
= 0
$$

that is, $(x, y, z) = (\frac{1}{2}, 3, -1)$.

The preceding systems of three equations are equivalent in the sense that the solutions of each system are also solutions of every other system. Equivalent systems of linear equations are obtained if any of the three elementary row transformations are applied to the coefficient matrix.

<table>
<tr><td>For Equations</td><td>For the Coefficient Array,
or Matrix</td></tr>
<tr><td>1. The order in which the equations are written is immaterial.</td><td>1. Any two rows of the matrix may be interchanged.</td></tr>
<tr><td>2. Both members of an equation may be multiplied by the same nonzero constant.</td><td>2. The elements of a row of the coefficient matrix may be multiplied by a nonzero constant.</td></tr>
<tr><td>3. K times the members of one equation may be added to the corresponding members of another equation.</td><td>3. K times the elements of a row of the matrix may be added to the corresponding elements of a different row.</td></tr>
</table>

Let us test our understanding by showing that an elementary transformation does yield an equivalent system in the case where M has three rows and four columns (three equations in three unknowns). We must show that, if $[x, y, z] = [r_x, r_y, r_z]$ is a solution of the system

$$
[M] \cdot \begin{bmatrix} x \\ y \\ z \\ -1 \end{bmatrix} = \begin{bmatrix} 0 \\ 0 \\ 0 \end{bmatrix}
$$

then $[r_x, r_y, r_z]$ is also a solution of $[E \cdot M] \cdot \begin{bmatrix} x \\ y \\ z \\ -1 \end{bmatrix} = \begin{bmatrix} 0 \\ 0 \\ 0 \end{bmatrix}$, where E is any

elementary matrix[1] and conversely. (Query: What are the dimensions of E?)

Given that

$$[M] \cdot \begin{bmatrix} r_x \\ r_y \\ r_z \\ -1 \end{bmatrix} = \begin{bmatrix} 0 \\ 0 \\ 0 \end{bmatrix}$$

it follows from the associative law that

$$[E \cdot M] \cdot \begin{bmatrix} r_x \\ r_y \\ r_z \\ -1 \end{bmatrix} = [E] \cdot \left[M \cdot \begin{bmatrix} r_x \\ r_y \\ r_z \\ -1 \end{bmatrix} \right] = [E] \cdot \begin{bmatrix} 0 \\ 0 \\ 0 \end{bmatrix} = \begin{bmatrix} 0 \\ 0 \\ 0 \end{bmatrix}$$

as desired.

The converse is only slightly more difficult to prove. Given that

$$[E \cdot M] \cdot \begin{bmatrix} r_x \\ r_y \\ r_z \\ -1 \end{bmatrix} = \begin{bmatrix} 0 \\ 0 \\ 0 \end{bmatrix}$$

---

[1] That is, E is the matrix corresponding to an elementary (row) operation. Actually we need only that E has an inverse.

it follows that

$$[E] \cdot \left[M \cdot \begin{bmatrix} r_x \\ r_y \\ r_z \\ -1 \end{bmatrix} \right] = \begin{bmatrix} 0 \\ 0 \\ 0 \end{bmatrix}$$

Since E is an elementary matrix, E^{-1} exists. Then,

$$[E^{-1}E] \cdot \left[M \cdot \begin{bmatrix} r_x \\ r_y \\ r_z \\ -1 \end{bmatrix} \right] = E^{-1} \cdot \begin{bmatrix} 0 \\ 0 \\ 0 \end{bmatrix}$$

or

$$M \cdot \begin{bmatrix} r_x \\ r_y \\ r_z \\ -1 \end{bmatrix} = \begin{bmatrix} 0 \\ 0 \\ 0 \end{bmatrix}$$

as desired.

There is nothing in this proof which need restrict M to a 3×4 matrix, nor $[r_x, r_y, r_z]$ to a numerical solution. Parametric solutions often occur, for example, in linear systems of five equations involving eight unknowns. In the next problem set, the reader is asked to carry through this proof in the case of a specific matrix.

It is convenient to note here that the identical equation $0 = 0$ may be annexed to any system of K equations to obtain a system of $K + 1$ equations which is equivalent to the original system. The reader should prove this. The detached coefficient (matrix) method, applied by an efficient operator, usually yields a solution more rapidly than other methods.

EXAMPLE 2

Apply the matrix method to determine solutions of

$$\begin{cases} x + y - 3z = -2 \\ 2x - y + 4z = 3 \\ x + y + z = 0 \end{cases}$$

The coefficient matrix is
$\begin{bmatrix} 1 & 1 & -3 & -2 \\ 2 & -1 & 4 & 3 \\ 1 & 1 & 1 & 0 \end{bmatrix}$

Add -2 times row one to row two.

Add -1 times row one to row three.
$\begin{bmatrix} 1 & 1 & -3 & -2 \\ 0 & -3 & 10 & 7 \\ 0 & 0 & 4 & 2 \end{bmatrix}$

Multiply row two by $-\frac{1}{3}$.

Multiply row three by $\frac{1}{4}$, giving $z = \frac{1}{2}$.
$\begin{bmatrix} 1 & 1 & -3 & -2 \\ 0 & 1 & -\frac{10}{3} & -\frac{7}{3} \\ 0 & 0 & 1 & \frac{1}{2} \end{bmatrix}$

Add 3 times row three to row one.

Add $\frac{10}{3}$ times row three to row two, giving $y = -\frac{2}{3}$.
$\begin{bmatrix} 1 & 1 & 0 & -\frac{1}{2} \\ 0 & 1 & 0 & -\frac{2}{3} \\ 0 & 0 & 1 & \frac{1}{2} \end{bmatrix}$

Add -1 times row two to row one.
$\begin{bmatrix} 1 & 0 & 0 & \frac{1}{6} \\ 0 & 1 & 0 & -\frac{2}{3} \\ 0 & 0 & 1 & \frac{1}{2} \end{bmatrix} \cdot \begin{bmatrix} x \\ y \\ z \\ -1 \end{bmatrix} = 0$

The desired solution is

$$
\begin{aligned}
1x \quad\quad &= \tfrac{1}{6} \\
1y \quad &= -\tfrac{2}{3} \\
1z &= \tfrac{1}{2}
\end{aligned}
$$

The matrix method of solving systems of linear equations is well adapted to machine computation as well as longhand computation. The method may also be applied to dependent systems (infinitely many solutions) and inconsistent systems (no solution) of linear equations.

EXAMPLE 3

In the case of the dependent system

$$
\begin{cases}
x + 2y - 3z + 5t = -8 \\
2x + 2y + 4z - 2t = 0
\end{cases}
$$

the coefficient matrix becomes

$$
A = \begin{bmatrix} 1 & 2 & -3 & 5 & -8 \\ 2 & 2 & 4 & -2 & 0 \end{bmatrix}, \quad \text{where} \quad A \cdot \begin{bmatrix} x \\ y \\ z \\ t \\ -1 \end{bmatrix} = \begin{bmatrix} 0 \\ 0 \end{bmatrix}
$$

It is then possible to reduce A, by elementary row (*not column*) transformations, to a row-equivalent matrix B,

$$
B = \begin{bmatrix} 1 & 0 & 7 & -7 & 8 \\ 0 & 1 & -5 & 6 & -8 \end{bmatrix}, \quad \text{where} \quad B \cdot \begin{bmatrix} x \\ y \\ z \\ t \\ -1 \end{bmatrix} = \begin{bmatrix} 0 \\ 0 \end{bmatrix}
$$

or

$$x + 7z - 7t - 8 = 0$$

$$y - 5z + 6t + 8 = 0$$

i.e.,
$$\begin{cases} x = \quad 8 - 7z + 7t \\[2mm] y = -8 + 5z - 6t \\[2mm] z = z \\[2mm] t = t \end{cases}$$

is a solution no matter what values are given to z and t. For example, $(x, y, z, t) = (1, 0, -2, -3)$ is the particular numerical solution obtained by taking $z = -2$, $t = -3$. The general solution may also be expressed as $(x, y, z, t) = (8, -8, 0, 0) + (-7, 5, 1, 0)z + (7, -6, 0, 1)t$.

EXAMPLE 4

The matrix of a system of inconsistent equations will reduce to a matrix having at least one row containing all zeros except the last element, which is nonzero. For example, the system

$$\begin{cases} 3x + 2y - 5z = 3 \\[2mm] 2x - 6y + 4z = 9 \\[2mm] 5x - 4y - \cdot z = 5 \end{cases}$$

has coefficient matrix

$$A = \begin{bmatrix} 3 & 2 & -5 & 3 \\ 2 & -6 & 4 & 9 \\ 5 & -4 & -1 & 5 \end{bmatrix}, \quad \text{where} \quad A \cdot \begin{bmatrix} x \\ y \\ z \\ -1 \end{bmatrix} = \begin{bmatrix} 0 \\ 0 \\ 0 \\ 0 \end{bmatrix}$$

and A is row equivalent to

$$B = \begin{bmatrix} 3 & 2 & -5 & 3 \\ 2 & -6 & 4 & 9 \\ 0 & 0 & 0 & -7 \end{bmatrix}$$

In forming

$$\begin{bmatrix} 3 & 2 & -5 & 3 \\ 2 & -6 & 4 & 9 \\ 0 & 0 & 0 & -7 \end{bmatrix} \cdot \begin{bmatrix} x \\ y \\ z \\ -1 \end{bmatrix} = \begin{bmatrix} 0 \\ 0 \\ 0 \end{bmatrix}$$

the bottom row leads to the statement $7 = 0$, which is not satisfied for any value of (x, y, z). Hence, the system has no solution.

In a system of dependent equations (many solutions), it is possible to express, x, y, and z in terms of one or more *parameters*. The *general solution* may be specialized to obtain *particular solutions* by substituting numerical values for the parameter k.

EXAMPLE 5

$$\begin{cases} x + 2y - z = 3 \\ 2x - y - 2z = 5 \end{cases}$$

The General Solution	*Two Particular Solutions*	
$x = k$	$x = 0$	$x = 1$
$y = \frac{1}{5}$	$y = \frac{1}{5}$	$y = \frac{1}{5}$
$z = -\frac{13}{5}$	$z = -\frac{13}{5}$	$z = -\frac{8}{5}$

Every particular (no parameter) solution may be obtained by giving the proper value to the parameter k and every set of values obtained by giving a value to k is a particular solution of the original system. It is possible (and even likely) that the general solution of a system of linear equations in several unknowns may involve more than one independent parameter.

It is convenient to write a general solution as the sum of two or more vectors:

$$(x, y, z) = (0, \tfrac{1}{5}, -\tfrac{13}{5}) + k(1, 0, 1)$$

where the vector $(0, \frac{1}{5}, -\frac{13}{5})$ is a particular solution of the original set of equations, while $k(1, 0, 1) = (k, 0, k)$ is a general solution of the related homogeneous system obtained by replacing the constant (right hand) members of the original equations by zeros. This may strike a responsive chord in those who have studied differential equations.

EXAMPLE 6

Solve the system

$$\begin{cases} x + 2y - z = 3 \\ 2x - y - 2z = 5 \\ 3x + y - 3z = 8 \end{cases}$$

In matrix notation this becomes

$$\begin{bmatrix} 1 & 2 & -1 & 3 \\ 2 & -1 & -2 & 5 \\ 3 & 1 & -3 & 8 \end{bmatrix} \cdot \begin{bmatrix} x \\ y \\ z \\ -1 \end{bmatrix} = \begin{bmatrix} 0 \\ 0 \\ 0 \end{bmatrix}$$

Adding -2 times row one to row two, then adding -3 times row one to row three and, finally, adding -1 times (new) row two to (new) row three, we obtain

$$\begin{bmatrix} 1 & 2 & -1 & 3 \\ 0 & -5 & 0 & -1 \\ 0 & 0 & 0 & 0 \end{bmatrix}$$

If $\frac{2}{5}$ row two is added to row one, and row two is multiplied by $-\frac{1}{5}$, then the reduction has been carried as far as is feasible, yielding

$$\begin{bmatrix} 1 & 0 & -1 & \frac{13}{5} \\ 0 & 1 & 0 & \frac{1}{5} \\ 0 & 0 & 0 & 0 \end{bmatrix} \cdot \begin{bmatrix} x \\ y \\ z \\ -1 \end{bmatrix} = \begin{bmatrix} 0 \\ 0 \\ 0 \end{bmatrix}$$

which is the same as

$$x - z - \tfrac{13}{5} = 0$$

$$y - \tfrac{1}{5} = 0$$

$$0 \cdot z = 0$$

Clearly if this system is to be satisfied we must have $y = \tfrac{1}{5}$. Furthermore, since $0 \cdot z = 0$ is satisfied by every value z, we select z to be an arbitrary value $z = c$ then it follows that the system will be satisfied if and only if $x = \tfrac{13}{5} + c$.

We have as solution:

$$x = \tfrac{13}{5} + c$$

$$y = \tfrac{1}{5}$$

$$z = c, \text{ an arbitrary value}$$

or $(x, y, z) = (\tfrac{13}{5}, \tfrac{1}{5}, 0) + c(1, 0, 1)$.

It would also have been possible to select x as the arbitrary value, say $x = k$ and then a solution may be obtained by setting $z = k - \tfrac{13}{5}$ to obtain a parametric form for the "single infinitude of solutions." In general, the coefficient matrix reduces, by row transformations, to a matrix having one or more rows of zeros if, and only if, the equations are dependent. In this case infinitely many solutions exist.

The *general solution* of the original system is expressed as a sum of a *particular solution* of the *original* system and the *general solution of the homogeneous system*. It is important that the reader understand why the result so obtained is a general solution. Consider $\alpha_1 = (r_1, r_2, r_3)$ and $\alpha_2 = (s_1, s_2, s_3)$ as two distinct, particular solutions of the original system. Show that $\beta = (\alpha_1 - \alpha_2)$ is a root of the homogeneous system. Let α be a root of the original system and β a root of the homogeneous system. Show that $(\alpha + \beta)$ is a root of the original system since it satisfies the equations of the original system upon substitution therein.

In more general notation, it is usual to use $x_1, x_2, x_3, \cdots, x_n$ as variables in place of $x, y, z, \cdots, w$.

In this case, the linear system

$$
\begin{cases}
a_{11}x_1 + a_{12}x_2 + a_{13}x_3 + \cdots + a_{1n}x_n = k_1 \\[1em]
a_{21}x_1 + a_{22}x_2 + a_{23}x_3 + \cdots + a_{2n}x_n = k_2 \\[1em]
a_{31}x_1 + a_{32}x_2 + \phantom{a_{33}x_3} \cdots + a_{3n}x_n = k_3 \\
\phantom{a_{31}x_1}\vdots \vdots \vdots \vdots \\
a_{m1}x_1 + a_{m2}x_2 + \cdots + a_{mn}x_n = k_m
\end{cases}
$$

may be represented symbolically as

$$\sum_{i=1}^{n} a_{ti}x_i = k_t \qquad (t = 1, 2, \cdots, m)$$

while the corresponding homogeneous system (constant terms zero) is represented as

$$\sum_{i=1}^{n} a_{ti}x_i = 0 \qquad (t = 1, 2, \cdots, m)$$

Two distinct solutions of the original nonhomogeneous system are represented as

$$\alpha_1 = (r_1, r_2, r_3, \cdots, r_n)$$

$$\alpha_2 = (s_1, s_2, s_3, \cdots, s_n)$$

The statement that α_1 and α_2 are solutions of the system

$$\sum_{i=1}^{n} a_{ti}x_i = k_t \qquad (t = 1, 2, \cdots, m) \text{ is equivalent to writing}$$

$$\sum_{i=1}^{n} a_{ti}r_i = k_t, \qquad \sum_{i=1}^{n} a_{ti}s_i = k_t \qquad (t = 1, 2, \cdots, m)$$

(Why?)

We now note that, if $\beta = \alpha_1 - \alpha_2$, then, upon substituting β for x in the homogeneous system, we obtain

$$\sum_{i=1}^{n} a_{ti}(r_i - s_i) = \sum_{i=1}^{n} a_{ti}r_i - \sum_{i=1}^{n} a_{ti}s_i$$

$$= k_t - k_t$$

$$= 0$$

Hence, $\beta = \alpha_1 - \alpha_2$ is a solution of the homogeneous system. Furthermore, if $\alpha = (r_1, r_2, \cdots, r_n)$ is any solution of the original (nonhomogeneous) system and $\beta = (h_1, h_2, \cdots, h_n)$ is any solution of the homogeneous system, then

$$\alpha + \beta = (r_1 + h_1, r_2 + h_2, \cdots, r_n + h_n)$$

is also a solution of the nonhomogeneous system. This follows from

$$\sum_{i=1}^{n} a_{ti}(r_i + h_i) = \sum_{i=1}^{n} a_{ti}r_i + \sum_{i=1}^{n} a_{ti}h_i$$

$$= k_t + 0$$

$$= k_t$$

EXAMPLE 7

In the case of an inconsistent system such as

$$\begin{cases} x + 2y - z = 5 \\ x + 2y - z = 2 \\ 2x + y - z = 1 \end{cases}$$

the matrix reduces to a form in which every element excepting the last (constant term) element is zero in some row. Show that the matrix of the above system reduces to

$$\begin{bmatrix} 1 & 2 & -1 & 5 \\ 1 & 2 & -1 & 2 \\ 2 & 1 & -1 & 1 \end{bmatrix} \rightarrow \begin{bmatrix} 1 & 2 & -1 & 5 \\ 0 & 0 & 0 & -3 \\ 2 & 1 & -1 & 1 \end{bmatrix}$$

and explain why this implies that the system is inconsistent by considering the meaning of the matrix equation

$$\begin{bmatrix} 1 & 2 & -1 & 5 \\ 0 & 0 & 0 & -3 \\ 2 & 1 & -1 & 1 \end{bmatrix} \cdot \begin{bmatrix} x \\ y \\ z \\ -1 \end{bmatrix} = \begin{bmatrix} 0 \\ 0 \\ 0 \end{bmatrix}$$

In general, the solution of a system of linear equations in n unknowns

$$A \cdot \xi = 0$$

where ξ represents a column vector containing the n unknowns and -1, is the result of transforming the matrix A into a matrix $(T \cdot A)$, by use of a transformation matrix, T, which is obtained by performing a suitable series of elementary row operations on an identity matrix, I. (See Section 5.12.) In short, the system of equations is mapped into an equivalent (but more desirable) system having the same solutions.

If $n = 2$, the system of lines

$$\begin{cases} x + y = 5 \\ x - y = -1 \end{cases}$$

is mapped into the system

$$\begin{cases} x = 2 \\ y = 3 \end{cases}$$

which represents two lines parallel to the coordinate axes, and whose intersection $(2, 3)$ is also the point of intersection of the original system.

Problem Set 6.1

Use the matrix (detached coefficient) method to solve the following systems of equations. Check your results by substitutions:

1. $$\begin{cases} x + 3y + 2z = 3/2 \\ 12x + 2y - z = -5 \\ 2x + y + z = 3 \end{cases}$$

2. $$\begin{cases} 3x + 6y + z = 0 \\ 2x - y + 5z = 3 \\ 5x + 4y + 7z = 4 \end{cases}$$

3. $$\begin{cases} A - 3B + 2Z = 0 \\ A - 2B - Z = 0 \\ 2A - B + 3Z = 0 \end{cases}$$

4. $$\begin{cases} A + B + C - D = 2 \\ 2A + 3B - 4C + D = 0 \\ 5A + 8B - 10C + 3D = 3 \\ 7A + 2B - 3C + D = 6 \end{cases}$$

5. $$\begin{cases} x + y + q = 7 \\ 2x - y + 3q = 0 \\ 3x + 4q = 7 \end{cases}$$

6. $$\begin{cases} 2R + V + A = 24 \\ R - V + 4A = 37 \end{cases}$$

7. $$\begin{cases} J + 2P + 4A = 13 \\ J + P = 1 \end{cases}$$

8. $$\begin{cases} P + 2E + T = 1 \\ 4P - E + T = 0 \end{cases}$$

9. $$\begin{cases} x + 2y - z = 4 \\ 3x - 2y + 5z = 1 \\ 5x + 2y + 3z = 10 \end{cases}$$

10. $$\begin{cases} 3A + Q + 2X = 9 \\ A - 3Q + 5X = 4 \\ 2A + 4Q - 3X = 7 \end{cases}$$

11. Does the system composed of the four equations given in Problems 7 and 8 have a solution?

12. Does the system composed of the equations given in Problems 3 and 7 have a solution?

13. Joel Robertson has \$2.88 in pennies, nickels, and dimes. There are 144 coins, with more dimes than nickels. Can you tell how many coins of each type Joel has, if you know that he has at least one of each type?

14. Make up a system of four equations in four unknowns which is inconsistent, and trade problems with a classmate, each solving the other's problem using matrix methods.

15. Same as Problem 14 with "dependent" substituted for "inconsistent."

16. Show, as stated in this section, that the general solution of a system of linear equations may be expressed as the sum of a particular solution of the system and the general solution of the corresponding homogeneous system.

17. Compare the systems of Examples 5 and 6.

18. Give the explanation requested in Example 7.

19. Carry through the proof that an elementary row transformation E on the coefficient matrix M of the system of linear equations below yields the coefficient matrix of an equivalent system of linear equations, where two systems of linear equations are said to be equivalent if, and only if, each solution of either system is also a solution of the other system.

    Let the linear system be

$$\begin{cases} 3x + 4y - 2z + w = 0 \\ 3x - y - z - w = 3 \\ x + y + z - w = 3 \\ 2x + y - z + w = -4 \end{cases}$$

and let (a) $E = \begin{bmatrix} 1 & 0 & 0 & 0 \\ 4 & 1 & 0 & 0 \\ 0 & 0 & 1 & 0 \\ 0 & 0 & 0 & 1 \end{bmatrix}$ (b) $E = \begin{bmatrix} 1 & 0 & 0 & 0 \\ 0 & 1 & 0 & 0 \\ 0 & -6 & 1 & 0 \\ 0 & 0 & 0 & 1 \end{bmatrix}$

20. *Replace* the last equation in the system of Problem 19 by the equation $0 = 0$, and carry through the proof.

21. Show, that, if a system of linear equations has another equation *which is an identity* annexed thereto, the resulting "enlarged" system is equivalent to the original system. Note that this is *not* what was done in Problem 20.

22. Read and report on one of the following:

>   Forsythe, G. E., "Solving Linear Algebraic Equations Can be Interesting," *Bulletin Amer. Math. Soc.*, **59** (1953), pp. 299–329.
>   Forsythe, G. E. and C. B. Moler, *Computer Solution of Linear Algebraic Systems* (Englewood Cliffs, New Jersey: Prentice-Hall, 1967).
>   Lanczos, C., "Linear Systems in Self Adjoint Form," *American Mathematical Monthly*, **65** (1958), p. 665.

Selected Reading List, Chapter 6
LINEAR SYSTEMS

The following books and journals will provide additional reading for students interested in term projects or independent study.

Books

Any text on college algebra or on linear algebra.

Articles

See Problem 22 above.

Givins, C., "An Exercise in Vector Identities," *Mathematics Magazine* **43** (May 1970), p. 153.

Gibson, P., "Solutions of $A^k + B^k = C^k$ in Nonsingular Integral Matrices," *Mathematics Magazine* **43** (Nov. 1970), p. 276.

Notes on Chapter 6

Notes on Chapter 6

DETERMINANTS

7

The reader who is familiar with determinant theory may wish to omit Chapter 7, or to read it rapidly. Either Chapter 8 or Chapter 9 may be studied next.

7.1 DETERMINANTS

Every complex number, $a + bi$, has a number associated with it which is called its absolute value or modulus, $\sqrt{a^2 + b^2}$. In a similar manner, we can associate a number with each square matrix. This number is called the *determinant* of the square matrix.

Let us denote a matrix as

$$M = \begin{bmatrix} a_{11} & a_{12} & a_{13} & a_{14} & \cdots & a_{1n} \\ a_{21} & a_{22} & a_{23} & a_{24} & \cdots & a_{2n} \\ a_{31} & a_{32} & a_{33} & a_{34} & \cdots & a_{3n} \\ a_{41} & a_{42} & a_{43} & a_{44} & \cdots & a_{4n} \\ \vdots & & & & & \\ a_{n1} & a_{n2} & a_{n3} & a_{n4} & \cdots & a_{nn} \end{bmatrix}$$

where the first subscript indicates the row, and the second subscript indicates the column in which each element appears. The *determinant* of the matrix M is indicated by placing a vertical bar on each side, $|M|$. It is convenient, for certain theoretical considerations, to define the determinant M as the sum

$$|M| = \sum(-1)^k a_{1i_1} \cdot a_{2i_2} \cdot a_{3i_3} \cdot a_{4i_4} \cdots a_{ni_n}$$

where the second subscripts are so chosen that *each column is represented exactly once in each term of the sum.* The first subscripts insure that each row is represented exactly once in each term. The exponent, k, is the number of interchanges[1] necessary to bring the second subscripts $i_1, i_2, i_3, i_4, \cdots, i_n$ into the so-called *normal order* $1, 2, 3, \cdots, n$. Note that every row of M and every column of M is represented exactly once in each term of the sum. Thus,

$$\begin{vmatrix} \begin{bmatrix} a_{11} & a_{12} & a_{13} \\ a_{21} & a_{22} & a_{23} \\ a_{31} & a_{32} & a_{33} \end{bmatrix} \end{vmatrix} = a_{11}a_{22}a_{33} - a_{13}a_{22}a_{31} + a_{13}a_{21}a_{32} - a_{12}a_{21}a_{33} + a_{12}a_{23}a_{31} - a_{11}a_{23}a_{32}$$

The determinant $|M|$ is rarely computed from this definition when $n > 3$. Instead, we study the effect upon the determinant $|M|$ of applying the elementary (row) transformations to the matrix M.

It is important to note that *a determinant is a single number*, an element of the system from which the elements of the matrix are selected, and hence obeys the postulates of this system.

THEOREM 7.1   If *the matrix M has an entire row of zeros*, then *its determinant is zero,* $|M| = 0$.

By definition,

$$|M| \equiv \sum(-1)^k a_{1i_1} \cdot a_{2i_2} \cdot a_{3i_3} \cdot a_{4i_4} \cdots a_{ni_n}$$

Since each term of this sum contains an element from the row of zeros, each term is zero and M is a sum of zeros. Hence, $|M| = 0$.

The converse of this theorem is *not* valid. (Show this.)

---

[1] An important theorem of group theory states that, while the number of interchanges is not fixed, its oddness or evenness is independent of the way in which the normal order is obtained.

THEOREM 7.2   If I *is an identity matrix,* $|I| = 1$. *If* E *is a matrix producing an elementary row transformation,* *then* $|E| = -1, c,$ *or* $+1$ *according as* E *produces an elementary row transformation of the first (interchange of two rows), second (multiply a row by* $c \neq 0$), *or third (add* k *times the elements of one row to another row) type.*

In expanding $|I|$, the only nonzero term in the sum is the product of the elements (all ones) in the principal diagonal. Hence, since the subscripts on the diagonal elements already are in normal order, $|I| = 1$.

An elementary row transformation of the first type (interchange of two rows) on I produces a matrix E_1 having only one nonzero term in its determinant,

$$|E_1| = \sum (-1)^l a_{1i_1} \cdot a_{2i_2} \cdot a_{3i_3} \cdots a_{ni_n}$$

namely, that term of the sum which is the product of all the ones. Since one interchange of subscripts is needed to restore them to the normal order, it follows from the definition that $|E_1| = -1$.

An elementary row transformation of the second type (multiply a row by $c \neq 0$) on I produces a matrix E_2 having ones down the main diagonal with the exception of one element which is c, and zeros elsewhere. In expanding

$$|E_2| = \sum (-1)^l a_{1i_1} \cdot a_{2i_2} \cdot a_{3i_3} \cdots a_{ni_n}$$

there will be only one nonzero term, namely $1 \cdot 1 \cdots 1 \cdot c \cdot 1 \cdots 1 = c$. Since the subscripts already appear in normal order, $|E_2| = c$.

An elementary row transformation of the third type (add k times the elements of one row to another) on I produces a matrix E_3 having ones down the main diagonal and a k in some one other position, and zeros elsewhere. In expanding

$$|E_3| = \sum (-1)^l a_{1i_1} \cdot a_{2i_2} \cdot a_{3i_3} \cdots a_{ni_n}$$

the term containing the elements (all ones) from the main diagonal is a nonzero term. Moreover, since each row and each column is represented exactly once in every term of the sum, each term containing the nondiagonal element k must have another nondiagonal element as a factor. All other nondiagonal elements are zero; thus, each term containing a nondiagonal element is zero. (Try it and see for yourself.)

Hence, $|E_3| = 1$.

This completes the proof of Theorem 7.2. Theorem 7.3 follows from a similar consideration of the effect of an elementary transformation, E, upon the rows of A. The proof is essentially no more difficult than that just given, but repetitious.

THEOREM 7.3 *If E is an elementary matrix and A is a square matrix of the same size,* then

$$| E \cdot A | = | E | \cdot | A | = | A | \cdot | E | = | A \cdot E |$$

A square matrix is said to be in triangular form when every element below the main diagonal is zero. Every square matrix can be transformed into triangular form by elementary row transformations. (See Section 5.7.)

THEOREM 7.4 *A matrix T in triangular form has determinant equal to the product of the elements in its main diagonal.*

Let T be a matrix in triangular form (zeros below the main diagonal). If $a_{11} = 0$, then each element of the first column is zero. Since each term of $| T |$ must contain a factor from the first column, it follows that $| T | = 0$. (Why?) If $a_{11} \neq 0$, then it follows that each nonzero term of $| T |$ must contain a_{11} as a factor. (Why?) Then a_{12} and a_{22} are the only possible nonzero factors from the second column, and a_{12} cannot appear,

$$T = \begin{bmatrix} a_{11} & a_{12} & a_{13} & a_{14} & a_{15} & a_{16} & \cdots & a_{1n} \\ 0 & a_{22} & a_{23} & a_{24} & a_{25} & a_{26} & & a_{2n} \\ 0 & 0 & a_{33} & a_{34} & a_{35} & a_{36} & & a_{3n} \\ 0 & 0 & 0 & a_{44} & a_{45} & a_{46} & & a_{4n} \\ 0 & 0 & 0 & 0 & a_{55} & a_{56} & & a_{5n} \\ 0 & 0 & 0 & 0 & 0 & a_{66} & & a_{6n} \\ \vdots & & & & & & \ddots & \vdots \\ 0 & 0 & 0 & 0 & 0 & \cdots & 0 & a_{nn} \end{bmatrix}$$

since a_{11} has already been selected from the first row. Thus, if $| T |$ has any nonzero terms in its expansion, they must contain a_{11} and a_{22} as the representatives of the first two rows. Since rows one and two are already represented, the only possible nonzero factor of the third column is a_{33}. (Why?)

Proceeding in this fashion we see that $| T |$ is a sum of terms, each of which is zero, with the possible exception of the term consisting of the product of the elements in the main diagonal of T.

$$\begin{vmatrix} 7 & 9 & 4 & 8 \\ 0 & 3 & 2 & 1 \\ 0 & 0 & 5 & 6 \\ 0 & 0 & 0 & 2 \end{vmatrix} = 7 \cdot 3 \cdot 5 \cdot 2 + (23 \text{ other terms, each of which is zero}) =$$

$210 + (\text{twenty-three zeros}) = 210.$

THEOREM 7.5   *A square matrix M has a multiplicative inverse M^{-1} such that $M^{-1} \cdot M = I$ if, and only if, $|M| \neq 0$.*

If M^{-1} exists, then there is a series of elementary row transformations $E_k, E_{k-1}, \cdots, E_3, E_2, E_1$ which carry M into I (See Section 5.6), and $M^{-1} = E_k \cdot E_{k-1} \cdots E_3 \cdot E_2 \cdot E_1$. Furthermore, by repeated applications of Theorem 7.3 and the associative law, we find that $|M^{-1}| \cdot |M| = |I|$. By Theorem 7.2, $|I| = 1$ and it follows that $|M| \neq 0$. (Why?)

It has now been shown that M^{-1} exists *only if* $|M| \neq 0$. We still need to show that *if* $|M| \neq 0$, then M^{-1} exists. There exist elementary row transformations which will transform M into a triangular matrix $T, R \cdot M = T$, where T has only ones and zeros as elements of its principal diagonal. By Theorem 7.4, $|T| = 0$ or 1, depending upon whether or not any zeros appear in the main diagonal. Since $M \cong T$ by a series of elementary row transformations, each of which has a nonzero determinant (Theorems 7.2, 7.3), it follows that $|M| \neq 0$ if, and only if, $|T| \neq 0$. However, if $|T| \neq 0$, then T is triangular with ones down the main diagonal and there exists a further set of elementary row transformations, S, which will carry T into I. (See Problem 24, Set 5.6.) $S \cdot T = I$. Then $I = S \cdot (R \cdot M) = (S \cdot R) \cdot M$, or $S \cdot R = M^{-1}$. Thus, if $|M| \neq 0$, M^{-1} exists. This completes the proof of Theorem 7.5.

THEOREM 7.6   *If A and B are square matrices of the same size*, then $|(A \cdot B)| = |A| \cdot |B|$.

The nonmatrix proof of this theorem, often given in courses on determinant theory, is quite complicated. An easier derivation is available for our consideration. Two cases, $|A| \neq 0$ and $|A| = 0$, will be considered.

If $|A| \neq 0$, then there exists a series of elementary row transformations $E_1, E_2, \cdots, E_k$ which transforms A into I.

$$E_k \cdot E_{k-1} \cdot \cdots \cdot E_2 \cdot E_1 \cdot A = I$$

Hence,
$$A = E_1^{-1} \cdot E_2^{-1} \cdot \cdots \cdot E_k^{-1}$$

$$(A \cdot B) = (E_1^{-1} \cdot E_2^{-1} \cdot \cdots \cdot E_k^{-1}) \cdot B$$

By repeated applications of Theorem 7.3 and the associative law,

$$| (A \cdot B) | = | E_1^{-1} | \cdot | E_2^{-1} | \cdot \cdots \cdot | E_k^{-1} | \cdot | B |$$
$$= | A | \cdot | B |$$

If $| A | = 0$, then A can be reduced to a diagonal matrix T having ones and zeros as elements of the main diagonal, with zeros elsewhere. Hence, some row of T must be all zeros. (Why?)

$$A = (E_1^{-1} \cdot E_2^{-1} \cdot \cdots \cdot E_k^{-1}) \, T$$

and[2] thus,

$$AB = (E_1^{-1} E_2^{-1} \cdots E_k^{-1}) \, T \cdot B = (E_1^{-1} E_2^{-1} \cdots E_k^{-1}) \cdot (T \cdot B)$$

where $(T \cdot B)$ must also have a row of zeros. (Why?) Then $| (T \cdot B) | = 0$ by Theorem 7.1, and by Theorem 7.3 it follows that

$$| A \cdot B | = | E_1^{-1} | \cdot | E_2^{-1} | \cdot \cdots \cdot | E_k^{-1} | \cdot | T \cdot B | = 0$$

The proof of Theorem 7.6 is complete.

Problem Set 7.1

In Problems 1 to 4, express the given matrix as a product of elementary (row transformation) matrices as was done in the proof of Theorem 7.6. If this is impossible, give reasons.

1.
$$A = \begin{bmatrix} 1 & 0 & 0 \\ 2 & 4 & 1 \\ 9 & 2 & 6 \end{bmatrix}$$

2.
$$B = \begin{bmatrix} 1 & 2 & 3 \\ 1 & 5 & 9 \\ 0 & 2 & 6 \end{bmatrix}$$

3.
$$C = \begin{bmatrix} 1 & 2 & 3 & 5 \\ 0 & 1 & 4 & 9 \\ 0 & 0 & 1 & 7 \\ 0 & 0 & 0 & 1 \end{bmatrix}$$

4. (a)
$$D = \begin{bmatrix} -19 & -4 \\ 3 & 11 \end{bmatrix}$$

4. (b)
$$H = \begin{bmatrix} 6 & 4 & 9 \\ 0 & 1 & 0 \\ 2 & 7 & 3 \end{bmatrix}$$

[2] The actual elementary matrices E_i involved will, of course, depend upon the choice of matrix A.

5. How many terms are there in the expansion of the determinant of an $n \times n$ matrix?

6.–10. Find the determinants of the matrices given in Problems 1–4, respectively, and show that each determinant is actually the product of the determinants of the elementary matrices whose product equals the given matrix. Note that while your classmates may have found a different sequence of elementary matrices, this product property is still valid.

11. Prove that a matrix having (a) two identical rows or (b) one row a constant times another row, has determinant zero by using Theorems 7.1, 7.2, and 7.3, and:
    (a) An elementary row transformation of type 1.
    (b) Elementary row transformations of types 2 and 3.

12. Let W be a 7×7 matrix with complex elements. Let W have the additional property that $a_{ij} = 0$ if $i < j$; that is, let the elements above the main diagonal be zeros. Prove that $|W|$ is the product of the diagonal elements of A.

13. The proof of Theorem 7.4 essentially showed that each term of the sum $\sum (-1)^l a_{1i_1} a_{2i_2} a_{3i_3} \cdots a_{ni_n}$, except the term containing all the elements of the main diagonal, must contain an element from below the main diagonal. Demonstrate using a 5×5 matrix.

14. The proof of Theorem 7.2 for a row transformation of type 3 shows that the term of the sum $\sum (-1)^l a_{1i_1} a_{2i_2} a_{3i_3} \cdots a_{ni_n}$, containing the nondiagonal element 1, must also contain a zero element as a factor. Demonstrate using a 5×5 matrix.

$\star$15. Show that, if $|A| \neq 0$, then $|A^{-1}| = 1/|A|$.

Problems 1–4 contain matrices A, B, C, D, H. In Problems 16–25, show that Theorem 7.6 holds for the given matrix products by evaluating $|(X \cdot Y)|$ and $|X| \cdot |Y|$.

16. $X = A,\ Y = B$	21. $X = C^2,\ Y = C$
17. $X = B,\ Y = A$	22. $X = A,\ Y = H$
18. $X = B,\ Y = H$	23. $X = H,\ Y = A$
19. $X = H,\ Y = B$	24. $X = D,\ Y = D$
20. $X = C,\ Y = C$	25. $X = D^2,\ Y = D$

7.2 MINORS AND COFACTORS

Let M_n be an $n \times n$ matrix.
For example,

$$M_3 = \begin{bmatrix} a_{11} & a_{12} & a_{13} \\ a_{21} & a_{22} & a_{23} \\ a_{31} & a_{32} & a_{33} \end{bmatrix}$$

A *minor* of an *element* a_{ij} in the matrix M_k, $k \geq 2$, is the determinant of the matrix remaining after removing the row and the column containing the element a_{ij}. For example, the minor of a_{23} in M_3 is

$$\begin{vmatrix} \begin{bmatrix} a_{11} & a_{12} & a_{13} \\ a_{21} & a_{22} & a_{23} \\ a_{31} & a_{32} & a_{33} \end{bmatrix} \end{vmatrix} = \begin{vmatrix} a_{11} & a_{12} \\ a_{31} & a_{32} \end{vmatrix}$$

The *cofactor of an element* of a matrix is the product of the minor of the element and $(-1)^{i+j}$, where i is the row subscript and j is the column subscript of the element a_{ij}. The corresponding capital letter A_{ij} is used to indicate the cofactor of an element a_{ij}. The cofactor of a_{23} in M_3 is

$$A_{23} = (-1)^{2+3} \cdot \begin{vmatrix} a_{11} & a_{12} \\ a_{31} & a_{32} \end{vmatrix} = - \begin{vmatrix} a_{11} & a_{12} \\ a_{31} & a_{32} \end{vmatrix} = -a_{11}a_{32} + a_{12}a_{31}$$

The *determinants* $|M_k|$ *of the square matrices* M_k could have been defined by the following expansions: (see Theorem 7.7).

$$|M_2| = \begin{vmatrix} a_{11} & a_{12} \\ a_{21} & a_{22} \end{vmatrix} = a_{11}A_{11} + a_{21}A_{21} = a_{11}a_{22} + a_{21}(-a_{12})$$

$$|M_3| = \begin{vmatrix} a_{11} & a_{12} & a_{13} \\ a_{21} & a_{22} & a_{23} \\ a_{31} & a_{32} & a_{33} \end{vmatrix} = a_{11}A_{11} + a_{21}A_{21} + a_{31}A_{31}$$

$$| M_n | = \sum_{i=1}^{n} a_{ij} \cdot A_{ij}, \text{ for any (constant) } j, \text{ where } A_{ij} \text{ is the cofactor of}$$

element a_{ij}. Several specific examples are discussed before stating Theorem 7.7.

EXAMPLE 1

$$\begin{vmatrix} 3 & -4 \\ 5 & 6 \end{vmatrix} = 3(-1)^{1+1} | 6 | + 5(-1)^{2+1} | -4 | = 3(6) - 5(-4) = 38\,^3$$

EXAMPLE[3] 2

$$| M_2 | = \begin{vmatrix} a_1 & b_1 \\ a_2 & b_2 \end{vmatrix} = a_1(-1)^{1+1} | b_2 | + a_2(-1)^{2+1} | b_1 | = a_1 b_2 - a_2 b_1$$

EXAMPLE 3

$$\begin{vmatrix} 1 & 6 & 7 \\ -2 & -5 & 8 \\ 3 & 4 & -9 \end{vmatrix} = 1(-1)^{1+1} \begin{vmatrix} -5 & 8 \\ 4 & -9 \end{vmatrix} + (-2)(-1)^{2+1} \begin{vmatrix} 6 & 7 \\ 4 & -9 \end{vmatrix}$$

$$+ 3(-1)^{3+1} \begin{vmatrix} 6 & 7 \\ -5 & 8 \end{vmatrix}$$

$$= 1[(-5)(-9) - 4(8)] + 2[6(-9) - 4(7)]$$

$$+ 3[6(8) - (-5)(7)]$$

$$= 1[13] + 2[-82] + 3[83] = 98$$

EXAMPLE 4

Show that $| M_3 |$ can also be expanded as

$$| M_3 | = a_{11}A_{11} + a_{12}A_{12} + a_{13}A_{13}$$

---

[3] The student should be cautious not to confuse "determinant" and "absolute value" notations in the case of a 1×1 matrix.

Let us expand $|M_3|$ as $a_{11}A_{11} + a_{21}A_{21} + a_{31}A_{31}$ and recombine to obtain the desired expansion.

$$|M_3| = \begin{vmatrix} a_{11} & a_{12} & a_{13} \\ a_{21} & a_{22} & a_{23} \\ a_{31} & a_{32} & a_{33} \end{vmatrix} = a_{11}(-1)^2 \cdot \begin{vmatrix} a_{22} & a_{23} \\ a_{32} & a_{33} \end{vmatrix} + a_{21}(-1)^3 \cdot \begin{vmatrix} a_{12} & a_{13} \\ a_{32} & a_{33} \end{vmatrix}$$

$$+ a_{31}(-1)^4 \cdot \begin{vmatrix} a_{12} & a_{13} \\ a_{22} & a_{23} \end{vmatrix}$$

$$= a_{11}(a_{22}a_{33} - a_{32}a_{23}) - a_{21}(a_{12}a_{33} - a_{32}a_{13}) + a_{31}(a_{12}a_{23} - a_{22}a_{13})$$

$$= a_{11}a_{22}a_{33} - a_{11}a_{32}a_{23} - a_{21}a_{12}a_{33} + a_{21}a_{32}a_{13} + a_{31}a_{12}a_{23} - a_{31}a_{22}a_{13}$$

If the terms containing a_{11}, a_{12}, a_{13}, respectively, are collected, then

$$|M_3| = a_{11}(a_{22}a_{33} - a_{32}a_{23}) - a_{12}(a_{21}a_{33} - a_{31}a_{23}) + a_{13}(a_{21}a_{32} - a_{31}a_{22})$$

$$= a_{11}(-1)^2 \cdot \begin{vmatrix} a_{22} & a_{23} \\ a_{32} & a_{33} \end{vmatrix} + a_{12}(-1)^3 \cdot \begin{vmatrix} a_{21} & a_{23} \\ a_{31} & a_{33} \end{vmatrix}$$

$$+ a_{13}(-1)^4 \cdot \begin{vmatrix} a_{21} & a_{22} \\ a_{31} & a_{32} \end{vmatrix}$$

$$= a_{11}A_{11} + a_{12}A_{12} + a_{13}A_{13}$$

It follows that $|M_3|$ may be obtained by using cofactors of elements either of the first column or of the first row. Similarly, it can be shown that the expansion of every determinant may be obtained by summing all products obtained on multiplying each element of the first row (rather than column) by its cofactor. A more general theorem follows:

THEOREM 7.7 *A determinant may be expanded by summing all products obtained by multiplying each element of a fixed column (or row) by its cofactor.*

The proof consists of examining the products so formed. They consist of the $n!$ products, each having n factors selected in such a manner that each row and each column is represented exactly once in each product. The signs of these terms are such that the desired determinant is actually obtained as their sum. (Show this.)

In the summation notation Theorem 7.7 states that:

$$|M_n| = \sum_{i=1}^{n} a_{i1}A_{i1} = \sum_{i=1}^{n} a_{i2}A_{i2} = \sum_{i=1}^{n} a_{i3}A_{i3} = \cdots = \sum_{i=1}^{n} a_{in}A_{in}$$

$$= \sum_{i=1}^{n} a_{1i}A_{1i} = \sum_{i=1}^{n} a_{2i}A_{2i} = \sum_{i=1}^{n} a_{3i}A_{3i} = \cdots = \sum_{i=1}^{n} a_{ni}A_{ni}.$$

EXAMPLE 5

Expand the determinant of Example 3 using the cofactors of the elements of (a) the second column, (b) the third row.

(a)
$$\begin{vmatrix} 1 & 6 & 7 \\ -2 & -5 & 8 \\ 3 & 4 & -9 \end{vmatrix} = 6(-1)^{1+2} \cdot \begin{vmatrix} -2 & 8 \\ 3 & -9 \end{vmatrix}$$

$$+ (-5)(-1)^{2+2} \cdot \begin{vmatrix} 1 & 7 \\ 3 & -9 \end{vmatrix} + 4(-1)^{3+2} \cdot \begin{vmatrix} 1 & 7 \\ -2 & 8 \end{vmatrix}$$

$$= -6[(-2)(-9) - (3)(8)] - 5[1(-9) - 3(7)]$$

$$- 4[1(8) - (-2)(7)]$$

$$= -6[-6] - 5[-30] - 4[22] = 98$$

(b)
$$\begin{vmatrix} 1 & 6 & 7 \\ -2 & -5 & 8 \\ 3 & 4 & -9 \end{vmatrix} = 3(-1)^{3+1} \cdot \begin{vmatrix} 6 & 7 \\ -5 & 8 \end{vmatrix} + 4(-1)^{3+2} \cdot \begin{vmatrix} 1 & 7 \\ -2 & 8 \end{vmatrix}$$

$$+ (-9)(-1)^{3+3} \cdot \begin{vmatrix} 1 & 6 \\ -2 & -5 \end{vmatrix}$$

$$= 3[6(8) - (-5)(7)] - 4[1(8) - (-2)(7)]$$

$$- 9[1(-5) - (-2)(6)]$$

$$= 3[83] - 4[22] - 9[7] = 98$$

Problem Set 7.2

Expand the determinants given in Problems 1–9, using the cofactors of the elements of (a) the first column, (b) the second row.

1. $\begin{vmatrix} -21 & 6 \\ 4 & -5 \end{vmatrix}$

2. $\begin{vmatrix} -2-x & 5 \\ 2 & -5-x \end{vmatrix}$

3. $\begin{vmatrix} 4-x & -5 \\ -5 & -3-x \end{vmatrix}$

4. $\begin{vmatrix} -0 & -2 & 3 \\ -2 & 1 & 4 \\ 3 & 4 & 2 \end{vmatrix}$

5. $\begin{vmatrix} -2 & 4 & 1 \\ 3 & -6 & 2 \\ 4 & -8 & 5 \end{vmatrix}$

6. $\begin{vmatrix} 3 & -4 & -1 \\ 0 & 2 & 1 \\ -15 & 20 & 5 \end{vmatrix}$

7. $\begin{vmatrix} 1-x & 2 & -3 \\ 2 & -1-x & -2 \\ 3 & -2 & 2-x \end{vmatrix}$

8. $\begin{vmatrix} 2-x & 0 & 2 \\ 0 & 1-x & 0 \\ 2 & 0 & 3-x \end{vmatrix}$

9. $\begin{vmatrix} 1 & x & x^2 \\ 1 & y & y^2 \\ 1 & z & z^2 \end{vmatrix}$

Solve the equations given in Problems 10 and 11.

10. $\begin{vmatrix} -2-x & -3 \\ 3 & -5-x \end{vmatrix} = 0$

11. $\begin{vmatrix} 1-x & 2 & 0 \\ 2 & -1-x & -2 \\ 0 & -2 & 1-x \end{vmatrix} = 0$

7.3 THE TRANSPOSE OF A MATRIX

Section 5.6, where the elementary row operations were defined, pointed out that similar operations could be defined for columns. It is also possible to prove theorems similar to those of the last section for column transformations. Instead, we shall prove that the determinant of a matrix M is equal to the determinant of the matrix M^t obtained from M by interchanging its rows and columns. M^t is called the *transpose* of M.

$$
\text{If} \quad M = \begin{bmatrix} a_{11} & a_{12} & a_{13} \\ a_{21} & a_{22} & a_{23} \\ a_{31} & a_{32} & a_{33} \end{bmatrix}, \quad \text{then} \quad M^t = \begin{bmatrix} a_{11} & a_{21} & a_{31} \\ a_{12} & a_{22} & a_{32} \\ a_{13} & a_{23} & a_{33} \end{bmatrix}
$$

THEOREM 7.8 *The determinant of the transpose of a matrix, M^t, is equal to the determinant of M, that is, $|\,M^t\,| = |\,M\,|$.*

Since each determinant is the sum of terms made up of exactly one factor from each row and each column, the terms of the sum $|\,M\,|$ and the sum $|\,M^t\,|$ must be identical, except possibly for signs and for the order in which they occur. The order of terms is immaterial. The sign of each term of $|\,M\,|$ is determined by the number, k, of interchanges needed to bring the second subscripts of $(-1)^k a_{1i_1} a_{2i_2} a_{3i_3} \cdots a_{ni_n}$ into normal order, $1, 2, 3, \cdots, n$; that is, into the same order as the first subscripts. The term of $|\,M^t\,|$ which contains the same factors as the term of $|\,M\,|$ described above has $(-1)^l$ as its sign, determined by the number of interchanges needed to bring the first subscripts into the same order as the second. (Why?) If the interchanges of the second subscripts used in determining the sign of a term in $|\,M\,|$ are applied in reverse order to the first subscripts, the latter will then be in the same order as the second subscripts originally were. (Why?) The oddness or evenness of the number of interchanges of subscripts needed to restore normal order is the same for each term of $|\,M\,|$ as for the corresponding term of $|\,M^t\,|$. Hence, the sign of each term of $|\,M\,|$ is the same as that of the corresponding term of $|\,M^t\,|$. Thus, $|\,M\,| = |\,M^t\,|$.

From Theorem 7.8, it follows that theorems about determinants stated for rows may be restated for columns.

EXAMPLE 6

$$
\begin{vmatrix} 5 & 6 & 9 \\ 3 & -4 & -6 \\ -2 & 2 & 3 \end{vmatrix} = \begin{vmatrix} 5 & 2(3) & 3(3) \\ 3 & 2(-2) & 3(-2) \\ -2 & 2(1) & 3(1) \end{vmatrix} = 0
$$

since the second and third columns are proportional. (Why does proportionality of columns or rows assure that the determinant is zero? State and prove a related theorem.)

EXAMPLE 7

Find the value of

$$D = \begin{vmatrix} 2 & 4 & -8 \\ 5 & 2 & 1 \\ 3 & 1 & 4 \end{vmatrix}$$

(1) Divide the first row by 2 and multiply the resulting determinant by 2 (Theorem 7.3) to obtain

$$D = 2 \begin{vmatrix} 1 & 2 & -4 \\ 5 & 2 & 1 \\ 3 & 1 & 4 \end{vmatrix}$$

(2) Now, add -2 times the elements of the first column to the corresponding elements of the second column (Theorem 7.3).
(3) Add 4 times the elements of the first column to the corresponding elements of the third column.
(4) Finally, expand using cofactors of elements of the first row, obtaining $D = -46$.

The reader should carry out these steps to obtain $D = -46$, and then should find a simpler reduction.

EXAMPLE 8

Evaluate:

$$D = \begin{vmatrix} 2 & 3 & -5 & 4 \\ 4 & 2 & 3 & 1 \\ -7 & 6 & 4 & 3 \\ 5 & 6 & 1 & 2 \end{vmatrix}$$

To find the value of D, one may obtain zeros in the first, third, and fourth rows of the fourth column, and expand as follows:

(1) Add -4 times the elements of the second row to the corresponding elements of the first row.

(2) Add -3 times the elements of the second row to the corresponding elements of the third row.

(3) Add -2 times the elements of the second row to the corresponding elements of the fourth row.

$$D = \begin{vmatrix} 2-4(4) & 3-4(2) & -5-4(3) & 4-4(1) \\ 4 & 2 & 3 & 1 \\ -7-3(4) & 6-3(2) & 4-3(3) & 3-3(1) \\ 5-2(4) & 6-2(2) & 1-2(3) & 2-2(1) \end{vmatrix}$$

$$= \begin{vmatrix} -14 & -5 & -17 & 0 \\ 4 & 2 & 3 & 1 \\ -19 & 0 & -5 & 0 \\ -3 & 2 & -5 & 0 \end{vmatrix}$$

(4) Expand, using cofactors of elements of the fourth column.

$$D = 1(-1)^{2+4} \cdot \begin{vmatrix} -14 & -5 & -17 \\ -19 & 0 & -5 \\ -3 & 2 & -5 \end{vmatrix}$$

(5) To expand the resulting determinant, add $(5/2)$ times the elements of the third row to the corresponding elements of the first row.

(6) Expand, using cofactors of elements of the second column.

$$D = \begin{vmatrix} -14 + \left(\dfrac{5}{2}\right)(-3) & -5 + \left(\dfrac{5}{2}\right)(2) & -17 + \left(\dfrac{5}{2}\right)(-5) \\ -19 & 0 & -5 \\ -3 & 2 & -5 \end{vmatrix}$$

$$= \begin{vmatrix} -\left(\dfrac{43}{2}\right) & 0 & -\left(\dfrac{59}{2}\right) \\ -19 & 0 & -5 \\ -3 & 2 & -5 \end{vmatrix} = 2(-1)^{3+2} \begin{vmatrix} -\left(\dfrac{43}{2}\right) & -\left(\dfrac{59}{2}\right) \\ -19 & -5 \end{vmatrix}$$

$$= -2(215/2 - 1121/2) = 906$$

Problem Set 7.3

Verify Problems 1–5, without expanding, if possible.

1. $$\begin{vmatrix} 3 & -6 & 9 \\ 0 & 2 & 5 \\ -2 & 4 & -6 \end{vmatrix} = 0$$

2. $$\begin{vmatrix} 1 & 3 & -5 \\ 2 & 4 & 7 \\ 8 & -3 & 2 \end{vmatrix} = - \begin{vmatrix} 8 & -3 & 2 \\ 2 & 4 & 7 \\ 1 & 3 & -5 \end{vmatrix}$$

3. $$\begin{vmatrix} 1 & x^2 \\ 1 & y^2 \end{vmatrix} = (x + y)(y - x)$$

4. $$\begin{vmatrix} x^2 & x & 1 \\ y^2 & y & 1 \\ z^2 & z & 1 \end{vmatrix} = -(x - y)(y - z)(z - x)$$

5.
$$\begin{vmatrix} 2 & -4 & 6 \\ 3 & 2 & -5 \\ -2 & 3 & 7 \end{vmatrix} = 2 \cdot \begin{vmatrix} -1 & 1 & 10 \\ 3 & 2 & -5 \\ 1 & 5 & 2 \end{vmatrix}$$

Use elementary row and column transformations to place zeros in as many positions as possible in some row or column before expanding, then expand and find the results for Problems 6–11.

6.
$$\begin{vmatrix} 20 & 17 & 2 \\ 15 & 12 & 8 \\ 25 & 22 & -6 \end{vmatrix}$$

7.
$$\begin{vmatrix} -1 & 6 & 5 \\ -2 & 3 & 3 \\ -10 & 8 & 10 \end{vmatrix}$$

8.
$$\begin{vmatrix} 2 & 2 & -2 \\ 3 & -2 & 4 \\ 2 & 2 & 8 \end{vmatrix}$$

9.
$$\begin{vmatrix} 4 & -3 & -1 & 2 \\ -1 & 2 & 2 & 4 \\ 2 & -1 & 3 & 1 \\ 3 & 0 & 10 & 5 \end{vmatrix}$$

10.
$$\begin{vmatrix} 3 & 5 & 4 & 2 \\ 2 & 6 & 2 & 3 \\ 1 & 4 & 1 & 2 \\ 2 & 9 & 2 & 5 \end{vmatrix}$$

11.
$$\begin{vmatrix} 6 & 5 & 7 & 4 \\ 0 & -2 & 1 & -1 \\ 3 & 3 & 6 & 2 \\ 3 & 1 & 7 & 1 \end{vmatrix}$$

12. Show that

$$\begin{vmatrix} x & y & 1 \\ x_1 & y_1 & 1 \\ x_2 & y_2 & 1 \end{vmatrix} = 0$$

is the equation of the line through the points (x_1, y_1), and (x_2, y_2). Discuss any necessary restrictions.

13. Write the equation of the line through $(2, -3)$ and $(-4, 5)$ in determinant form.

14. Show that the area of the triangle whose vertices are (x_1, y_1), (x_2, y_2) and (x_3, y_3) is the positive value of

$$\pm \left(\tfrac{1}{2}\right) \cdot \begin{vmatrix} x_1 & y_1 & 1 \\ x_2 & y_2 & 1 \\ x_3 & y_3 & 1 \end{vmatrix}$$

Discuss.

[HINT: Drop perpendiculars from each vertex to the x axis and consider the areas of the resulting trapezoids.]

15. Use the determinant form of Problem 14 to find the area of the triangle whose vertices are

(a) $(2, -3), (-4, 5), (3, 2)$;
(b) $(-1, 3), (2, 15), (-2, -1)$. Discuss.

16. Verify that

$$\begin{vmatrix} x^2 + y^2 & x & y & 1 \\ x_1^2 + y_1^2 & x_1 & y_1 & 1 \\ x_2^2 + y_2^2 & x_2 & y_2 & 1 \\ x_3^2 + y_3^2 & x_3 & y_3 & 1 \end{vmatrix} = 0$$

is the equation of a circle through the points (x_1, y_1), (x_2, y_2), and (x_3, y_3). Discuss any necessary restrictions. Note that it is unnecessary to expand the given determinant. Be sure you prove both aspects of this problem.

(1) The equation is the equation of a circle.
(2) The graph of the equation passes through the three given points.

17. Write the equation of the circle through $(2, -3)$, $(-4, 5)$, $(3, 2)$ in determinant form. Expand and simplify. Discuss.

18. Use Theorem 7.7 to obtain a simple proof of Theorem 7.1.

★7.4 THE ADJOINT MATRIX

We begin by proving the following.

THEOREM 7.9 *The sum of the products of each element of one row of square matrix A, with the cofactors of the corresponding elements of a different row, is zero.*

$$\sum_{i=1}^{n} a_{hi}A_{ki} = a_{h1}A_{k1} + a_{h2}A_{k2} + \cdots + a_{hn}A_{kn} = 0, \quad \text{if} \quad h \neq k$$

The proof is immediate, since such expansions are equivalent to expansions of determinants having two identical rows.

A similar theorem is obtained for columns by using transposes.

EXAMPLE 9

$$\sum_{i=1}^{3} a_{1i}A_{3i} = a_{11}A_{31} + a_{12}A_{32} + a_{13}A_{33} = \begin{vmatrix} a_{11} & a_{12} & a_{13} \\ a_{11} & a_{12} & a_{13} \\ a_{21} & a_{22} & a_{23} \end{vmatrix} = 0$$

In later work, the adjoint matrix of a given square matrix A, written (adj. A), is used. The *adjoint* of a given $n \times n$ matrix A is another $n \times n$ matrix obtained from A in the following fashion:

(1) Replace each element a_{ij} of matrix A by its cofactor A_{ij}.
(2) Take the transpose of the resulting matrix.

In short,

$$(\text{adj. } A) = \begin{bmatrix} A_{11} & A_{12} & \cdots & A_{1n} \\ A_{21} & A_{22} & \cdots & A_{2n} \\ A_{31} & A_{32} & \cdots & A_{3n} \\ \vdots & \vdots & & \vdots \\ A_{n1} & A_{n2} & \cdots & A_{nn} \end{bmatrix}^{t} = \begin{bmatrix} A_{11} & A_{21} & A_{31} & \cdots & A_{n1} \\ A_{12} & A_{22} & A_{32} & \cdots & A_{n2} \\ \vdots & \vdots & \vdots & & \vdots \\ A_{1n} & A_{2n} & A_{3n} & \cdots & A_{nn} \end{bmatrix}$$

where A_{ij} is the cofactor[4] of element a_{ij} of A.

---

[4] It may be unnecessary to point out that, since each cofactor A_{ij} is a determinant, not a matrix, the A_{ij} are single elements of (adj. A).

EXAMPLE 10

$$\text{Find (adj. } A) \text{ for } A = \begin{bmatrix} -2 & 0 & 4 \\ 2 & 1 & 0 \\ 3 & 1 & 2 \end{bmatrix}$$

$$(\text{adj. } A) = \begin{bmatrix} A_{11} & A_{12} & A_{13} \\ A_{21} & A_{22} & A_{23} \\ A_{31} & A_{32} & A_{33} \end{bmatrix}^{t}$$

$$= \begin{bmatrix} +\begin{vmatrix} 1 & 0 \\ 1 & 2 \end{vmatrix} & -\begin{vmatrix} 2 & 0 \\ 3 & 2 \end{vmatrix} & +\begin{vmatrix} 2 & 1 \\ 3 & 1 \end{vmatrix} \\ -\begin{vmatrix} 0 & 4 \\ 1 & 2 \end{vmatrix} & +\begin{vmatrix} -2 & 4 \\ 3 & 2 \end{vmatrix} & -\begin{vmatrix} -2 & 0 \\ 3 & 1 \end{vmatrix} \\ +\begin{vmatrix} 0 & 4 \\ 1 & 0 \end{vmatrix} & -\begin{vmatrix} -2 & 4 \\ 2 & 0 \end{vmatrix} & +\begin{vmatrix} -2 & 0 \\ 2 & 1 \end{vmatrix} \end{bmatrix}^{t}$$

$$= \begin{bmatrix} 2 & -4 & -1 \\ 4 & -16 & 2 \\ -4 & 8 & -2 \end{bmatrix}^{t} = \begin{bmatrix} 2 & 4 & -4 \\ -4 & -16 & 8 \\ -1 & 2 & -2 \end{bmatrix}$$

Note also that

$$A \cdot (\text{adj. } A) = \begin{bmatrix} -2 & 0 & 4 \\ 2 & 1 & 0 \\ 3 & 1 & 2 \end{bmatrix} \cdot \begin{bmatrix} 2 & 4 & -4 \\ -4 & -16 & 8 \\ -1 & 2 & -2 \end{bmatrix}$$

$$= \begin{bmatrix} -8 & 0 & 0 \\ 0 & -8 & 0 \\ 0 & 0 & -8 \end{bmatrix} = |A| \cdot I$$

as Theorem 7.10 indicates.

THEOREM 7.10   If A *is a square matrix, then*

$$A \cdot (\text{adj. } A) = (\text{adj. } A) \cdot A = |A| \cdot I$$

$$\begin{bmatrix} |A| & 0 & 0 & 0 & \cdots & 0 \\ 0 & |A| & 0 & & & 0 \\ 0 & 0 & |A| & 0 & & 0 \\ 0 & & 0 & \ddots & & \vdots \\ \vdots & & & & \ddots & 0 \\ 0 & 0 & 0 & \cdots & 0 & |A| \end{bmatrix}$$

This theorem states that, if a matrix A and its adjoint are multiplied together in either order, the result is the scalar matrix having diagonal elements each equal to the determinant of A, and zeros elsewhere.

The proof of this interesting theorem may be based on Theorems 7.7 and 7.9. Theorem 7.7 states that

$$\sum_{i=1}^{n} a_{hi}A_{hi} = |A|,$$ while Theorem 7.9 states that

$$\sum_{i=1}^{n} a_{hi}A_{ki} = 0,$$ if $h \neq k$. Thus, using the notation of Section 5.5

$$(A) \cdot (\text{adj. } A) = (a_{rc})(A_{cr}) = \left[\left(\sum_{i=1}^{n} a_{ri}A_{hi} \right)_{rc} \right]$$

$$\begin{bmatrix} |A| & 0 & 0 & 0 & \cdots & 0 \\ 0 & |A| & 0 & & & 0 \\ 0 & 0 & |A| & 0 & & 0 \\ 0 & & 0 & \ddots & & \vdots \\ \vdots & & & & \ddots & 0 \\ 0 & 0 & 0 & \cdots & 0 & |A| \end{bmatrix}$$

The immediate corollary of Theorem 7.10 is left as an exercise for the reader.

THEOREM 7.11   *If A is a square matrix having an inverse, then*

$$A^{-1} = \frac{1}{|A|} \cdot (\text{adj. } A)$$

Theorem 7.11 is *not* a convenient method of computing A^{-1}, since it will require the evaluation of n^2 different determinants of $(n-1)$ rows each in addition to $|A|$, but it has important applications in further theory.

Problem Set 7.4

Compute (adj. A) for the matrix A given in Problems 1–5. Then check the result using Theorem 7.10.

1. $\begin{bmatrix} 7 & 6 & 2 \\ 1 & 4 & 9 \\ 3 & 1 & 8 \end{bmatrix}$

2. $\begin{bmatrix} 2 & 1 & 3 \\ 0 & 1 & 2 \\ 0 & 0 & 0 \end{bmatrix}$

3. $\begin{bmatrix} 2 & 1 & 3 \\ 0 & 1 & 2 \\ 0 & 0 & 1 \end{bmatrix}$

4. $\begin{bmatrix} 7 & 6 & 2 & 9 \\ 4 & 1 & 3 & 8 \\ 0 & 0 & 1 & 0 \\ 1 & 0 & 0 & 0 \end{bmatrix}$

5. $\begin{bmatrix} 7 & 0 & 0 & 0 \\ 0 & 4 & 0 & 0 \\ 0 & 0 & 2 & 0 \\ 0 & 0 & 0 & 1 \end{bmatrix}$

6. Prove or disprove: The adjoint of a diagonal matrix is also a diagonal matrix. A diagonal matrix has zero for each element which is not on the main diagonal. The identity matrix is a diagonal matrix, so is $\begin{bmatrix} 3 & 0 \\ 0 & 7 \end{bmatrix}$.

7. Prove or disprove: The adjoint of a scalar matrix is also a scalar matrix.

8. Prove or disprove: The adjoint of a triangular matrix (either zeros above or zeros below the main diagonal) is also a triangular matrix.

9. Prove or disprove: The adjoint of a singular matrix is necessarily the zero matrix. A matrix is called *singular* if it has no multiplicative inverse.

10. Prove or disprove: The only matrix having the zero matrix as its adjoint is the zero matrix.

11. Prove Theorem 7.11.

12. (a) Prove that, if $|A| \neq 0$, then

$$A \cdot B = \begin{bmatrix} |A| & 0 & 0 & 0 \cdots 0 \\ 0 & |A| & 0 & & 0 \\ 0 & 0 & |A| & 0 & 0 \\ 0 & & 0 & \ddots & \vdots \\ \vdots & & & \ddots & 0 \\ 0 & 0 & 0 \cdots & 0 & |A| \end{bmatrix}$$

implies $B = (\text{adj. } A)$.

(b) Attempt a similar proof, or disproof, if $|A| = 0$.

★7.5 DETERMINANTS AND LINEAR SYSTEMS

Most high school mathematics students learn to solve systems of linear equations in several unknowns using determinants (Cramer's Rule). The subject is discussed in this book mainly because it is so widely *misunderstood*. Determinants are important—very important—but their importance does *not* lie in the solution of systems of linear equations. Indeed, this is one of the poorest ways of solving linear systems. The detached coefficient (matrix) method discussed in Chapter 6 is usually both more efficient and more accurate, even when high-speed computers are used. (Errors in "rounding" accumulate rapidly in evaluating large determinants on computers, as elsewhere.)

The proof of validity for the determinant method of solving simultaneous equations makes use of the adjoint matrix.

The linear system

$$\begin{cases} a_{11}x_1 + a_{12}x_2 + a_{13}x_3 + \cdots + a_{1n}x_n = k_1 \\ a_{21}x_1 + a_{22}x_2 + \cdots\cdots\cdots + a_{2n}x_n = k_2 \\ \vdots \qquad \vdots \qquad\qquad\qquad \vdots \qquad \vdots \\ a_{n1}x_1 + a_{n2}x_2 + \cdots\cdots\cdots + a_{nn}x_n = k_n \end{cases}$$

may be represented in terms of matrices either as

$$\begin{bmatrix} a_{11} & a_{12} & a_{13} & \cdots & a_{1n} & k_1 \\ a_{21} & a_{22} & \cdots\cdots & & a_{2n} & k_2 \\ \vdots & & & & \vdots & \\ a_{n1} & \cdots\cdots\cdots\cdots & a_{nn} & k_n \end{bmatrix} \cdot \begin{bmatrix} x_1 \\ x_2 \\ \vdots \\ x_n \\ -1 \end{bmatrix} = \begin{bmatrix} 0 \\ 0 \\ \vdots \\ 0 \end{bmatrix}$$

as was done in Chapter 6, or as

$$
\begin{bmatrix} a_{11} & a_{12} & a_{13} & \cdots & a_{1n} \\ a_{21} & a_{22} & \cdots\cdots\cdots & a_{2n} \\ \vdots & & & \vdots \\ a_{n1} & \cdots\cdots\cdots\cdots & a_{nn} \end{bmatrix} \cdot \begin{bmatrix} x_1 \\ x_2 \\ \vdots \\ x_n \end{bmatrix} = \begin{bmatrix} k_1 \\ k_2 \\ \vdots \\ k_n \end{bmatrix}.
$$

We prefer the latter method here, since the coefficient matrix $A = (a_{ij})$ is square in this case. We use the notation

$$A \cdot X = K$$

where A is the large square coefficient matrix, X the column matrix containing the variables or unknowns, and K the column matrix containing the constants k_i.

Then, if A^{-1} exists,

$$A \cdot X = K$$

after left multiplication by A^{-1} yields

$$X = A^{-1} \cdot K$$

showing that the solution is unique. By Theorem 7.11 the above equation is equivalent to

$$X = \frac{1}{|A|} \cdot (\text{adj. } A) \cdot K$$

or

$$
\begin{bmatrix} x_1 \\ x_2 \\ \vdots \\ x_h \\ \vdots \\ x_n \end{bmatrix} = \frac{1}{|A|} \cdot (\text{adj. } A) \cdot \begin{bmatrix} k_1 \\ k_2 \\ \vdots \\ k_h \\ \vdots \\ k_n \end{bmatrix}
$$

Let us examine row h of the left member. It is the unknown x_h. Since row h of (adj. A) contains the cofactors $A_{1h}, A_{2h}, \cdots, A_{nh}$, it follows that, for each h,

$$x_h = \frac{A_{1h}k_1 + A_{2h}k_2 + \cdots + A_{nh}k_n}{|A|} \qquad \text{(Why?)}$$

The numerator is the expansion (using column h) of the determinant of a matrix, obtained from A by replacing its column h with the column of constants K. Thus, "Cramer's Rule" is proved.

If A^{-1} does not exist, then $|A| = 0$. (Why? Quote a theorem or problem number, or prove it.) In this case, either no solution may exist or infinitely many solutions may exist. Recall the difference between $5/0$ (meaningless), and $0/0$ (indeterminant). This becomes geometrically meaningful as follows. Consider

$$\begin{cases} a_{11}x_1 + a_{12}x_2 = k_1 \\ \\ a_{21}x_1 + a_{22}x_2 = k_2 \end{cases} \qquad \text{two lines in the plane.}$$

If $|A| \neq 0$, the lines intersect in a single point.
If $|A| = 0$,

$$\text{the lines are parallel if} \quad \frac{a_{11}}{a_{21}} = \frac{a_{12}}{a_{22}} \neq \frac{k_1}{k_2}$$

$$\text{the lines are coincident if} \quad \frac{a_{11}}{a_{21}} = \frac{a_{12}}{a_{22}} = \frac{k_1}{k_2}$$

Students who have studied solid analytic geometry will be able to extend this to three- and higher-dimensional space.

The method of solving simultaneous linear equations by determinants is often referred to as Cramer's Rule in honor of the Swiss mathematician G. Cramer (1704–1752). In fairness, however, it should be noted that as early as 1100 B.C. the Chinese solved two linear equations in two unknowns by a rule equivalent to Cramer's Rule. The present notation for determinants was introduced during the nineteenth century by A. Cayley (1821–1895, English). The importance of determinant theory does *not* lie in determining solutions of linear equations. The powerful tensor algebra, which is important in the modern theory of relativity, has its roots in the theory of determinants. Modern geometry and the theory of transformations use determinants. The Jacobian and the Wronskian are determinants. The very definitions of volume in fourth- and higher-dimensional space are often phrased in terms of determinants.

Problem Set 7.5

1.–5. Use determinants to solve Problems 1–5 of Set 6.1.

6. Solve for x and y: $2z + 3iz = -7 - 4i$, where $z = x + iy$ and $i^2 = -1$.

7. Solve: $2^{x+y} = 3^{2x-y}$, $3^{2x-y} = 2^{3y}$.

8. Find K such that the lines $Kx - 2y - 7 = 0$, $2x - Ky - 8 = 0$, and $x + 2y + K = 0$ intersect in a point.

9. The following puzzle recently appeared in a national magazine: The owner of a small dog-and-bird shop claims that his pets have a total of 36 heads and 100 feet. How many of each are there?

10. Two particles of mass M_1 and M_2 collide. The particle of mass M_1 has a velocity U_1 before collision and V_1 after collision. The particle of mass M_2 is stationary before collision and has a velocity of V_2 after collision. The law of conservation of momentum states that $M_1 U_1 = M_2 V_2 + M_1 V_1$. The law of conservation of kinetic energy states that

$$M_1 U_1^2 = M_2 V_2^2 + M_1 V_1^2.$$

   (a) Assuming that M_1, M_2, and U_1 are known, determine the velocity of each particle after collision.
   (b) What happens to each particle after collision, if $M_1 = M_2$?
   (c) What happens to each particle after collision, if M_1 is very small in comparison to M_2? [HINT: Take $M_2 = 1$ and let $M_1 \to 0$.]
   (d) What happens to each particle if M_2 is very small in comparison to M_1?

11. Use the matrix method to solve the system of equations:

   $$3x + 4y - 2z = 15, \quad x - y + 3z = -7, \quad x + 2y - z = 7.$$

12. Evaluate
$$\begin{vmatrix} 1 & 3 & 2 & 5 \\ 1 & 9 & 3 & 6 \\ 2 & 1 & 4 & 1 \\ 0 & 1 & 3 & 2 \end{vmatrix}$$

13. Find z such that
$$\begin{vmatrix} 1 & 4 & 3 \\ z & 5 & 8 \\ 2 & 1 & 6 \end{vmatrix} = 0.$$

14. Solve Problems 12 and 13 in the mod 7 system.

*15. The construction of nomograms and alignment charts is often expedited by the use of determinants. Read the chapter on this subject in the text *Nomography* by A. S. Levins (Wiley).

16. Read the article "Determinants as Area of a Triangle and Volume of a Tetrahedron" by G. Petersen in the *American Mathematical Monthly*, **62** (1955), p. 249.

17. Reexamine the work in Section 5.12 on transformations. What is the determinant of each transformation matrix? You may wish to investigate this subject further for a term project.

Selected Reading List, Chapter 7
DETERMINANTS

The following books and journals will provide additional reading for students interested in term projects or independent study.

Books

Any text on college algebra.

Articles

Robinson, S. M., "A Short Proof of Cramer's Rule," *Mathematics Magazine*, **43,** No. 2 (Mar. 1970), pp. 94–95.

"A Determinant Evaluation," Problem E 2064, *The American Mathematical Monthly*, **76,** No. 6 (June–July 1968), p. 692.

"Divisor of a Determinant," Problem E 2046, *The American Mathematical Monthly*, **76,** No. 1 (Jan. 1969), p. 92.

"Putnam Competition Problem A 2, *The American Mathematical Monthly*, **77,** No. 7 (Sept. 1970), p. 723.

Notes on Chapter 7

8

8.1 FIELD

The symbol $\in$ is used to mean "is an element of" or "are elements of." An integral domain was defined in Chapter 1 as follows:

Integral Domain

A set of elements, called D, an equals relation, two well-defined operations $+, \cdot$ which satisfy the following postulates:

For all $a, b, c, \in D$

Under $+$
(commutative group)

1. *Closure*: $a + b \in D$
2. *Associativity*:
   $(a + b) + c = a + (b + c)$
3. *Additive identity*: There exists $z \in D$ such that
   $b + z = z + b = b$

Under $\cdot$

1. *Closure*: $a \cdot b \in D$
2. *Associativity*:
   $(a \cdot b) \cdot c = a \cdot (b \cdot c)$
3. *Multiplicative identity*: There exists $u \in D$ such that
   $b \cdot u = u \cdot b = b$

4. *Additive inverse*: For each a, there | 4. *Cancellation law*: $a \neq z$ and
exists a corresponding a^* such | $a \cdot (b) = a \cdot (c)$ implies
that $a + a^* = a^* + a = z$ | $b = c$
5. *Commutativity*: $a + b = b + a$ | 5. *Commutativity*: $a \cdot b = b \cdot a$

Distributive laws:

$$a \cdot (b + c) = a \cdot b + a \cdot c$$

$$(b + c) \cdot a = b \cdot a + c \cdot a$$

The elements of an integral domain form a commutative group under addition. However, they need not form a commutative group under multiplication, since no multiplicative inverse is postulated. Indeed, even if zero (which cannot have an inverse, since, for all x, $z \cdot x \neq u$) is excluded, the remaining elements need not have inverses in the system. Examples are the integers and the set of all polynomials in x with real (or, if you prefer, with complex) coefficients.

There are algebraic systems in which the elements form a commutative group under addition and the elements with z(zero) excluded form a commutative group under multiplication as well. Such a system which also obeys the distributive law is called a *field*. The rational numbers form a convenient example of a field. The real numbers, the complex numbers and the mod 7 system are three other examples of fields. It is unnecessary to postulate the cancellation law in a field, since it can be deduced easily using the multiplicative inverse.

Field Postulates

Elements, equals relation, two well-defined operations $+$, $\cdot$.
For all elements $a, b, c \in D$.

Under $+$ | Under $\cdot$
(commutative group) | (with zero excluded, commutative group)

1. *Closure*: $a + b \in D$. | 1. *Closure*: $a \cdot b \in D$.
2. *Associativity*: | 2. *Associativity*:
$(a + b) + c = a + (b + c)$. | $(a \cdot b) \cdot c = a \cdot (b \cdot c)$.
3. *Additive identity*: | 3. *Multiplicative identity*:
There exists $z \in D$ such that | There exists $u \in D$ such that
$b + z = z + b = b$. | $b \cdot u = u \cdot b = b$.

4. *Additive inverse*: For each a, there exists a corresponding a^* such that
$$a + a^* = a^* + a = z.$$

4. *Multiplicative inverse*: For each $b \neq z$, there exists a corresponding element b^{-1} such that $b \cdot b^{-1} = b^{-1} \cdot b = u$.

5. *Commutativity*: $a + b = b + a$.

5. *Commutativity*: $a \cdot b = b \cdot a$.

Distributive laws:

$$a \cdot (b + c) = a \cdot b + a \cdot c$$

$$(b + c) \cdot a = b \cdot a + c \cdot a.$$

Problem 17 Set 8.1 shows that some of these postulates are redundant (deducible from the remaining postulates). An easily understood presentation is more important here than a minimal set of postulates.

It is the field postulates which are implied by the often heard phrase, "the rules of ordinary algebra." Any system satisfying these postulates is almost certain to be of mathematical interest.

Problem Set 8.1

1. Show that the complex numbers form a field.

2. Show that the mod 7 system is a field.

3. Prove that every field is also an integral domain.

4. Show that the converse (see Section 1.6) of Problem 3 is not valid.

5. Which of the systems of Problem Set 1.4, Problems 1–12, are fields?

6. Prove that the set of residue classes mod p, a prime, form a field. (See Sections 2.2, 2.3.)

7. (a) Prove that division, except by zero, is always possible and unique in a field.
   (b) Prove the cancellation law holds in a field.

8. Prove that the set of all polynomials with real coefficients forms an integral domain. Is this domain also a field?

9. Prove that a subset S of a field F is also a field, providing that S is closed under addition and multiplication (Postulate 1) and that for each $a \in S$, the additive inverse a^* and (if $a \neq$ zero) the multiplicative inverse a^{-1} are also in S (Postulate 4), assuming S is nonvoid.

10. Prove that the only subfield of the field R of rational numbers is R itself. [HINT: A subfield must contain 1 (Why?), hence, the subfield contains $1 + 1 + 1 + \cdots + 1 = n$ for all $n > 0$. (Why?) It also contains zero and solutions of $x + n = 0$ for all $n > 0$, hence (you supply conclusion). Finally, S contains all m/n, where m, n are integers, $n \neq 0$. (Why?)]

11. Show that, if S_1 and S_2 are subfields of field F, then the elements common to S_1 and S_2; that is, $S_1 \cap S_2$, in the notation of Section 3.10, is also a subfield.

12. Is the following system a field?

$+$	x	y		$\cdot$	x	y
x	x	y		x	x	x
y	y	x		y	x	y

Can you find a modular system which is isomorphic to the above system? (Isomorphism of a system containing two operations must preserve both operations.)

13. Does the system of 2×2 square matrices $\begin{bmatrix} a & b \\ c & d \end{bmatrix}$ having nonzero determinant $ad - bc \neq 0$ form a field?

14. Does the Boolean algebra $(0, 1)$ form a field?

15. (a) Does the mod 12 system form a field?
    (b) Does the mod 13 system form a field? Explain, in terms of equivalence classes, what the elements of the mod 13 system are.

*16. Does the mod m system, where m is composite, form a field?

17. (a) Show that the second portion of the distributive law $(b + c) \cdot a = b \cdot a + c \cdot a$ may be deduced from the remaining field postulates and hence is redundant.
    (b) Discover parts of two other postulates which are also redundant.

*18. Attempt to obtain as small a set of field postulates as possible. Consider the conjecture, "If any two of x, y, z are given in the equation $x + y = z$ (or $x \cdot y = z$) the corresponding third element exists and is unique."

$\star$19. Prove that the additive group F^+ of a field F cannot be isomorphic to the multiplicative group $F^\times$ of the same field. Consult Advanced Problem 4644 in *American Mathematical Monthly* (Oct., 1956) for the solution.

8.2 RINGS

The last section discussed a mathematical system (field) which is a special case of the integral domain. We next discuss a system (ring) which is *more general* than the integral domain, i.e., the integral domain is a special case of a ring.

Ring Postulates

Elements, equals relation, two well-defined operations $+$, $\cdot$.

For all $a, b, c \in R$,

Under $+$
(Commutative group)

Under $\cdot$

1. *Closure*: $a + b \in R$
2. *Associativity*:
   $(a + b) + c = a + (b + c)$
3. *Additive identity*: There exists
   $z \in R$ such that
   $b + z = z + b = b$
4. *Additive inverse*: For each a, there
   exists a corresponding a^* such
   that $a + a^* = a^* + a = z$
5. *Commutativity*: $a + b = b + a$

1. *Closure*: $a \cdot b \in R$
2. *Associativity*:
   $(a \cdot b) \cdot c = a \cdot (b \cdot c)$

Distributive Laws:

$$a \cdot (b + c) = a \cdot b + a \cdot c$$

$$(b + c) \cdot a = b \cdot a + c \cdot a$$

The set of all $n \times n$ matrices with real elements (or with elements in any field) form a ring. A ring may have proper divisors of zero; i.e., $A \neq 0$, $B \neq 0$, but $A \cdot B = 0$. (See Section 5.10.)

The mod 6 system is a ring having commutative multiplication and a multiplicative identity, but is not a field or an integral domain. (Why?) The mod 7 system is a field. The set of polynomial functions with coefficients in the mod 7 system (including constant polynomials) satisfies all the postulates for integral domain *except* the cancellation law, which fails,

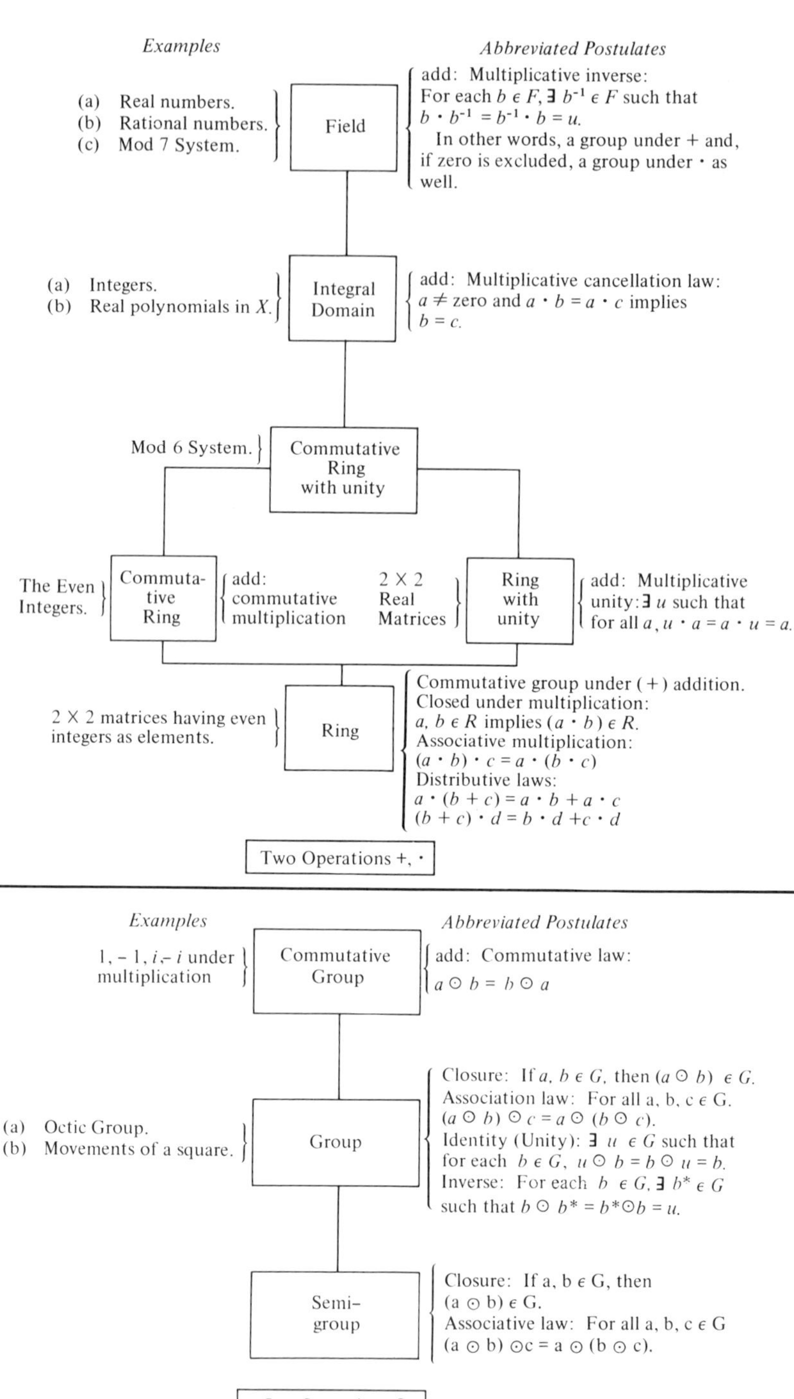

Examples
Abbreviated Postulates
(a)   Real numbers.
(b)   Rational numbers.
(c)   Mod 7 System.
Field
add: Multiplicative inverse:
For each $b \in F$, $\exists\, b^{-1} \in F$ such that
$b \cdot b^{-1} = b^{-1} \cdot b = u$.
In other words, a group under $+$ and, if zero is excluded, a group under $\cdot$ as well.
(a)   Integers.
(b)   Real polynomials in X.
Integral Domain
add: Multiplicative cancellation law:
$a \neq$ zero and $a \cdot b = a \cdot c$ implies $b = c$.
Mod 6 System.
Commutative Ring with unity
The Even Integers.
Commutative Ring
add:
commutative multiplication
2×2 Real Matrices
Ring with unity
add: Multiplicative unity: $\exists\, u$ such that for all a, $u \cdot a = a \cdot u = a$.
2×2 matrices having even integers as elements.
Ring
Commutative group under $(+)$ addition.
Closed under multiplication:
$a, b \in R$ implies $(a \cdot b) \in R$.
Associative multiplication:
$(a \cdot b) \cdot c = a \cdot (b \cdot c)$
Distributive laws:
$a \cdot (b + c) = a \cdot b + a \cdot c$
$(b + c) \cdot d = b \cdot d + c \cdot d$
Two Operations $+, \cdot$
Examples
Abbreviated Postulates
$1, -1, i, -i$ under multiplication
Commutative Group
add: Commutative law:
$a \odot b = b \odot a$
(a)   Octic Group.
(b)   Movements of a square.
Group
Closure: If $a, b \in G$, then $(a \odot b) \in G$.
Association law: For all $a, b, c \in G$.
$(a \odot b) \odot c = a \odot (b \odot c)$.
Identity (Unity): $\exists\, u \in G$ such that for each $b \in G$, $u \odot b = b \odot u = b$.
Inverse: For each $b \in G$, $\exists\, b^* \in G$ such that $b \odot b^* = b^* \odot b = u$.
Semi-group
Closure: If $a, b \in G$, then $(a \odot b) \in G$.
Associative law: For all $a, b, c \in G$
$(a \odot b) \odot c = a \odot (b \odot c)$.
One Operation $\odot$

since the product $(x) \cdot (x - 1) \cdot (x - 2) \cdot (x - 3) \cdot (x - 4) \cdot (x - 5) \cdot (x - 6)$ mod 7 is zero for all values of x, but is a product of nonzero functions; i.e., proper divisors of zero exist in this function space.[1] The system is, however, a ring. It is a *commutative ring*. (Since all rings are commutative with respect to addition, the term *commutative ring* means a ring with commutative multiplication.) This particular example also contains a multiplicative identity (namely, the constant polynomial 1) and, hence, is properly called a *commutative ring with unity*. It is true, but not obvious, that the set of *all functions* having coefficients in a given commutative ring with unity is itself a commutative ring with unity. The polynomial functions form a commutative subring with unity.

The charts (p. 276) may help you visualize the interrelations among algebraic varieties under consideration in this text. Further study will extend the chart in various directions.

Problem Set 8.2

1. Which of the following systems are rings? In each case also discuss commutativity of multiplication and existence of a multiplicative identity (unity).

   (a) Mod 12 system.
   (b) All 2×2 matrices with complex elements.
   (c) All functions with rational coefficients.
   (d) All polynomials f with rational coefficients and such that $f(0) \neq 0$.
   (e) The even integers.

2. Prove that the mod m system is a ring for every integral value of m.

3. Prove that the set of all polynomials (including constant polynomials) with coefficients in a commutative ring is a ring. Is it necessarily commutative, too?

4. Show that the system of all $n \times n$ real matrices is a ring. Is it a commutative ring? Does it have a unity?

---

[1] For sophisticated readers we comment that the set of *polynomial functions* is *not* the same as the set of *polynomials*. Two polynomials belong to the same equivalence class mod 7 if, and only if, corresponding coefficients are congruent mod 7. Two polynomial *functions* $f(x)$, $g(x)$ belong to the same equivalence classes mod 7 if, and only if, $f(i) \equiv g(i)$ mod 7 for each $i = 0, 1, 2, 3, 4, 5, 6$. It is the set of *polynomial functions* which is currently under discussion. A polynomial function $P(x)$ is defined to be zero if, and only if, the value zero corresponds to each permissible value of x.

5. Consider the set of all ordered pairs $\left(a, \begin{bmatrix} w & x \\ y & z \end{bmatrix}\right)$, where a is a complex number and w, x, y, z are elements of the mod 6 system. Define addition and multiplication as follows:

$$(a_1, M_1) + (a_2, M_2) = (a_1 + a_2, M_1 + M_2)$$

$$(a_1, M_1) \cdot (a_2, M_2) = (a_1 \cdot a_2, M_1 \cdot M_2),$$

where $M_1 \cdot M_2$ is the usual matrix product reduced mod 6. Is this system a ring? Remember, the elements are the ordered pairs (x, M).

6. Prove that, in any ring, if a^* and b^* are the additive inverses of a and b respectively, then

$$(a^*) \cdot (b^*) = (ab); \text{ i.e., } (-a) \cdot (-b) = (ab). \quad (\textit{Use the postulates!})$$

7. Prove that, in any ring, $(a^*)^* = a$; i.e., $-(-a) = a$.

8. (a) Show that every integral domain is a ring.
   (b) Does (a) also show that every field is a ring, in view of Problem 3 Section 8.1?

9. Does the Boolean algebra $(0, 1)$ form a ring?

$\star$10. Consider the rational points in the unit interval (i.e., all rational x, $0 \le x \le 1$). Let the elements of a mathematical system consist of all possible *sets* of these points. Some sample elements are

$A = \{\text{the set } \frac{1}{3} < x \le 1\}$,
$B = \{\frac{5}{6}, \frac{1}{10}, \text{and } \frac{1}{2} \le x < \frac{3}{4}\}$,
$C = \{\text{the entire interval } 0 \le x \le 1\}$,
$D = \{\text{the empty or null set}\}$,
$E = \{\text{the set consisting of all points } 1/n \text{ with numerator 1}\}$, and
$F = \{\text{the point } \frac{7}{13}\}$. There are, of course, infinitely many such elements. Define multiplication as set intersection; i.e., $X \cdot Y = X \cap Y = $ all points common to both X and Y if any, or the null set otherwise. Define addition as the set of all points in X or Y but not in both; i.e., $X + Y = (X \cup Y) \cap (X \cap Y)'$. Show that this system forms a ring. Does it have a unity?

11. Are the following sets rings? Are they rings with unit elements?

   (a) All integral multiples of 11.
   (b) All integral multiples of 12.

(c) All integral multiples of an integer n.

(d) All 2×2 matrices $\begin{bmatrix} 0 & r \\ 0 & 0 \end{bmatrix}$ where r is a rational number.

12. Can you deduce either part of the distributive law from the other ring postulates? If not, consider a possible way of showing they are not redundant, but do not bother to carry out your scheme; merely get it to the stage of, "If I could find a system which $\cdots$, this would show that the second part of the distributive law is not redundant."

13. Show that a Boolean algebra (defined in Section 3.12) is a ring $\tilde{\mathbf{B}}$ with unit element and operations $\cup$ and $\cap$ in which for each $x \in \tilde{\mathbf{B}}$, $x \cup x = x$ and $x \cap x = x$ and conversely that a ring with unit element in which for each x it follows that $x \cup x = x$ and $x \cap x = x$ is a Boolean algebra.

8.3 IDEALS

Let R be a ring having a nonvoid subset M such that M is closed under subtraction[2] ($x \in M$ and $y \in M$ implies $x - y \in M$) and, furthermore, M is closed under multiplication not only by elements of M but by elements of R ($m \in M$ and $r \in R$ implies $m \cdot r \in M$ and $r \cdot m \in M$.) Then M is called an *ideal* of R. Reiterating, M is an ideal of a ring R if $1 \neq M \subset R$, and M is closed under subtraction from within, and M is also closed under multiplication by elements of R. An example of such a system is

$$R = \text{the ring (actually a domain) of all integers}$$

$$M = \text{the even integers.}$$

The even integers are closed under subtraction (the difference of two even integers is even). The even integers also have the ideal properties, the product

$$r \cdot m = (\text{integer}) \cdot (\text{even integer}) = \text{even integer} \in M, \quad \text{and}$$

$$m \cdot r = (\text{even integer}) \cdot (\text{integer}) = \text{even integer} \in M.$$

[2] We use the common word "subtraction" in place of the phrase "addition of the additive inverse of the second element."

Thus the set of even integers, $M = \cdots, -6, -4, -2, 0, 2, 4, 6, 8, \cdots$ is an *ideal* of the set of all integers $\cdots, -3, -2, -1, 0, 1, 2, 3, 4, 5, \cdots$.

Since a ring need not have commutative multiplication, it is possible to consider one-sided ideals; i.e., for all $x, y \in M$, $r \in R$, $x - y \in M$, and $x \cdot r \in M$. We shall, however, not do so here.

It is not difficult to see that the integral multiples of an integer k, $\cdots, -3k, -2k, -k, 0, k, 2k, \cdots$ form an ideal in the ring of integers. Actually, in the ring of integers every ideal is of this type. Such an ideal is said to be *generated* by k. Ideals, each element of which is a multiple of an element of the (commutative) ring R, are called *principal ideals*. Principal ideals are of great importance in more advanced work.

In more general rings, it is possible to find examples of ideals which are *not principal*. Indeed, in the ring of all possible sets of rational points in the unit interval (See Problem 10 Set 8.2), one ideal M is the set of all subsets containing only a finite number of points. This ideal is not only nonprincipal, it cannot even be generated by a finite number of elements. It is within your power to prove that this set is an ideal, if you wish, but we shall not do so here.

Problem Set 8.3

1. Prove that all multiples of 5 form an ideal in the ring of all integers. Is the ideal a subring of the ring of integers?

2. Let R be the ring of all integral multiples of 3. Show that all integral multiples of 12 form an ideal of R. Is the ideal principal? Is it a ring? Does it have a unit element?

3. Consider the set M of all 3×3 matrices with integral elements. Do they form a ring? If so, is the set of all multiples of elements of

$$M \text{ by } \begin{pmatrix} 3 & 0 & 0 \\ 0 & 3 & 0 \\ 0 & 0 & 3 \end{pmatrix} \text{ an ideal of } M?$$

4. Show that, if R is the ring of integers, then the set of all integers of the form $k = 6m + 8n$, where m and n are in R, form an ideal of R.

5. Does the set of all integers which are congruent to 1 (mod 5) form an ideal of the set of integers?

6. Does the set of all integers which are congruent to 1 (mod 6) form an ideal of the set of integers?

7. (a) For "1" in Problem 5 read "0."
   (b) For "1" in Problem 6 read "3."

8. In the ring of the mod 12 system, does (a) 0, 2, 4, 6, 8, 10 form an ideal? (b) 1, 3, 5, 7, 9, 11? (c) 0, 4, 8?

9. In the ring of the mod 15 system, does (a) 0, 5, 10 form an ideal? (b) 1, 6, 11?

10. Let R be the ring of all polynomials with integral coefficients. Let D be the set of all polynomials $f(x)$ for which $f(3) = 0$. Is D an ideal of R?

11. Let R be the ring of integers. Let M be an ideal containing, among other elements, the integers 12 and 17. Show that $M = R$.

12. Prove that an ideal M of a ring R is always a subring, but that the converse is not valid.

8.4 RESIDUE CLASS RINGS

Sections 2.1 and 2.2 introduced notions of equivalence relations and equivalence classes. The reader will be well advised to reread these sections before continuing.

Let R be a ring and M an ideal of R. First, let R be separated into cosets in the following manner: If x and y are two elements of R, then x and y belong to the same coset (relative to M) if, and only if, there is some $m \in M$ such that $x = y + m$. This permits us to separate R into cosets, much as groups (see Section 4.6) were separated into cosets. This process also establishes residue classes (see Chapter 2 on congruences). In fact, it is usual to write

$x \equiv y \bmod M$, meaning there exists an $m \in M$ such that $x = y + m$.

The reader is already familiar with this notation from Chapter 2, where the ring of integers was separated into cosets (residue or equivalence classes) by the ideal consisting of all multiples of a constant m. Section 4.6 also dealt with related material and should be reviewed if difficulty is encountered here. The phrase *residue class ring* is used to designate the ring whose elements are the cosets. The term *quotient ring* is also used because of the similarity to the concept of quotient groups (see Section 4.8), and the notations R mod M, or simply R/M, are frequently seen. The reader is asked in the next problem set to prove that congruence modulo an ideal is an equivalence relation and that the cosets do form a ring. The proofs parallel closely similar proofs given in Chapters 2 and 4.

Problem Set 8.4

1. Prove that, in a ring, congruence modulo an ideal has the following properties of a well-defined equivalence relation:

    (a) $a \equiv a \bmod M$.
    (b) If $a \equiv b \bmod M$, then $b \equiv a \bmod M$.
    (c) If $a \equiv b \bmod M$ and $b \equiv c \bmod M$, then $a \equiv c \bmod M$.
    (d) If $a \equiv b \bmod M$ and $c \equiv d \bmod M$, then $a + c \equiv b + d \bmod M$ and $a \cdot c \equiv b \cdot d \bmod M$.

2. Show that the cosets of a ring R modulo an ideal M form a ring under the type of coset operations discussed in Chapter 4.

3. Let R be the ring of integers and let M be the ideal consisting of all multiples of 5. Find the cosets of R mod M. Form several products in the quotient group R/M.

4. Same as Problem 3 with M the ideal consisting of all multiples of 8.

5. Let R be the ring of all integral multiples of 6.

    (a) Show that the set of all multiples of 18 is an ideal M of R.
    (b) Find the cosets of R mod M.
    (c) How many elements has the quotient ring R/M?
    (d) Find a familiar ring which is isomorphic to R/M.

6. Let R be the ring of integers. Let M be the ideal of R consisting of all multiples of m. Show that R/M is an integral domain *if, and only if, m* is prime. Note that two proofs are required for an *if, and only if,* theorem. See Section 1.6 for comments on necessary and sufficient conditions.

7. Find all the proper divisors of zero $(A \cdot B = 0$, but $A \neq 0$, $B \neq 0)$ in R/M, where R is the ring of integers and M is the ideal of all integral multiples (a) of 28, (b) of 24, (c) of 23.

8. Let R be the ring of real numbers. Let C be the set of all integral multiples of 2π. (C may or may not be an ideal—you may investigate the question if you wish.) Show that the group cosets relative to C of the *additive group* of R form a valid congruence set. Discuss the (im)possibility of multiplication of these cosets. Discover a relation between this problem and periodic trigonometric functions.

9. Let R be the ring of all polynomials in x with real coefficients. Let M be the set of all elements of R for which $f(1) = f(2) = 0$, i.e., M contains polynomials of the form $k(x - 1)(x - 2)g(x)$. Show that M is an ideal of R.

★10. Find the cosets into which M separates R for the M and R of Problem 9.

8.5 POLYNOMIALS MODULO $(x^2 + 1)$—COMPLEX NUMBERS

The real numbers form a field. The polynomials in one variable, x, with real coefficients form a ring which contains the real numbers (constant polynomials) as a subset. Let us denote the ring of polynomials with real coefficients by the symbol $R[x]$. The set of all polynomials of the form $(x^2 + 1)g(x)$ is an ideal M of $R[x]$.

Upon separating the elements of $R[x]$ into cosets modulo M, one finds that two typical cosets are:

$$
\begin{bmatrix}
17 \\[1ex]
17 + 4(x^2 + 1) \\[1ex]
17 + 7(x^2 + 1)(x^7 - 4x + 3) \\[1ex]
17 + (x^2 + 1)(x^5 - 3x^2 + 6x + 1) \\[1ex]
\vdots \\[1ex]
\text{infinitely many polynomials of} \\
\text{the form } 17 + (x^2 + 1)q(x)
\end{bmatrix}
\quad\text{and}\quad
\begin{bmatrix}
4x - 3 \\[1ex]
4x - 3 + (x^2 + 1)(9x^2 + 1) \\[1ex]
4x - 3 + (x^2 + 1)(x^{81})(x^8 - 3x + 2) \\[1ex]
\vdots \\[1ex]
\text{infinitely many polynomials of} \\
\text{the form } 4x - 3 + (x^2 + 1)q(x)
\end{bmatrix}
$$

In general, each coset contains infinitely many elements of the form

$$a + bx + (x^2 + 1)\cdot q(x), \text{ for fixed } a \text{ and } b.$$

Using the polynomial generalization of Archimedes' Division Axiom (Section 1.10), we see that every polynomial (i.e., every element of $R[x]$ may be expressed in the form

$$f(x) = (x^2 + 1) \cdot q(x) + bx + a$$

and that elements of $R[x]$ belong to the same coset of $R[x]/M$ if, and only if, exactly the same values of b and a are so obtained.

In the modular algebra discussed in Chapters 1 and 2, it was convenient to use the smallest non-negative number in each residue class to represent the entire class (coset). A similar device may be used here by noting that each residue class (coset) contains either exactly one polynomial of degree one or exactly one constant polynomial, but not both. [This occurs when $q(x)$ is zero.] By selecting the polynomials of the form $b_1x + a_1$ to represent the coset (equivalence class) containing all elements of the form

$$(x^2 + 1) \cdot q(x) + b_1x + a_1$$

and reducing sums and products modulo $(x^2 + 1)$, we have an algebra for the ring $R[x]/M$, where M is the ideal containing all polynomials having $(x^2 + 1)$ as a factor. The reader may wish to review Chapter 2 at this point.

EXAMPLE 1

$$A = (7 + 3x)$$

$$B = (-5 + 2x)$$

$$A \cdot B = (7 + 3x)(-5 + 2x) = 6x^2 - x - 35$$

$$= (x^2 + 1) \cdot 6 + (-x - 41)$$

Hence,

$$A \cdot B \equiv -x - 41 \quad \mod (x^2 + 1)$$

More generally,

$$(a + bx) \cdot (c + dx) = ac + (bc + ad)x + bdx^2$$

$$= \underline{ac - bd + (bc + ad)x} + bd(x^2 + 1)$$

$$(a + bx) \cdot (c + dx) \equiv ac - bd + (bc + ad)x \quad \mod (x^2 + 1)$$

Note the similarity to the rules for multiplying complex numbers

$$(a + bi) \cdot (c + di) = ac - bd + (bc + ad)i$$

Actually, the quotient ring $R[x]$ mod ($x^2 + 1$) *is* isomorphic to the field of complex numbers!

We shall not prove the following theorem, but its statement may give the reader some understanding of the importance of the concept of residue class rings.

THEOREM 8.1   *Every finite field is isomorphic to a residue class ring $R[x]/M$ for some suitable choice of M, where R is a subset of the real field.*

Actually the theorem remains valid when $R[x]$ is replaced by the set of polynomials in x with *integral* coefficients in place of real coefficients. It is also valid for many nonfinite fields.

Problem Set 8.5

1. Determine the "constant or first degree representative" of the coset class of $R[x]/M$ for

   (a) $x^2 + 3x + 6$
   (b) $27x^4 - 3x^2 + 6x - 11$
   (c) $4x^3 + 5x - 3$
   (d) $7x + 28$
   (e) $\frac{4}{5}x^2 - 6x + \frac{5}{6}$
   (f) $17x^5 - 4$

2. Form the following sums and products of polynomials in Problem 1 and reduce them mod ($x^2 + 1$). Show that the same results are obtained if the reduced representatives are used in place of the given polynomials. The notation 3(c) means 3 times polynomial (c) in Problem 1.

   $(a) + 3(b), \quad 3(c) + (b), \quad (d) + (f), \quad 2(b) + (e), \quad (a) \cdot (d)$

3. Show that, in the mod ($x^2 + 1$) system, the polynomial x^2 and the polynomial -1 are equal; that is, belong to the same coset (equivalence class) of $R[x]/M$.

4. In the complex numbers, $a + bi$ with $i^2 = -1$, reduce the following expressions to the form $a + bi$:

   (a) $i^2 + 3i + 6$
   (b) $27i^4 - 3i^2 + 6i - 11$
   (c) $4i^3 + 5i - 3$
   (d) $7i + 28$
   (e) $\frac{4}{5}i^2 - 6i + \frac{5}{6}$
   (f) $17i^5 - 4$

5. Form the following sums and product of complex numbers in Problem 4. Show that the same results are obtained if reduced $(a + bi)$ representatives are used in place of the given expressions

$$(a) + 3\,(b), \quad 3\,(c) + b, \quad (d) + (f), \quad 2\,(b) + (e), \quad (a)\cdot(d)$$

6. Compare Problem 1 and Problem 4.

7. Compare Problem 2 and Problem 5.

8. Show that Problem 3 helps explain the similarities noted in Problems 6 and 7.

9. (a) Show that, in the mod $(x^2 + 1)$ system, the four elements 1, $-1, x, -x$ form a group under multiplication.
   (b) To which of the groups of four elements (Section 4.5) is the above group isomorphic?

10. Consider the following system: Let $D[x]$ consist of the ring of all polynomials with integral coefficients. Let M be the ideal of $D[x]$ consisting of all multiples $(x^2 + 1)\cdot q(x)$. Is $D[x]$ mod M a field? Prove your answer by showing that the field postulates are or are not satisfied by the system of residue classes.

11. Prove *one* of the following statements
    S1   If B and K are elements of a ring R, then the equation $B + x = K$ has a unique solution in R.
    S2   If B and K are elements of ring R, then the equation $B\cdot x = K$ has a unique solution in R.

12. If R is a commutative ring, does this imply that R is an integral domain? If not, what additional postulates must be satisfied? Would your answer change if R was known to contain only a finite number of elements?

13. If W is a set of elements that form a commutative group under some operation $+$ and also form a commutative group under a different operation $\cdot$ what additional postulates are needed to guarantee that under these operations W is a

    (a)   commutative ring with unity
    (b)   field
    (c)   integral domain
    (d)   all three of the above.

14. Let I_1 and I_2 be two distinct ideals of a ring R. Either prove that $I_1 \cap I_2$ is also an ideal of R, or give a counterexample.

15. Let R be a ring having only a *finite* number of elements. Show that there exists a positive integer N such that the sum $b + b + b + \cdots + b$ equals z, the zero or additive identity of R *no matter which elements $b \in R$ is used.*

16. Prove that if R is a ring such that for each $x \in R$ it is true that $x \cdot x = x$, then R is commutative.

17. Let P be the set of all polynomials with real coefficients. Is P a ring under the operation of addition and multiplication of polynomials? Is it an integral domain? Why or why not?

Selected Reading List, Chapter 8
FIELDS, RINGS, AND IDEALS

The following books and journals will provide additional reading for students interested in term projects or independent study.

Books

McCoy, N. H., *Rings and Ideals.* Carus Monograph No. 8 (Mathematical Association of America.)

Hartley, B. and T. O. Hawkes, *Rings, Modules, and Linear Algebra* (London: Chapman & Hall, 1970).

Articles

Hooi-Tong, Loh, "Notes on Semi-rings," *Mathematics Magazine,* **40,** No. 3 (May 1967), pp. 150–152.

Johnsen, E. C., D. L. Outcalt and Adil Yaqub, "An Elementary Commutativity Theorem for Rings," *The American Mathematical Monthly,* **75,** No. 3 (Mar. 1968), p. 288.

Peinado, R. E., "On Finite Rings," *Mathematics Magazine,* **40,** No. 2 (Mar. 1967), pp. 83–85.

Notes on Chapter 8

MORE MATRIX THEORY

9

9.1 CHARACTERISTIC EQUATIONS

Certain elementary properties of vectors and matrices were studied in Chapter 5. Let $f(X) = a_n X^n + a_{n-1} X^{n-1} + \cdots a_1 X + a_0 I$ be a matrix polynomial with scalar coefficients a_i. The matrix A will be called a root of the equation $f(X) = 0 \cdot I$, if and only if, $f(A) = 0 \cdot I$.

EXAMPLE 1

Show that the matrix $A = \begin{bmatrix} 3 & 0 & 6 \\ 3 & 4 & 0 \\ 0 & 0 & 2 \end{bmatrix}$ satisfies the equation

$f(X) = X^3 - 9X^2 + 26X - 24I = 0 \cdot I.$

Substituting $A^2 = \begin{bmatrix} 9 & 0 & 30 \\ 21 & 16 & 18 \\ 0 & 0 & 4 \end{bmatrix}$ and $A^3 = \begin{bmatrix} 27 & 0 & 114 \\ 111 & 64 & 162 \\ 0 & 0 & 8 \end{bmatrix}$ into

$f(A) = A^3 - 9A^2 + 26A - 24I$, one obtains $f(A) = \begin{bmatrix} 0 & 0 & 0 \\ 0 & 0 & 0 \\ 0 & 0 & 0 \end{bmatrix}$.

Hence, the given matrix A is a root of the equation $f(X) = 0 \cdot I$.

Where no confusion can arise, it is customary to drop the identity matrix, writing a_0 in place of $a_0 \cdot I$ and 0 in place of $0 \cdot I$.

Each square matrix has associated with it an equation of special importance called its *characteristic equation*. The characteristic equation of the square matrix A is $|I \cdot x - A| = 0$. The determinant $|I \cdot x - A|$ is called the *characteristic polynomial*, $f(x)$, of the square matrix A.

EXAMPLE 2

Find the characteristic equation of the matrix

$$A = \begin{bmatrix} 5 & 3 & 6 \\ 0 & 1 & 0 \\ 0 & 4 & -2 \end{bmatrix}$$

$$|Ix - A| = \left| \begin{bmatrix} x & 0 & 0 \\ 0 & x & 0 \\ 0 & 0 & x \end{bmatrix} - \begin{bmatrix} 5 & 3 & 6 \\ 0 & 1 & 0 \\ 0 & 4 & -2 \end{bmatrix} \right| = 0$$

$$\begin{vmatrix} x - 5 & -3 & -6 \\ 0 & x - 1 & 0 \\ 0 & -4 & x + 2 \end{vmatrix} = 0$$

The reader should expand the given determinant to show that the desired characteristic equation is:

$$x^3 - 4x^2 - 7x + 10 = 0.$$

9.2 HAMILTON–CAYLEY THEOREM

This powerful theorem enables one to produce a polynomial equation $f(x) = 0$ with scalar coefficients, which a given square matrix will satisfy.

THEOREM 9.1   (Hamilton–Cayley). *Every square matrix, A, satisfies its characteristic equation, $| Ix - A | = 0$.*

It is, however, not the matrix solutions, but the *ordinary complex numbers* which satisfy the characteristic equation $| Ix - A | = 0$ that are of special importance. If both you and your instructor are willing to postpone the proof of the Hamilton–Cayley Theorem until a later course, you may now skip to Section 9.3.

The proof uses the adjoint matrix that is discussed in Section 7.4, through extensions of the axiom of Archimedes and the Remainder Theorem (Chapter 1) either to polynomials with matrix coefficients or to matrices with polynomial elements. The two concepts are closely related. A polynomial with matrix coefficients may always be expressed as a matrix with polynomial elements and conversely:

$$\begin{bmatrix} 1 & 3 \\ 2 & 5 \end{bmatrix} x^3 + \begin{bmatrix} 0 & -1 \\ 4 & 2 \end{bmatrix} x^2 + \begin{bmatrix} 7 & 0 \\ 0 & 1 \end{bmatrix} x + \begin{bmatrix} 3 & -5 \\ 1 & 2 \end{bmatrix}$$

$$= \begin{bmatrix} x^3 + 7x + 3 & 3x^3 - x^2 - 5 \\ 2x^3 + 4x^2 + 1 & 5x^3 + 2x^2 + x + 2 \end{bmatrix}$$

We say the above matrix (or polynomial) is of degree three. In general, the degree of a matrix is the same as that of its element of highest degree.

Since matrix multiplication is not commutative, care must be taken. We shall restrict our attention to left divisors. If $R = M \cdot Q$, then M is said to be a *left divisor* of R, or briefly, "M left divides R to give Q." Actually, it is possible to define a "greatest common left divisor" of two compatible matrices and to determine a Euclidean Algorithm which is an extension of Section 1.10. We do not need to do so here, but it should make an interesting class report or student project if one is desired. The two extensions we shall need are:

LEMMA 1:   If P and B are $n \times n$ matrices with polynomial elements (actually, with elements in an integral domain), then $P = B \cdot Q + R$, where either $R = (0)$ or the degree of R is less than the degree of B. The Q and R so determined are unique for a given P and B.

LEMMA 2:   If P is a polynomial with $n \times n$ matrix coefficients, $P(x) = A_m x^m + \cdots + A_1 x + A_0$, and P is left divided by the matrix $(Ix - A)$ to obtain $P = (Ix - A)Q + R$, then $R = P(A)$.

To prove the Hamilton–Cayley Theorem we employ Theorem 7.10, obtaining:

$$(x \cdot I - A) \cdot [\text{adj. } (x \cdot I - A)] = |\,(x \cdot I - A)\,| \cdot I = f(x) \cdot I,$$

where $f(x)$ is the characteristic function of A.

Letting $Q = [\text{adj. } (x \cdot I - A)]$ and $P(x) = f(x) \cdot I$, we have Lemma 1 with $R = (0)$,

$$P(x) = (x \cdot I - A) \cdot Q + (0).$$

From Lemma 2 it then follows that $(0) = P(A)$. Since $P(x) = f(x) \cdot I$, it then follows that $f(A) = 0$, as desired.

The student should show that the matrix of Example 2 satisfies its characteristic equation.

9.3 CHARACTERISTIC ROOTS AND CHARACTERISTIC VECTORS

The complex numbers (or other field elements, if the elements of the matrix are not taken from a subset of the complex field) which are roots of the characteristic equation $|\,Ix - A\,| = 0$ are called the *characteristic roots* of the matrix A. In quantum mechanics and elsewhere, the terms *latent roots, proper value, eigenvalue,* and *eigenwerte* are often used in place of characteristic root. Much computer time is currently spent in determining the characteristic roots of large matrices which arise in applications of mathematics to physics and engineering as well as to biological and sociological problems.

EXAMPLE 3

Find the characteristic roots of the matrix $A = \begin{bmatrix} 1 & 3 \\ -1 & 5 \end{bmatrix}$.

$$| Ix - A | = 0$$

$$\begin{vmatrix} x - 1 & -3 \\ 1 & x - 5 \end{vmatrix} = 0$$

$$x^2 - 6x + 8 = 0$$

Hence, $x = 2, 4$ are the characteristic roots of A.

In this case, both characteristic roots are real, but this is not always true. The roots (real or complex) of the equation $| Ix - A | = 0$ are of vital importance in applications of matrix theory to differential equations in physics, aeronautical and electrical engineering, and elsewhere. In general, an $n \times n$ matrix has n characteristic roots (Why?), although they need not all be distinct. The squares of the frequencies of the vibrations of a mechanical system near equilibrium (an air foil, for example) are obtained as the characteristic roots of a matrix.

Associated with each characteristic root r of a matrix A there is a set of one or more nonzero column vectors V such that

$$A \cdot V = r \cdot V, \quad \text{with} \quad V \neq \begin{bmatrix} 0 \\ 0 \\ \vdots \\ 0 \end{bmatrix}$$

Such a vector V is called a *characteristic vector* (*eigenvector*)[1] of A, corresponding to the root r. The equation $A \cdot V = r \cdot V$ may be written as

$$(r \cdot I - A) \cdot V = \begin{bmatrix} 0 \\ 0 \\ \vdots \\ 0 \end{bmatrix} \quad \text{or} \quad B \cdot \begin{bmatrix} x_1 \\ x_2 \\ \vdots \\ x_n \end{bmatrix} = \begin{bmatrix} 0 \\ 0 \\ \vdots \\ 0 \end{bmatrix}$$

[1] Actually, the characteristic vector defined here is a *right* characteristic vector. A *left* characteristic vector is obtained from the equation $V \cdot A = V \cdot r$ (or, since r is a scalar, $= r \cdot V$). The most usual application is when A is symmetric (that is, $A^t = A$). In this case, there is no distinction between right and left characteristic vectors for a given characteristic root r.

which is a system of n homogeneous (constant terms zero) linear equations in the n unknowns $x_1, x_2, \cdots, x_n$. (See Chapters 6 and 7.) The coefficient determinant is $|B|$, and since r is a root of the characteristic equation, $|B| \equiv |r \cdot I - A| = 0$ and there exists a solution

$$V = (x_1, x_2, \cdots, x_n)^t \neq (0, 0, 0, \cdots, 0)^t \text{ of the system}$$

$$(r \cdot I - A) \cdot V = \begin{bmatrix} 0 \\ 0 \\ 0 \\ \vdots \\ 0 \end{bmatrix}$$

Hence, for each characteristic root of a square matrix A, at least one characteristic vector will be determined.

If c is some number (scalar) which is *not* a characteristic root of A, then the only vector V which satisfies the equation $A \cdot V = c \cdot V$ is the zero vector since, in this case, $|B| = |cI - A| \neq 0$ and the system has a unique solution which can only be $(0, 0, 0, \cdots, 0)^t$ since the constant term in each equation is zero.

EXAMPLE 4

Find the characteristic vectors of the matrix A given in Example 1:

$$A \cdot V = r \cdot V$$

Using the root $r = 2$,

$$\begin{bmatrix} 1 & 3 \\ -1 & 5 \end{bmatrix} \cdot \begin{bmatrix} x_1 \\ x_2 \end{bmatrix} = 2 \cdot \begin{bmatrix} x_1 \\ x_2 \end{bmatrix}$$

$$\begin{bmatrix} x_1 + 3x_2 \\ -x_1 + 5x_2 \end{bmatrix} = \begin{bmatrix} 2x_1 \\ 2x_2 \end{bmatrix}$$

We thus seek the nonzero solution of the equations:

$$\begin{cases} x_1 + 3x_2 = 2x_1 \\ -x_1 + 5x_2 = 2x_2 \end{cases} \quad \text{or} \quad \begin{cases} x_1 = 3x_2 \\ x_1 = 3x_2 \end{cases}$$

The obvious solution is

$$\begin{cases} x_1 = 3t \\ x_2 = t \end{cases} \quad \text{or} \quad (x_1, x_2) = (3t, t) = (3, 1)t$$

Any multiple of $(3, 1)$ is a characteristic vector of A. Another characteristic vector corresponding to the root $r = 4$ may be determined.

If the matrix is larger than 2×2, the detached coefficient (matrix) method of solving systems of linear equations which was discussed in Chapter 6 is useful.

An example of a matrix having nonreal characteristic roots follows.

EXAMPLE 5

Determine the characteristic roots and characteristic vectors of the matrix $\begin{bmatrix} 1 & 2 \\ -2 & 1 \end{bmatrix}$

$$\begin{vmatrix} x - 1 & -2 \\ 2 & x - 1 \end{vmatrix} = 0$$

$$x^2 - 2x + 5 = 0$$

$$x = \frac{2 \pm \sqrt{4 - 20}}{2} = 1 \pm 2i$$

The characteristic vector corresponding to $1 + 2i$ is

$$\begin{bmatrix} 1 & 2 \\ -2 & 1 \end{bmatrix} \cdot \begin{bmatrix} x_1 \\ x_2 \end{bmatrix} = (1 + 2i) \cdot \begin{bmatrix} x_1 \\ x_2 \end{bmatrix}$$

$$\begin{bmatrix} x_1 + 2x_2 \\ -2x_1 + x_2 \end{bmatrix} = \begin{bmatrix} (1 + 2i)x_1 \\ (1 + 2i)x_2 \end{bmatrix}$$

$$x_1 + 2x_2 = (1 + 2i)x_1 \qquad -2ix_1 + 2x_2 = 0$$
$$\text{or}$$
$$-2x_1 + x_2 = (1 + 2i)x_2 \qquad -2x_1 - 2ix_2 = 0$$

each of which reduces to $x_1 = -ix_2$ giving a characteristic vector corresponding to the root $r = 1 + 2i$ as $(-i, 1)$, or any (complex) multiple $(-i, 1)t$.

The reader should verify that a vector corresponding to the characteristic root $r = 1 - 2i$ is $(1, -i)$. Is $(i, 1)$ a characteristic vector corresponding to either root?

Problem Set 9.3

Form the characteristic equation and find the characteristic roots and characteristic vectors of the following matrices.

1. $A = \begin{bmatrix} 5 & 3 \\ -1 & 1 \end{bmatrix}$

2. $B = \begin{bmatrix} 2 & 7 \\ 0 & 4 \end{bmatrix}$

3. $C = \begin{bmatrix} 1 & 9 \\ 5 & 7 \end{bmatrix}$

4. $D = \begin{bmatrix} 4 & -2 \\ 3 & 5 \end{bmatrix}$

5. $E = \begin{bmatrix} 5 & 1 \\ 9 & 8 \end{bmatrix}$

6. $F = \begin{bmatrix} 2 & 3 + i \\ -4 + 2i & 6 \end{bmatrix}$

7. $G = \begin{bmatrix} 3i & -7 \\ 4 & 2 + 4i \end{bmatrix}$

8. $H = \begin{bmatrix} 3 & 2 & 2 \\ 1 & 4 & 1 \\ -2 & -4 & -1 \end{bmatrix}$

9. $K = \begin{bmatrix} 4 & 0 & 6 & 0 \\ 4 & 7 & 0 & 5 \\ 2 & 0 & 6 & 0 \\ 1 & 0 & 3 & 5 \end{bmatrix}$

10. $L = \begin{bmatrix} 7 & 0 & 0 & 0 & 0 \\ 0 & 9 & 0 & 0 & 0 \\ 0 & 0 & 6 & 0 & 0 \\ 0 & 0 & 0 & 4 & 0 \\ 0 & 0 & 0 & 0 & 1 \end{bmatrix}$

11. Show that the characteristic equation of the matrix $\begin{bmatrix} 3 & 0 & 6 \\ 3 & 4 & 0 \\ 0 & 0 & 2 \end{bmatrix}$ is that obtained in Example 1, Section 9.1.

*12. Show that, if $B = PAP^{-1}$, then B has the same characteristic equation as does A, where A is a 3×3 matrix.

13. Discover and state a rule for determining the characteristic roots of a diagonal matrix; i.e., of a matrix having zero for each element which is not on the main diagonal.

14. (a) Show that, if a 4×4 matrix has determinant zero, then one of the characteristic roots is zero.
    (b) Generalize to an $n \times n$ matrix.

*15. Prove that a square matrix A has nonzero determinant (that is, is nonsingular) *if, and only if*, all the characteristic roots of A are nonzero.

16. Prove the Hamilton–Cayley Theorem for 2×2 matrices by direct computation.

17. Consult the excellent paper "A Recurring Theorem on Determinants" by O. Taussky (Todd) in *American Mathematical Monthly*, **56** (1949), pp. 672–76 where the following theorem is discussed. If A is a matrix with real elements and $|a_{ii}| > \sum_{j \neq i} |a_{ij}|$ for all i, then $|A| \neq 0$. In words, if the absolute value of the element on the main diagonal is greater than the sum of the absolute values of the remaining elements on the same row, for each row of the matrix A, then the determinant of matrix A is not zero (that is, A is nonsingular and has an inverse).

18. Another important paper by Dr. Taussky (Now Mrs. Todd) is the National Bureau of Standards Report "Bibliography on Bounds for Characteristic Roots of Finite Matrices" (Sept. 1951).

19. Chapters 6 and 8 in *A Survey of Numerical Analysis* by J. Todd (New York: McGraw-Hill, 1962) contain additional material on eigenvalues and their computation using computers which may be of interest to some readers. If you are interested in computers, spend an hour reading parts of the above book and write a brief report on it.

9.4 MINIMUM FUNCTIONS

The Hamilton–Cayley Theorem assures us that every $n \times n$ matrix A satisfies a polynomial equation with coefficients of degree n. Some $n \times n$ matrices satisfy polynomial equations of degree less than n. For example, the 4×4 matrix $\begin{bmatrix} 7 & 0 & 0 & 0 \\ 0 & 7 & 0 & 0 \\ 0 & 0 & 7 & 0 \\ 0 & 0 & 0 & 7 \end{bmatrix}$ satisfies the first degree equation $x - 7 \cdot I = 0 \cdot I$.

Let A be a square matrix and let S be the set of all polynomials such that $f(A) = 0$. The set S is not empty (null), since the characteristic function is in it. S is closed under addition from within, and furthermore the product of an element of S with any polynomial is again an element of S. In short, S *is an ideal of the set of all polynomial functions* with scalar coefficients. Since S is not empty, it contains a subset of polynomials of lowest degree. Among this subset there is at least one (actually exactly one) polynomial $m(x)$ of lowest degree having leading coefficient 1 and such that $m(A) = 0$. Such a polynomial $m(x)$ is called a (the) *minimum function of the matrix A*. If B is a nonsquare matrix, then B has no minimum function, since powers of B are not defined.

THEOREM 9.2   *The minimum function of the matrix A is unique.*

It has already been established that there exists at least one polynomial $m(x)$ of lowest degree such that $m(A) = 0$ and $m(x)$ has leading coefficient 1. Assume there are two such minimum polynomials.

$$m_1(x) = x^k + a_{k-1}x^{k-1} + a_{k-2}x^{k-2} + \cdots + a_1 x + a_0$$

$$m_2(x) = x^k + b_{k-1}x^{k-1} + b_{k-2}x^{k-2} + \cdots + b_1 x + b_0.$$

Either $m_1(x) - m_2(x)$ is identically zero [that is, $m_1(x) = m_2(x)$] or else $g(x) = m_1(x) - m_2(x)$ is a polynomial of degree less than the degree of the minimum function such that $g(A) = m_1(A) - m_2(A) = 0 - 0 = 0$. Since the latter alternative is contrary to the assumption that $m_1(x)$ is a minimum function (Why?), we conclude $m_1(x) \equiv m_2(x)$ and that the minimum function of a square matrix A is unique.

THEOREM 9.3   If $f(A) = 0$, then $m(x)$ *divides* $f(x)$, *where* $m(x)$ *is the minimum function of* A.

Problem 12 of the next set asks for a proof of this theorem.

Just as the complex roots of the characteristic function are of great importance in modern physics, quantum mechanics, aeronautical design, and structure analysis, so are the roots of the minimum function.

Since the minimum function $m(x)$ divides the characteristic function $| x \cdot I - A |$, it follows immediately that each root of the minimum function is also a characteristic root. The reader may be interested in consulting some texts on matrix theory in his school library to determine whether or not the characteristic equation may have roots which do not satisfy the minimum equation.

Problem Set 9.4

1. (a) Prove or disprove: The minimum function of a scalar matrix (See Section 5.10.) is linear.
   (b) Prove or disprove: If the minimum function of a matrix A is linear, then A is a scalar matrix.

2. In Problem 1 for "scalar" read "diagonal."

Find the minimum functions of the matrices given in Problems 3–8.

3. $\begin{bmatrix} 7 & 0 & 0 \\ 0 & 7 & 0 \\ 0 & 0 & 7 \end{bmatrix}$

4. $\begin{bmatrix} 3 & 2 \\ 1 & 4 \end{bmatrix}$

5. $\begin{bmatrix} 1 & 0 \\ 0 & 2 \end{bmatrix}$

6. $\begin{bmatrix} 1 & 0 & 3 \\ 2 & 0 & 4 \\ 0 & 1 & 0 \end{bmatrix}$

7. $\begin{bmatrix} 4 & 7 \\ 2 & 9 \end{bmatrix}$

8. $\begin{bmatrix} 4 & 3 & 2 & -1 \\ 3 & 2 & 1 & 0 \\ 2 & 1 & -1 & -2 \\ -1 & 0 & -2 & 5 \end{bmatrix}$

9. Prove that the minimum function of an $n \times n$ matrix is of degree n or less.

*10. Discuss the various possibilities for a characteristic function and for a minimum function of a singular matrix. Recall that a matrix A is called singular if $|A| = 0$.

*11. Consult recent issues of *Mathematical Reviews* under "Matrices" to determine what work is currently being done on characteristic roots, and prepare a one or two page report thereon.

12. Prove Theorem 9.3. {HINT: Using the division algorithm for polynomials, there exist polynomials $q(x)$ and $r(x)$ [where either $r(x) \equiv 0$ or the degree of $r(x)$ is less than the degree of $m(x)$] and such that $f(x) = q(x) \cdot m(x) + r(x)$.}

13. Show that the ideal of all polynomials $f(x)$ such that $f(A) = 0$ is a principal ideal (See Section 8.3.) consisting of all polynomial multiples of the minimum function of A.

14. Show that, if the constant term of the minimum function of A is zero, then A has no inverse. [HINT: If the minimum function is

$$m(x) = x^k + a_{k-1}x^{k-1} + \cdots + a_1 x + 0,$$

then $m(A) = A^k + a_{k-1}A^{k-1} + \cdots + a_1 A = 0$ may be expressed as

$$A(A^{k-1} + a_{k-1}A^{k-2} + \cdots + a_1) = 0]$$

If A has an inverse, then a contradiction of the type obtained in the proof of Theorem 9.2 follows. Also compare Problem 15 Set 9.3.

15. State and prove the converse of the theorem of Problem 14.

9.5 INFINITE SERIES WITH MATRIX ELEMENTS

It will be assumed that the reader is somewhat familiar with ordinary (Cauchy) convergence, as usually applied to infinite sequences and series of real (or complex) numbers and to power series. We now discuss the meaning of convergence of an infinite series of *matrices* of similar size.

$$A_0 + A_1 + A_2 + \cdots + A_n + \cdots.$$

Consider the matrix

$$S_n = A_0 + A_1 + A_2 + A_3 + \cdots + A_n \qquad \text{(a finite sum)}.$$

Each element of the matrix S_n is a (finite) series. As $n \to \infty$, each individual element of S_n becomes an infinite series. If *each* of these infinite series converges in the ordinary sense, then the matrix series

$$A_0 + A_1 + A_2 + A_3 + \cdots + A_n + \cdots$$

is said to converge to the matrix, $\lim_{n \to \infty} S_n$, composed of the corresponding limits of the elements of S_n. If any one of the "element series" diverges, then the entire matrix series is called divergent.

EXAMPLE 6

Let

$$A_0 = I, \quad A_1 = I, \quad A_2 = \frac{1}{2!} \cdot I, \quad A_3 = \frac{1}{3!} \cdot I, \cdots, A_n = \frac{1}{n!} \cdot I$$

Then

$$S_n = \begin{bmatrix} \sum\limits_{i=0}^{n} \dfrac{1}{i!} & 0 & 0 & 0 & 0 \\ 0 & \sum\limits_{i=0}^{n} \dfrac{1}{i!} & 0 & 0 & 0 \\ 0 & 0 & \sum\limits_{i=0}^{n} \dfrac{1}{i!} & 0 & 0 \\ 0 & 0 & 0 & \sum\limits_{i=0}^{n} \dfrac{1}{i!} & 0 \\ 0 & 0 & 0 & 0 & \sum\limits_{i=0}^{n} \dfrac{1}{i!} \end{bmatrix}$$

From calculus the reader may recall that

$$\sum_{i=0}^{\infty} \frac{1}{i!} = 1 + 1 + \frac{1}{2!} + \frac{1}{3!} + \frac{1}{4!} + \cdots + \frac{1}{n!} + \cdots = e$$

Hence,

$$\lim_{n \to \infty} S_n = \begin{bmatrix} e & 0 & 0 & 0 & 0 \\ 0 & e & 0 & 0 & 0 \\ 0 & 0 & e & 0 & 0 \\ 0 & 0 & 0 & e & 0 \\ 0 & 0 & 0 & 0 & e \end{bmatrix} = e \cdot I$$

We now restrict A to be a square matrix and consider a *power series* in powers of the square matrix A.

$$\sum_{i=0}^{\infty} a_i A^i = a_0 I + a_1 A + a_2 A^2 + a_3 A^3 + \cdots + a_n A^n + \cdots$$

where A is a square matrix and the a_i are scalars. It can be proved that, if A is a matrix having *all* of its *characteristic roots* (which may, of course, be complex numbers) *inside the unit circle*, then $\lim_{n \to \infty} A^n = (0)$. A matrix with non-negative real elements, in which the sum of each row (or of each column) is less than one, will have all its characteristic roots inside the unit circle. Such matrices occur frequently in statistics and probability theory. Much of modern engineering and physics is based on probability theory. Infinite series with matrix elements are of great importance in aeronautical engineering (flutter analysis, for example) and physics as well as in statistics and its applications.

It may come as somewhat of a shock to realize that sin A, cos A, and other trigonometric functions of a square matrix A are not only meaningful, but useful in practical applications. The definition is by means of power series.

$$\sin A = A - \frac{1}{3!} A^3 + \frac{1}{5!} A^5 - \frac{1}{7!} A^7 + \cdots + \cdots$$

$$\cos A = I - \frac{1}{2!} A^2 + \frac{1}{4!} A^4 - \frac{1}{6!} A^6 + \cdots + \cdots$$

The concept of e^A, where A is a square matrix, also has important applications in aeronautical engineering, electronics, physics, and elsewhere.

$$e^A = I + A + \frac{1}{2!} A^2 + \frac{1}{3!} A^3 + \cdots + \cdots$$

The power series expansion for e^A is convergent for all square matrices A.

In general, $A \cdot B \neq B \cdot A$, and $e^A \cdot e^B \neq e^B \cdot e^A$. One should not expect either to equal e^{A+B}. However, if A and B commute (i.e., $A \cdot B = B \cdot A$), then $e^A \cdot e^B = e^{A+B} = e^B \cdot e^A$. Furthermore, $e^A \cdot e^{-A} = I$. Thus, the inverse of e^A is e^{-A}. Just because the notation "looks right" for these statements does *not* prove they are true. Each requires additional proof.

9.6 DERIVATIVES AND INTEGRALS OF MATRICES

If $A(x)$ is a matrix having functions of x as elements, such as

$$A(x) = \begin{bmatrix} x^2 & x & 5 + 3x^6 \\ 2x & x^{17} & x \\ 13 - \sqrt{x} & \sqrt{\pi} & 14 - x \end{bmatrix}$$

we define the derivative and the integral of the matrix $A(x)$ to be the matrix obtained by differentiating or integrating elementwise the individual functions which are the elements of $A(x)$.

$$\frac{dA(x)}{dx} = \begin{bmatrix} 2x & 1 & 18x^5 \\ 2 & 17x^{16} & 1 \\ -\dfrac{1}{2\sqrt{x}} & 0 & -1 \end{bmatrix}$$

$$\int A(x)\,dx = \begin{bmatrix} \dfrac{x^3}{3} & \dfrac{x^2}{2} & 5x + \dfrac{3x^7}{7} \\ x^2 & \dfrac{x^{18}}{18} & \dfrac{x^2}{2} \\ 13x - \tfrac{2}{3}x^{3/2} & x\sqrt{\pi} & 14x - \dfrac{x^2}{2} \end{bmatrix} + C$$

where C is a matrix with constant elements.

Problem Set 9.6

1. Show that $\dfrac{d[A(x)\cdot B(x)]}{dx} = \dfrac{dA(x)}{dx}\cdot B(x) + A(x)\cdot\dfrac{dB(x)}{dx}$.

2. Show, using the result of Problem 1, that

$$\frac{dA^{-1}(x)}{dx} = -A^{-1}(x)\cdot\frac{dA(x)}{dx}\cdot A^{-1}(x).$$

3. Show that, if A is a constant matrix, then $\dfrac{d(e^{xA})}{dx} = A\cdot e^{xA} = e^{xA}\cdot A.$

4. (a) Determine a real 2×2 matrix A such that $A^2 = \begin{bmatrix} 3 & -4 \\ 1 & -1 \end{bmatrix}$. Show that there are exactly two such matrices, namely $\pm\begin{bmatrix} 2 & -2 \\ \frac{1}{2} & 0 \end{bmatrix}$.

   (b) Show that there are infinitely many real 2×2 matrices such that $B^2 = \begin{bmatrix} 1 & 0 \\ 0 & 1 \end{bmatrix}$. Find four examples.

⋆5. Let the "remainder after n terms" in the power series for e^A be R_n, where

$$e^A = I + A + \frac{1}{2!}A^2 + \frac{1}{3!}A^3 + \frac{1}{4!}A^4 + \cdots + \frac{1}{(n-1)!}A^{n-1} + R_n.$$

Show that $R_n = \dfrac{A^n}{n-1}\displaystyle\int_0^1 (1-z)^{n-1}e^{Az}dz$. Consult library texts for assistance if needed.

6. Let $A(x)$ be a square matrix having functions $f_{rs}(x)$ as elements. Form $\displaystyle\lim_{\Delta x\to 0}\dfrac{A(x_1+\Delta x) - A(x_1)}{\Delta x}$ and show that this is the same as the definition of $\dfrac{dA(x)}{dx}$ given (termwise differentiation).

7. Prove the theorem given in Problem 1 using the definition of Problem 6.

8. Let $A = \begin{bmatrix} \frac{1}{2} & 2 \\ 0 & 0 \end{bmatrix}$. Show that $\lim_{n \to \infty} A^n = \begin{bmatrix} 0 & 0 \\ 0 & 0 \end{bmatrix}$ both by considering its characteristic roots and by actual computation of A^n.

9. Let $B = \begin{bmatrix} \frac{1}{2} & 0 \\ 2 & 0 \end{bmatrix}$. Show that $\lim_{n \to \infty} B^n = \begin{bmatrix} 0 & 0 \\ 0 & 0 \end{bmatrix}$ by both methods.

10. Problems 8 and 9 provide examples of matrices A and B such that $\lim_{n \to \infty} A^n = 0$ and $\lim_{n \to \infty} B^n = 0$. Show, however, that $\lim_{n \to \infty} (A \cdot B)^n$ is *divergent*. (This will probably come as a surprise!)

11. Let $T = \begin{pmatrix} 1 & s \\ 0 & 1 - s - t \end{pmatrix}$, where $0 < s < 1, 0 < t < 1$.

    Show that

$$T^k = \begin{pmatrix} 1 & s + s(1 - s - t) + s(1 - s - t)^2 + \cdots + s(1 - s - t)^{k-1} \\ 0 & (1 - s - t)^y \end{pmatrix}$$

   and that $\lim_{n \to \infty} T^n = \begin{pmatrix} 1 & s/(s + t) \\ 0 & 0 \end{pmatrix}$.

12. Example 2, Section 5.11, discusses a chemical problem in which

$$\lim_{n \to \infty} \begin{pmatrix} \frac{3}{4} & \frac{1}{4} \\ \frac{1}{10} & \frac{9}{10} \end{pmatrix}^n$$

   is involved.

   (a) Show that, in general, if $P = \begin{pmatrix} 1 - s & s \\ t & 1 - t \end{pmatrix}$, with $0 < s < 1$, $0 < t < 1$, then, for $H = \begin{pmatrix} 1 & 0 \\ 1 & 1 \end{pmatrix}$, $H^{-1} \cdot P \cdot H = T$ and $H^{-1} \cdot P^n \cdot H = T^n$, where T is the matrix of Problem 11.

   (b) Show that, if $H^{-1}(\lim_{n \to \infty} P^n) H = \lim_{n \to \infty} T^n$

(a valid assumption, which Problem 10 might make us question), then, by Problem 11,

$$\lim_{n \to \infty} P^n = \begin{pmatrix} t/(s+t) & s/(s+t) \\ t/(s+t) & s/(s+t) \end{pmatrix}.$$

(c) Show that $(L_0, V_0)(\lim_{n \to \infty} P^n) = [t/(s+t), \ s/(s+t)]$, when

$L_0 + V_0 = 1$.

9.7 WHAT MATHEMATICS TO TAKE NEXT

Now that you have completed your introductory course in abstract algebra, it may be appropriate to discuss briefly what mathematics you should take next. Naturally, this is an individual matter which must be decided in cooperation with your advisor. Nevertheless, since many students are advised on what mathematics courses to take by nonmathematicians, it seems appropriate to make some general remarks here.

Mathematics is divided into several broad classifications. These divisions are not at all well defined in many cases, and, from certain viewpoints each classification is actually a subclassification of some other. (Geometry, for example, may be considered as a special application of matrix theory, although it is seldom taught in this manner.) One set of these broad classifications, along with some courses commonly placed in each category is as follows:

Analysis: (This usually deals with infinite processes, such as limits.)

Differential Equations
Advanced Calculus
Theory of a Complex Variable
Theory of a Real Variable
Partial Differential Equations
Integral Equations
Infinite Series
Elasticity
Calculus of Variations
Functional Analysis
Hilbert Space

Modern Algebra: (This usually deals with finite processes.)

> The course for which this book was written.
> Modern Abstract Algebra (beginning, intermediate, and advanced)
> Theory of Matrices
> Theory of Rings and Fields
> Theory of Numbers
> Boolean Algebra
> Theory of Groups
> Theory of Games
> Theory of Loops
> Algebraic Varieties

Geometry and Topology:

> College Geometry
> Analytic Geometry
> Projective Geometry (analytic and synthetic)
> Differential Geometry
> Tensor Calculus
> Topology (several types)
> Topological Groups

Mathematical Statistics: (A course in "statistics" which does not have calculus as a prerequisite *is probably not worthwhile* at your level of training.)

> Theory of Probability
> Mathematical Statistics (elementary, intermediate, advanced)
> Quality Control
> Monte Carlo Methods

Computing:

The areas of Information Science and Computing Science are new-comers to the academic field since the modern stored program computer is a "last half of the twentieth century" development. Being able to code computers in several languages is a marketable skill, and one you would be well advised to cultivate, but the sad truth is that many thousands of hours of valuable computer time are used (or misused) to obtain results which could be obtained with less effort and greater accuracy if the programmer had known a bit more about modern mathematics in general and abstract algebra (which deals with finite processes) in particular, before he wrote his instructions to the computer. If you are interested in computing, then by all means take as much mathematics as you can. It might be well to consult Chapter V in that fascinating compendium *Mathematics in the Modern World* (San Francisco: Freeman, 1968) as well as recent issues of *Scientific American*.

As was stated previously, these lines are not at all well drawn. Topology, for example, certainly deserves a separate classification. Vector analysis could well be considered as part of Modern Algebra. Many courses could be shifted from classification to classification depending upon the instructor. There is another broad classification of *applied mathematics* which selects from each of the above classifications any topics which seem to have current applications. This classification changes so fast it would be out of date between writing and printing, especially now that most industries are employing Ph.D. mathematicians as part of their regular staff.

The problem of what course to take next hinges upon what is offered at your institution. Most colleges and universities give several courses in modern abstract algebra, which you will now be prepared to take. If you tell your instructor which chapters of this text you particularly liked or disliked, he will be able to suggest appropriate courses being offered at *your* college.

Today modern algebra is in about the same relative position to mathematics as mathematics is to science and engineering. In addition to many fascinating problems in its own right, each provides the vocabulary and the general overall methods and concepts which are used in the larger body of knowledge. There is much to learn in each body of knowledge, but without a good foundation in modern abstract algebra your path will be appreciably rougher and more difficult. You may wish to read M. S. Klamkin's article on industrial mathematics in the January 1971 issue of *American Mathematical Monthly* (Vol. 78, p. 53).

Selected Reading List, Chapter 9
MORE MATRIX THEORY

The following books and journals will provide additional reading for students interested in term projects or independent study.

Books

MacDuffee, C. C., *Vectors and Matrices*. Carus Monograph No. 7 (Mathematical Association of America).

MacDuffee, C. C., *The Theory of Matrices* (One of the Ergebnisse Series, reprinted by Chelsea.)

Murdoch, D. C., *Linear Algebra for Undergraduates*. New York: Wiley, 1957.

Perlis, S., *Theory of Matrices*. Reading, Mass.: Addison-Wesley, 1952.

Articles

Bucher, R. and S. Godbole, "Cofactoral Matrices," *Mathematics Magazine*, **42,** No. 3 (May 1969), pp. 142–145.

Frame, J. S. "Matrix Functions and Applications," an excellent series of five papers appearing in *IEEE Spectrum* No. 3, pp. 208–220; No. 4, pp. 102–108; No. 5, pp. 100–109; No. 6, pp. 123–131; No. 7, pp. 103–109.

Hearon, J. Z., "Idempotent Matrices with Nilpotent Difference," *Mathematics Magazine*, **41,** No. 2 (Mar. 1968), pp. 80–84.

Kieffer, J. C. and F. M. Stein, "Solution of an Equation in a Linear Algebra by Means of the Minimal Polynomial," *Mathematics Magazine*, **42,** No. 3 (May 1969), pp. 114–121.

Laidacker, M. A., "Another Theorem Relating Sylvester's Matrix and the Greatest Common Divisor," *Mathematics Magazine*, **42,** No. 3 (May 1969), pp. 126–128.

Lightstone, A. H., "Two Methods of Inverting Matrices," *Mathematics Magazine*, **41,** No. 1 (Jan. 1968), pp. 1–7.

South, J. C., Jr., "Note on the Matrix Functions Sin πA and Cos πA," *Mathematics Magazine*, **39,** No. 5 (Nov. 1966), pp. 287–288.

"Square Root of a Matrix," Problem E 1942, *The American Mathematical Monthly*, **75,** No. 4 (Apr. 1968), p. 409 asks you to find the square root of the matrix

$$A = \begin{bmatrix} -91 & 28 & 9 \\ 47 & -14 & -4 \\ -1113 & 341 & 108 \end{bmatrix}.$$ You may wish to try your hand at this problem before consulting the solution given.

Notes on Chapter 9

ANSWERS AND HINTS

A *student* of mathematics will try to work problems on his own before consulting the hints given here, but should not hesitate to seek the assistance available if help is really needed. As always in textbook problems the real goal is *not* the answer to the problem but rather in the understanding of the methods involved so it is good to have answers available when needed to check your understanding. Good luck and I hope you have as much fun working through this book as I did in writing it for you.

CHAPTER 1

Set 1.2

1. $3 \pmod 7$.

3. $x \equiv 4 \pmod 7$.

5. $x \equiv 5 \pmod 7$. One method of working the problem is to notice that $7x \equiv 0 \pmod 7$ *for all* x. Thus $297x = 42 \cdot 7x + 3x \equiv 3x \pmod 7$.

7. $x \equiv 3 \pmod 7, x \equiv 4 \pmod 7$.

9. $x \equiv 3 \pmod 7$, $x \equiv 5 \pmod 7$, $x \equiv 6 \pmod 7$.

11. (a)

x	x^2	x^3	x^4	x^5	x^6	x^7	x^8
0	0	0	0	0	0	0	0
1	1	1	1	1	1	1	1
2	4	1	2	4	1	2	4
3	2	6	4	5	1	3	2
4	2	1	4	2	1	4	2
5	4	6	2	3	1	5	4
6	1	6	1	6	1	6	1

(b) $5^6 \equiv 1 \pmod 7$, thus $5^{236} \equiv 5^{39 \cdot 6} \cdot 5^2 \equiv 5^2 \equiv 4 \pmod 7$. Similarly, $3^{179} \equiv 3^5 \equiv 5 \pmod 7$.

(c) No.

(d) For all values of $b \pmod 7$, $x^5 \equiv b$ has a solution.

13.

$+$	0	1	2	3	4	5	6
0	0	1	2	3	4	5	6
1	1	2	3	4	5	6	0
2	2	3	4	5	6	0	1
3	3	4	5	6	0	1	2
4	4	5	6	0	1	2	3
5	5	6	0	1	2	3	4
6	6	0	1	2	3	4	5

$\cdot$	0	1	2	3	4	5	6
0	0	0	0	0	0	0	0
1	0	1	2	3	4	5	6
2	0	2	4	6	1	3	5
3	0	3	6	2	5	1	4
4	0	4	1	5	2	6	3
5	0	5	3	1	6	4	2
6	0	6	5	4	3	2	1

15. $b + c = (2 + 5) + (k + j)7 = 0 + (1 + k + j)7 = 0 + m \cdot 7$.

$b \cdot c = 2 \cdot 5 + (2j + 5k + 7jk)7$
$= 3 + (1 + 2j + 5k + 7jk)7 = 3 + n \cdot 7$.

17. There are 6 possible values of A and 7 possible values of B for each value of A; therefore, there are $6 \times 7 = 42$ distinct equations $Ax \equiv B \pmod 7$.

Set 1.3

1. $0 \pmod 6$.

3. No solution.

5. $x = 0, 2, 4 \pmod 6$. $297x = 49 \cdot 6x + 3x \equiv 3x$. $3x + 6 \equiv 0 \pmod 6$ has the solutions 0, 2, 4. Thus $297x + 6 \equiv 0 \pmod 6$ has solutions 0, 2, and 4.

7. No solution.

9. $x \equiv 0 \pmod 6$.

11. No solution.

13. $x \equiv 0, 3 \pmod 6$.

15. In contrast to the complex numbers system there are no solutions in the mod 6 system.

17. $a \equiv a \pmod 6$, since $a = a + c \cdot 6$, for $c = 0$.
$a \equiv b$ implies $b \equiv a \pmod 6$, since if $a = b + c_1 \cdot 6$, then $b = a + c_2 \cdot 6$, for $c_2 = -c_1$.
$a \equiv b$, $b \equiv c$ imply $a \equiv c \pmod 6$, since if $a = b + c_1 \cdot 6$ and $b = c + c_2 \cdot 6$, then $a = c + c_3 \cdot 6$, for $c_3 = c_1 + c_2$.

19. $Ax \equiv B \pmod 6$ has no solution if $A \equiv 2$ or 4 and $B \equiv 1$, 3, or 5, or if $A \equiv 3$ and $B \equiv 1, 2, 4$, or 5. In any other case with $A \not\equiv 0$ the congruence has solutions.

21. $Ax \equiv B \pmod 6$ has a unique solution if and only if $A \equiv 1$ or 5.

Set 1.4

1. All Postulates are satisfied:

3. (a) All Postulates.
(b) All except Postulate 7.

5. All Postulates.

7. All except Postulate 1. For example, $\frac{1}{2} \times \frac{1}{2} = \frac{1}{4}$. Closure for addition does hold.

9. All Postulates.

11. All but Postulates 5 and 7.

13. (1) $z + W = z$ (Hypothesis with $a = z$.)
    (2) $z + W = W$ (Postulate 4.)
    (3) $W = z$ (1, 2, symmetry and transitivity of $=$ relation.)

15. (1) $B + x = 0 = B + y$ (By assumption.)
    (2) $B^* + (B + x) = B^* + 0 = B^*$ (1, Post. 4.)
    (3) $(B^* + B) + x = 0 + x = x$ (2, Post. 3, Post. 6, Post. 4.)
    (4) $x = B^*$ (2, 3, Post. 3.)
    (5) $y = B^*$ (1, similarly.)
    (6) $x = y$ (4, 5.)
    Note: The three properties of the equals relation were used throughout.

Set 1.6

1. (a) None.
   (b) One.
   (c) One.
   (d) None.
   (e) One.

7. "All odd numbers are prime."

11. Opposite: "If two lines are not parallel, then they intersect." Valid in plane, invalid (because of skew lines) in solid geometry. Converse: "If two lines do not intersect, then they are parallel." Valid in plane, invalid in solid. Contrapositive: "If two lines intersect, then they are not parallel." Valid in plane, valid in solid.

13. Yes, Peter's statement is merely an inversion of the word order of the original sentence.

15. Not (not A) = A, so that "If not (not A), then not (not B)" is equivalent to "If A, then B."

21. (a) No.
    (b) No.
    (c) Yes.
    (d) No.
    (e) Answer depends upon the definition of "to the 21 Club."

23. (a) No.
    (b) No.

25. The conclusion is invalid, although its converse is valid.

Set 1.7

3. The analysis may be handled more easily by considering the system as a
    set of ordered pairs (a, b), where a and b are integers and $a \equiv b \pmod 2$.
    Then the operations are defined as $(a, b) + (c, d) = (a + c, b + d)$

    and $(a, b) \cdot (c, d) = \left(\dfrac{ac - 3bd}{2}, \dfrac{bc + ad}{2} \right)$. All integral domain postu-

    lates except closure are seen to be satisfied by noting that this system
    corresponds to a subset of the system of complex numbers (which is an
    integral domain) and that zero $(0, 0)$, unity $(2, 0)$, and inverses
    $(-a, -b)$ are in the subset. Since if $a \equiv b$ and $c \equiv d \pmod 2$, then
    $a + c \equiv b + d \pmod 2$, the system is closed under addition. Also,
    $ac - 3bd \equiv ac + 1bd \equiv ac + ac \equiv 2ac \pmod 2$ and $bc + ad \equiv$
    $ac + ac \equiv 2ac \pmod 2$, whenever $a \equiv b$ and $c \equiv d \pmod 2$, so that

    $\dfrac{ac - 3bd}{2}$ and $\dfrac{bc + ad}{2}$ are integers and are congruent $\pmod 2$, yielding

    closure under multiplication. The system is an integral domain.

5. Substitute $\cdot$ for $+$, u for z, and u' for z' in the proof of Problem 4.

7. (1) $ab + (a(-b) + (-a)(-b)) = (ab + a(-b)) + (-a)(-b)$
         (Postulate 3)
    (2) $ab + (a(-b) + (-a)(-b)) = ab + (a + (-a))(-b) =$
         $ab + z(-b)$ (Postulates 8 and 6)
    (3) $ab + z(-b) = ab$ (Example 8 on p. 30, Post. 4)
    (4) $ab + (a(-b) + (-a)(-b)) = ab$ $(2, 3)$
    (5) $(ab + a(-b)) + (-a)(-b) = a(b + (-b)) + (-a)(-b) =$
         $(-a)(-b)$ (Similarly)
    (6) $ab = (-a)(-b)$ $(4, 5, 1)$

Set 1.8

1. (a) $1, 2, 4, 13, 26, \underline{52}$.
    (b) $1, 3, 9, 27, \underline{91}$.
    (c) $1, \underline{103}$.

(d) $1, 3, 7, 11, 21, 33, 77, \underline{231}$.
One is not a proper divisor of itself, but is a proper divisor of 6. See definition.

3. (a) No. $3x = 1$ has no solution in the system.
   (b) Yes. $x = \frac{1}{3}$ is in the system.
   (c) No. $2 + 3i$ (much less its "inverse") has no meaning in the system (it is not included in the system of rationals).
   (d) No. $2 + 3i$ has no inverse in the system.
   (e) Yes. $(2 + 3i)\left(\dfrac{2}{13} - \dfrac{3}{13}i\right) = 1 + 0i,$ and $\dfrac{2}{13} - \dfrac{3}{13}i$ is in the system.

5. Prove the contrapositive: If $N = k_1 k_2$, for $1 < k_i < N$ (k_1, k_2 integers), then N is a multiple of a positive prime $\leq \sqrt{|N|}$. First show that either k_1 or $k_2 \leq \sqrt{|N|}$, then use induction on the k_i: Either k_i is prime or $k_i = k_{i_1} k_{i_2}$, etc.

9. (b) 2, 3, 5, 7, 11, 13, 17, 19, 23, 29, 31, 37, 41, 43, 47, 53, 59, 61, 67, 71, 73, 79, 83, 89, 97.

13. If N is even, $N(N + 1)(N + 2) = 2(4k^3 + 6k^2 + 2k)$, where $N = 2k$, and $4k^3 + 6k^2 + 2k \equiv k^3 + 2k \equiv k + 2k \equiv 0 \pmod{3}$, so that the product is divisible by 2 and by 3, and hence by 6. If N is odd, $N(N + 1)(N + 2) = 2(4k^3 + 12k^2 + 11k + 3)$, where $N = 2k + 1$, and $4k^3 + 12k^2 + 11k + 3 = k^3 + 2k = 0 \pmod{3}$, so that the product likewise is divisible by 6. (This is the hard way.)

15. The span $N, N + 1, \ldots, N + 4$ always includes a multiple of 5. If $N = 2k$, $N(N + 1)\ldots(N + 4) = 2^3 k(k + 1)(k + 2)(N + 1)(N + 3)$, and $k(k + 1)(k + 2)$ is a multiple of 3, by Problem 13. If $N = 2k + 1$, $N(N + 1)\ldots(N + 4) = 2^2(k + 1)(k + 2)N(N + 2)(N + 4) = 2^3 mN(N + 2)(N + 4)$, since either $k + 1$ or $k + 2$ is even. Also, $N(N + 2)(N + 4) \equiv 0 \pmod{3}$, hence is a multiple of 3. Therefore $N(N + 1)\ldots(N + 4)$ is divisible by $2^3 \cdot 3 \cdot 5 = 120$ and this is the maximum integer which always is a divisor, since $1 \cdot 2 \cdot 3 \cdot 4 \cdot 5 = 120$.

17. If $N > 24$ and every integer less than or equal to $\sqrt{N}$ divides N, then 1, 2, 3, 4, 5 divide N, which implies that $60 \mid N$. This in turn implies that $\sqrt{N} > 7$.
   If all integers $\leq \sqrt{N}$ divide N, then certainly the product of all primes $\leq \sqrt{N}$ divides N. Suppose $\sqrt{N} < 11$. Then $2 \cdot 3 \cdot 5 \cdot 7 = 420 \mid N$. But

$N < 11^2 = 121$, a contradiction. Thus for $k = 4$, $p_1 \cdot p_2 \ldots p_k > (p_{k+1})^2$. The process is completed by using induction to establish this statement for all $k \geq 4$.

$p_{k+1} < p_{k+2} < 2p_{k+1} - 2$. (This is Butrand's Postulate; it is mentioned on page 33.) Assuming that $p_1 \cdot p_2 \ldots p_k > (p_{k+1})^2$, then $p_1 \cdot p_2 \ldots p_k \cdot p_{k+1} > (p_{k+1})^2 p_{k+1}$. Since $p_{k+1} > 4$ for all $k > 1$, $p_1 \cdot p_2 \ldots p_{k+1} > 4(p_{k+1})^2 = (2p_{k+1})^2 > (p_{k+1})^2$. The proof follows by induction.

19. 3, 2; \$5.11.

21. Each number $n! + m \, (2 \leq m \leq n-1)$ in the sequence is at least divisible by m, for $n! + m = m(1 \cdot 2 \cdot \ldots (m-1)(m+1)(m+2) \ldots (n-1)(n) + 1)$

Set 1.10

1. $30 = (-11)1500 + (29)570 = 8(1500) + (-21)(570)$

3. $1 = (27)352 + (-43)221 = (-194)352 + (307)221$

5. $11 = (1)506 + (-15)33 = (-2) \cdot 506 + (31) \cdot 33$

7. $7 = (22)203 + (-13)343 = (-27) \cdot 203 + 16 \cdot 343$

9. $2 = (-79)418 + (24)1376 = (609)418 + (-185)1376$

11. (a) Let $A = nd$, $B = kd$: $nd = kdq + r$. Then $r = (n - kq)d$.
    (b) Since $r_{k-1} = r_k q_{k+1}$, $r_k \mid r_{k-1}$. But $r_{k-2} = r_{k-1}q_k + r_k$, so that $r_k \mid r_{k-2}$. etc.

13. Let $(ma, mb) = d$, $(a, b) = d'$. $d' \mid a$ and $d' \mid b$, so that $md' \mid ma$ and $md' \mid mb$. If $c \mid ma$ and $c \mid mb$, then $c \mid md'$, since $d' = ta + sb$ and so $md' = tma + smb$. Hence $md' = d$, q.e.d.!

15. $(a_1, a_2, \ldots, a_n) = d$ if and only if $d \mid a_i$ $(i = 1, \ldots, n)$ and $c \mid a_i$ $(i = 1, \ldots, n)$ implies $c \mid d$. $(405, 285, 495, 695) = 15(27, 19, 33, 45) = 15 \cdot 1 = 15$. (Since the proof of Problem 13 does not depend upon $n = 2$.)

CHAPTER 2

Set 2.1

1. All postulates are satisfied; *similar* is an equivalence relation.

3. The relation is:
Not reflexive: $x \mid 0$ but $0 \mid x$ is meaningless (reflexive for $x \neq 0$).
Not symmetric: $2 \mid 4$ but $4 \nmid 2$.
Transitive: $a \mid b, b \mid c$ imply $a \mid c$, unless a or $b = 0$.

5. The relation is:
Not reflexive: (The song "I'm My Own Gran'pa" to the contrary notwithstanding.)
Not symmetric: Abel is the descendent of Adam but not vice versa.
Transitive: For all X, Y, Z, if X is a descendent of Y, and Y a descendent of Z, then X is a descendent of Z.

7. All postulates are satisfied.

9. The relation is:
Not reflexive: Barring figurative speech, no person is married to himself.
Symmetric: A married to B implies B married to A.
Not transitive: "AmB, BmC implies AmC" is not true, for example, when $C = A$.

11. All postulates are satisfied.

13. The relation is:
Not reflexive: 117 does not have a larger integral divisor than 117.
Not symmetric: All relations involving "*x-er* than" are asymmetric.
Transitive: Since "greater than" is also transitive and an integer is always greater than or equal to each of its divisors.

15. All postulates are satisfied if reflexiveness is granted.

17. The conclusion aEa does not hold *for all* a. It holds only for those a such that there exists some b which satisfies aEb. Hence the assertion is still not true of zero.

19. All postulates are satisfied:
Reflexiveness: $a \equiv a \pmod 7$, since for some c (viz., $c = 0$), $a = a + c \cdot 7$.
Symmetry: $a \equiv b \pmod 7$ implies $b \equiv a \pmod 7$, since if for some c_1, $a = b + c_1 \cdot 7$, then for some c_2 (viz., $c_2 = -c_1$), $b = a + c_2 \cdot 7$.

Transitivity: $a \equiv b \; b \equiv c$ implies $a \equiv c \pmod{7}$, since if for some c_1 and c_2, $a = b + c_1 \cdot 7$ and $b + c + c_2 \cdot 7$, then for some c_3 (viz., $c_3 = c_2 + c_1$), $a = c + c_3 \cdot 7$.

Set 2.2

1. The equivalence relation partitions the set of all girls into equivalence classes of girls of the same hair color: the class of all redheads, the class of all blondes, etc.

3. The equivalence relation might be "falls in the same score bracket with," where the score brackets are defined in terms of E_1, E_2, E_3.

7. $\equiv \pmod{6}$ separates the integers into six equivalence classes: $E_1: \ldots, -6, 0, 6, 12, \ldots; E_2: \ldots, -5, 1, 7, 13, \ldots; \ldots;$ $E_6: \ldots, -1, 5, 11, 17, \ldots.$

Set 2.3

1. $x \equiv 9 \pmod{15}$

3. $x \equiv 4 \pmod{5}$

5. $x \equiv 22 \pmod{31}$

7. $x \equiv 6 \pmod{24}$

9. $x \equiv 4 \pmod{11}$

11. $-6, 24$, etc.

13. $-1, 9$, etc.

15. $-9, 53$, etc.

17. $-18, 30$, etc.

19. $-7, 15$, etc.

21. $a \equiv b \pmod{-|m|}$ if and only if $a \equiv b \pmod{|m|}$.
Proof: If $a = b + k_1 \cdot (-|m|)$, then for some k_2, viz., $k_2 = -k_1$, $a = b + k_2 \cdot |m|$.

23. If $a = b + k_1 m$, then for $k_2 = -k_1$, $-a = -b + k_2 m$. (Multiply through the first equation by -1.)

25. (a) 5, 7, 8, 4, 2, 1, 5, 7, 8, 4

    (b) $5^{237} \equiv 5^3 \equiv 8$, $5^{238} \equiv 4$.

    (c) $5^n \equiv 5^m \pmod 9$ if and only if $m \equiv n \pmod 6$.

27. Let $N = a_0 + a_1 \cdot 10^1 + a_2 \cdot 10^2 + \ldots + a_k \cdot 10^k = P(10)$. Then the sum of the digits is $S = a_0 + a_1 \cdot 1^1 + a_2 \cdot 1^2 + \ldots + a_k \cdot 1^k = P(1)$. Since $10 \equiv 1 \pmod 3$, each term in $N = P(10)$ is equivalent $\pmod 3$ to the corresponding term in $S = P(1)$, so that $N \equiv S \pmod 3$. But this means that the remainders after division by 3 are equal, q.e.d.

29. As in Ex. 27, let $N = P(10)$ and $T = P(-1)$. Since $10 \equiv -1 \pmod{11}$, $N \equiv T \pmod{11}$, q.e.d.

31. See Ex. 19, Set 2.1.

33. No solution.

35. $x \equiv 5 \pmod 7$

37. $x \equiv 3 \pmod{11}$

39. $x \equiv 13 \pmod{14}$

41. No solution. ($25x = 13 + k \cdot 60$ is impossible for integral x and k.)

43. No solution. ($21x = 16 + k \cdot 35$ is impossible for integral x and k.)

45. If $a = b + jm$ and $c = d + km$, then $ac = bd + lm$ for some integral l, viz., $l = bk + jd + jkm$.

47. No. e.g., $2 \cdot 4 = 2 \cdot 1 \pmod 6$ but $4 \not\equiv 1 \pmod 6$. See Ex. 46 and 49.

49. $ca \equiv cb \pmod{24}$ implies $a \equiv b \pmod{24}$ if $(c, 24) = 1$.

    Proof: Let $ca = cb + 24B$, i.e., $c(a - b) = 24k$. Consider $a - b = \dfrac{24k}{c}$ in conjunction with $(c, 24) = 1$, noting that $a - b$ must be integral.

Set 2.4

1. No solutions

3. $x \equiv 1 \pmod{12}$

5. No solutions.

7. $w \equiv 20 \pmod{35}$

9. No solutions.

11. $y \equiv 9 \pmod{17}$

13. No solutions.

15. $z \equiv 1 \pmod{285}$

17. Suppose two of the terms, say x^2b^2 and y^2b^2, left the *same* remainder when divided by p. Then, by definition $x^2b^2 = y^2b^2 \pmod{p}$. Since $(p, b) = 1$, we may cancel twice: $x^2 \equiv y^2 \pmod{p}$ or $x^2 - y^2 = (x + y)(x - y) = kp$, for some integral k. Since p is prime, either $p \mid (x + y)$ or $p \mid (x - y)$, by Ex. 16 (a), Set 1.10. It is easy to show that neither is possible as long as x and y stay smaller than $p/2$ and $x \neq y$, as required.

19. No solution. The equivalence classes $(0 + k \cdot 6)$ and $(4 + k \cdot 6)$ have no prime representatives.

21. Quotient: $3x^2 + x + 10 \pmod{11}$. Remainder: $3 \pmod{11}$.

23. Quotient: $20x + 14 \pmod{23}$. Remainder: $3x + 10 \pmod{23}$.

CHAPTER 3

Set 3.5

1. $0, 1, 1, 1$.

3. $0, 1, 1, 1$.

5. $0, 0, 1, 1$.

7. $0, 0, 1, 1$.

9. $1, 0, 1, 1$.

11. Proceed by the method of exhaustion of cases as in Section 3.3.

13. (a) $(GB + GS) + BS \neq (G + B) \cdot (G + S)$.
    (b) $G \neq (G + B) \cdot (G + S)$.
    (c) $G + BS = (G + B) \cdot (G + S)$.
    (d) $GB + GS \neq (G + B)(G + S)$

15. See figure under Theorem 3.5.

17. The premiums should be equal.

Set 3.6

5. The first and last: $A(B + CA)(C + B) = \ldots = AB + AC = A(B + C)$.

7. Theorem 3.5B for both 7a and 7b.

9. All but Postulate 6.

Set 3.7

1. $x = x + x \cdot y = x \cdot x + x \cdot y = x(x + y)$, by Theorems 3.5a, 3.4b, 3.3a.

3.

s	$0 + x$	$1 + x$	$1 + x$	$0 \cdot x$
0	0	0	1	0
1	1	1	1	0

, hence $0 + x = x$, and so forth.

5. See Problem 17.

7. (a) $1 + x = 1$

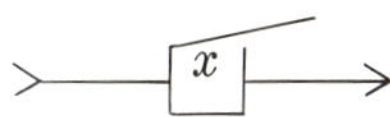

(b) $0 \cdot x = 0$

9. $(x + y)(x' + z) = xz + yx' + yz$, by Theorems 3.3a, 3.11a, and so forth. Multiply both sides times $(y + z)$ and reduce the result on the right to $xz + yx' + yz$. The dual is obtained by interchanging $\cdot$ and $+$.

11. The correspondence $+ \leftrightarrow \cdot$ yields $x \cdot y + x' \cdot z = (x + z) \cdot (z' + y)$. This is identical to Theorem 3.13, except that y and z are interchanged.

13. (a)　$(x) - (y)$

$$(x') - (z) \quad = \quad \begin{matrix} (x) - (y) \\ \\ (x') - (z) \end{matrix}$$

$$(y) - (z)$$

(b) $-(x)-(y)- = -(x)-\begin{bmatrix} (x') \\ \\ (y) \end{bmatrix}-.$

(c) Simplified: $(x + y') \cdot z$.

15. Since $A + B = (A' \cdot B')'$ for *all* A and B, replacing $(\ ) + (\ )$ by $[(\ )' \cdot (\ )']'$ in any formula results in an equivalent formula.

17. Let "x" mean "statement X is valid" and "x'" mean "statement X is not valid." Let 1 represent a statement that is always true (tautology) and 0 a statement which is always false (contradiction). Then use the ideas of isomorphism, which we've discussed in Section 3.4.

19. If Y is a tautology $(y = 1)$, then the statement "either Y or X" is also a tautology. If Z is a contradiction $(z = 0)$, then the statement "both Z and X" is also a contradiction.

21. (a) $(V + S)(P + V')$.

    (b) Theorem 3.13. The right side of Theorem 3.12a is also identical to (a), but does not simplify the problem.

    (c) "Both V and P, or S but not P" with V, S, P appropriately interpreted.

    (d) He may take Stevie by himself; otherwise he must take two or more.

    (e) $(1 + 1)(1 + 0) = 1$.

Set 3.9

1.

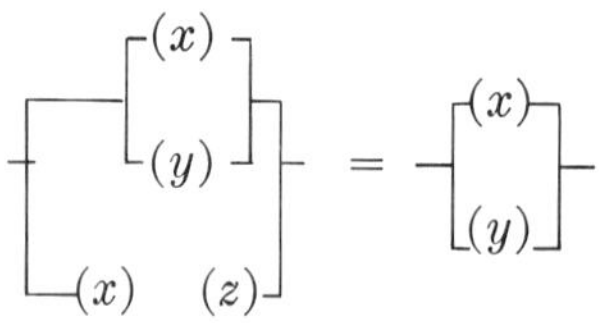

$$\left[\;\left[\begin{array}{c}(x)\\(y)\end{array}\right]\;(x)\;(z)\right] = \left[\begin{array}{c}(x)\\(y)\end{array}\right]$$

7. If $A = 0$, $A \cdot B = A \cdot C$ yields no information but $A + B = A + C$ yields $B = C$. If $A = 1$, $A \cdot B = A \cdot C$ says $B = C$ and $A + B = A + C$ says nothing. These are the only possibilities. (See Prob. 9, Set 3.4.)

9. $(A + B) \cdot (C + A') + B \cdot (B' + C') = AC + B$.

11. $(F + S') \cdot (F' + S) = (F + S') \cdot F' + (F + S') \cdot S = F \cdot F' + S' \cdot F' + F \cdot S + S' \cdot S = S' \cdot F' + F \cdot S = F \cdot S + F' \cdot S'$, by Theorems 3.3a, 3.1, and 3.8b.

13. $x \cdot (y + z) + y \cdot z$, with minimum equipment.

Set 3.10

1. $T = J \cdot M' \cdot (W + C) + J' \cdot M \cdot (W' + C')$ is one simplification. Four multiple pole, two throw switches can be used to build the "black box" to operate on a flash light battery. An additional on-off switch is convenient.

3. The problem (and its solution) is isomorphic to the previous one of Jeraldine. The correspondence is: Man—M, Wolf—W(or C), Cabbages—C(or W), Goat—J.

Set 3.11

1.

Theorem 3-1a
x
y
x + y
x
y
y + x
Theorem 3-1b
x
y
x · y
x
y
y · x
Theorem 3-2a
x
y
z
(x+y)+z
x
y
z
Theorem 3-3a
x
y
z
x·y+x·z
x
y
z
x · (y+z)
Theorem 3-4a
x
x + x
x
x
Theorem 3-5a
x + x · y
x
y
x

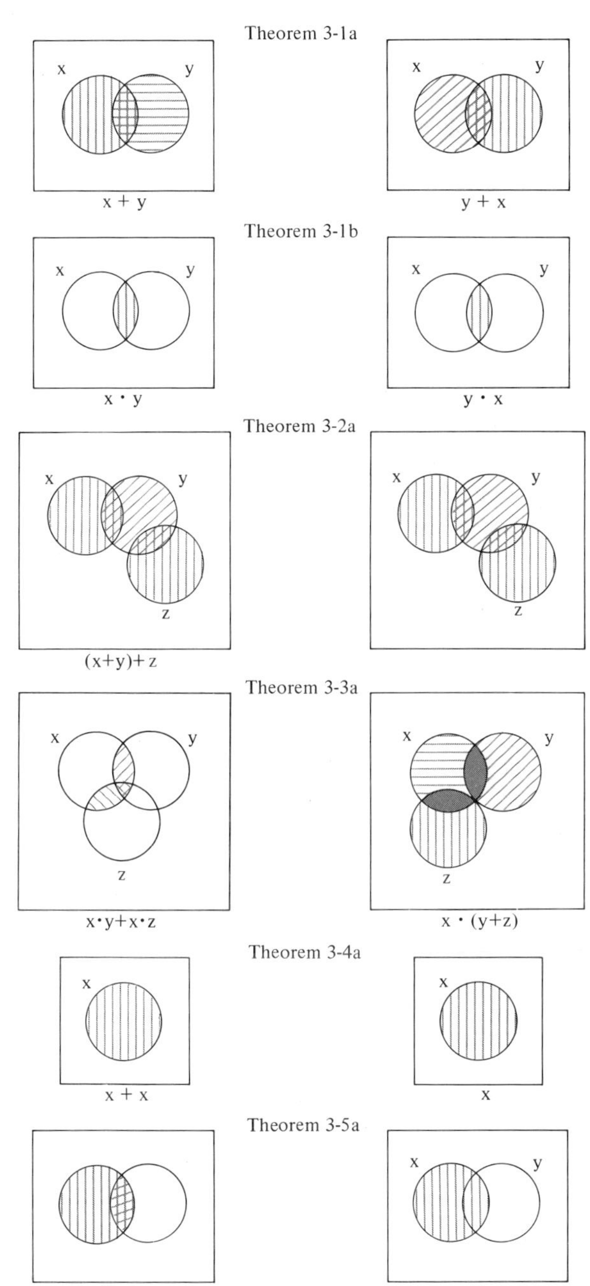

7. If $W \subset X$, then by Problem 6, $W \cdot X = W$. Then $(W \cdot X) \cdot X' = W \cdot (X \cdot X') = 0 = W \cdot X'$. If $W \cdot X' = 0$, then there exists no point of W which is not in X, i.e., all points of W belong to X: $W \subset X$.

9. $(Q \cdot Y)' = Q' + Y', Q' \cdot Y' = (Q + Y)'$

13. The set is closed under each operation: For all X and Y in the set, $X + Y = $ l.c.m. (X, Y), $X \cdot Y = $ g.c.d. (X, Y), and $X' = 30/X$ are in the set. Moreover, Theorems 3.1 through 3.13 are all satisfied by the system with $30 \leftrightarrow 1$ and $1 \leftrightarrow 0$. In the proofs the following algorithms are useful:

l.c.m. $(X, Y) = p_1^{\max(x_1, y_1)} \cdot p_2^{\max(x_2, y_2)} \ldots p_n^{\max(x_n, y_n)}$
and g.c.d. $(X, Y) = p_1^{\min(x_1, y_1)} \cdot p_2^{\min(x_2, y_2)} \ldots p_n^{\min(x_n, y_n)}$,
where $X = p_1^{x_1} \cdot p_2^{x_2} \ldots p_n^{x_n}$
and $Y = p_1^{y_1} \cdot p_2^{y_2} \ldots p_n^{y_n}$ (prime factorization).

15. That the set is closed and commutative under the operations is obvious. Using the algorithm mentioned above, associativity, the distributive laws, and idempotency $(X \cdot X = X + X)$ can be proven. For example, $X \cdot (Y \cdot Z) = (X \cdot Y) \cdot Z$ means l.c.m. $(X, $ l.c.m. $(Y, Z)) = $ l.c.m. (l.c.m. $(X, Y), Z)$, and follows from the fact that the algorithm yields terms involving $\max(x_i, \max(y_i, z_i))$, which is the same as $\max(\max(x_i, y_i), z_i)$. Similarly, $X \cdot (Y + Z) = X \cdot Y + X \cdot Z$ because $\max(x_i, \min(y_i, z_i)) = \min(\max(x_i, y_i), \max(x_i, z_i))$. Hence Theorems 3.1 through 3.4 can be established. Theorem 3.5 follows from the relation $\min(x_i, \max(x_i, y_i)) = x_i$ and its dual, $\max(x_i, \min(x_i, y_i)) = x_i$. If E is admitted as representing 0 (the integer 1 represents 1), then g.c.d. $(x, E) = x$ and l.c.m. $(x, E) = E$, so that the postulates on page 89 and theorems 3.9 and 3.10 now hold. Complementation is not defined.

CHAPTER 4

Set 4.3

1.

	i	a	b	c	d	e
i	i	a	b	c	d	e
a	a	d	c	e	i	b
b	b	e	i	d	c	a
c	c	b	a	i	e	d
d	d	i	e	b	a	c
e	e	c	d	a	b	i

3.

	I	R	H	V
I	I	R	H	V
R	R	I	V	H
H	H	V	I	R
V	V	H	R	I

Where $R_{360} = I$ and $R_{180} = R$.

5. Not a group: Postulate 4 violated.

7. A group: Closure, associativity, and existence of an identity ($u = 1$) follow from the definition of $\equiv$ (mod 13). Each element x has an inverse x^*, since the equation $xx^* \equiv 1$ (mod 13) has a solution x^* for each $x \not\equiv 0$.

9. A group: Associativity is satisfied since this set is a subset of the complex numbers.

	1	−1	i	$-i$
1	1	−1	i	$-i$
−1	−1	1	$-i$	i
i	i	$-i$	−1	1
$-i$	$-i$	i	1	−1

11. Not a group: Postulate 1 violated: $R_{90} \odot H$ is a foreign element.

13. A group:

	t	v	w	x	y	z
t	t	v	w	x	y	z
v	v	t	x	w	z	y
w	w	y	t	z	v	x
x	x	z	v	y	t	w
y	y	w	z	t	x	v
z	z	x	y	v	w	t,

where $v = \dfrac{1}{t}$, $w = 1 - t$, $x = \dfrac{1}{w}$, $y = \dfrac{t-1}{t}$, $z = \dfrac{1}{y}$.   (Associativity also satisfied.)

15. (a) See Problem 7.
    (b) Not a group; 0 (mod 13) is needed.

17. True: Integral Domain Postulates 1, 3, 4, and 6 are the Group Postulates.

19. Closure: $(x_1 + x_2 e^{-y_1}, y_1 + y_2)$ is a point with real coordinates if x_1, x_2, y_1, y_2 are real.
    Associativity: $(x_1, y_1) \odot ((x_2, y_2) \odot (x_3, y_3)) = (x_1 + x_2 e^{-y_1} + x_3 e^{-y_1-y_2}, y_1 + y_2 + y_3 = ((x_1, y_1) \odot (x_2, y_2)) \odot (x_3, y_3)$.
    Identity: $u = (0, 0)$ is a point.
    Inverse: $(x, y)^* = (-xe^y, -y)$ is a point if (x, y) is a point.

21. (a) c has no inverse. Even if c had an inverse, associativity would be violated.

    (b) b, c, and e have no inverses. With inverses, associativity might be violated without due consideration.

    (c) No inverse for b.

    (d) No inverses for y and 0, etc.

    (e) Associativity violated; e.g., $a_2(a_4a_3) = a_2a_5 = a_3$, but $(a_2a_4)a_3 = a_5a_3 = a_2$.

    (f) No identity; associativity violated. (For example, $b_1(b_3b_4) \neq (b_1b_3)b_4$.)

    (g) Associativity violated: $d_7(d_5d_4) = (d_7d_5)d_4$, but, $d_6(d_4d_7) \neq (d_6d_4)d_7$.

23. Define $(P, Q, \ldots)$ *equ* $(P', Q', \ldots)$ to mean $P' = P + (a, b)$, $Q' = Q + (a, b), \ldots$, where $(P, \ldots)$ and $(P', \ldots)$ are plane figures with points $P, Q, \ldots$ and $P', Q', \ldots$, respectively, and addition is by coordinates: If $P = (x, y)$, then $P + (a, b) = (x + a, y + b)$. Then $(P, \ldots)$ *equ* $(P, \ldots)$, with $a = b = 0$; $(P, \ldots)$ *equ* $(P', \ldots)$, with $X' = X + (a, b)$, implies $(P', \ldots)$ *equ* $(P, \ldots)$, with $X = X' + (-a, -b)$; and $(P, \ldots)$ *equ* $(P', \ldots)$, with $X' = X + (a, b)$, and $(P', \ldots)$ *equ* $(P'', \ldots)$, with $X'' = X' + (c, d)$, implies $(P, \ldots)$ *equ* $(P'', \ldots)$, with $X'' = X + (a + c, b + d)$. Translational equivalence divides the set of all plane figures into classes of congruent figures with corresponding sides parallel.

25. "For each$\ldots$, there exists$\ldots$" $\neq$ "There exists$\ldots$such that for each$\ldots$" (Consider also how these assertions could be negated.)

Set 4.4

1. The systems in Problems 7, 8, 9, 12, 13, 14b, 15a of Set 4.2 are finite groups.

3. (a) 17 is a prime, so every equation $AX \equiv B \pmod{17}$ has a solution X (by Theorem 2.4). See Problem 7, Set 4.2.

    (b)

	1	4	13	16
1	1	4	13	16
4	4	16	1	13
13	13	1	16	4
16	16	13	4	1

(c) $(1, 2, 4, 8, 9, 13, 15, 16)$.

(d) $S = (1, 16)$ is a subgroup of the subgroup in (b) and of the original group.

5. No: Closure is violated. For example, $\dfrac{1-z}{1+z} \odot \dfrac{z+1}{z-1} = -\dfrac{1}{z}$. With $-z$ and $-\dfrac{1}{z}$ added, the system is a group.

7. The integers (mod 14) form a group under addition but not under multiplication for some elements have no multiplicative inverses.

11. $b^m b^n = b^{m+n \ (mod \ 6)}$, $b^6 = u$, $(b^k)^* = b^{6-k}$.

13. The first two postulates in each postulate set are identical. To prove that existence of solutions of $a \cdot x = b$ and $y \cdot a = b$ is logically equivalent to existence of an identity, set $b = a$. To prove the existence of inverses, let $b = u$, where u is the identity element, whose existence has just been established.

15. For lack of inverses, Boolean algebras do not form groups under either operation.

17. The set of all points (x, y) is closed and associative under the operation defined (since so is the set of real numbers), $(0, 0)$ is the identity element, and $(-y, -x)$ is the inverse of (x, y).

19. See proof of Theorem 4.5.

21. $(R_{360}, R_{270}, R_{180}, R_{90})$ is contained in the set of the octic group. It is itself a group (a cyclic group of order four). Hence it is a sub-group.

Set 4.5

1. One-one correspondence is a reflexive, symmetric, and transitive relation. Preservation of group products may be traced through the various correspondences to show that isomorphism is also reflexive, symmetric, and transitive.

3. Let $a_i \leftrightarrow b_i$. Let $a_i^{-1} \leftrightarrow b'_i$. Then $a_i \textcircled{a} a_i^{-1} = u_a \leftrightarrow b_i \textcircled{b} b'_i$, since G_a and G_b are assumed isomorphic. But by Problem 2, $u_a \leftrightarrow u_b$. Hence $b_i \textcircled{b} b'_i = u_b$, so that $b'_i = b_i^{-1}$.

5. It is the subscripts which distinguish the a_j. a_i corresponds to $\begin{pmatrix} 1 & 2 & \ldots & n \\ i_1 & i_2 & \ldots & i_n \end{pmatrix}$ which corresponds to $\begin{pmatrix} a_1 & a_2 & \ldots & a_n \\ a_{i_1} & a_{i_2} & \ldots & a_{i_n} \end{pmatrix}$. In the multiplication table, $a_1, a_2, \ldots, a_n$ is just the first column and $a_{i_1}, a_{i_2}, \ldots, a_{i_n}$ is the ith, since a_{i_k} was defined as $(a_k \odot a_i)$.

7. Yes. They are both isomorphic to the abstract group given in Problem 4.

9. $(R_{360}, R_{270}, R_{180}, R_{90})$ is a cyclic four group whereas the other two (isomorphic) subgroups are non-cyclic.

11. (a) No.
    (b) Yes (Problem 7).
    (c) No (Part a).
    (d) $\left(\begin{pmatrix} 123 \\ 123 \end{pmatrix}, \begin{pmatrix} 123 \\ 231 \end{pmatrix}, \begin{pmatrix} 123 \\ 312 \end{pmatrix} \right)$ (Problem 4).

13. (a) One:

	u	a
u	u	a
a	a	u

    (b) One: The group of Problem 4.
    (c) Two: The cyclic and the non-cyclic four groups.
    (d) One: The cyclic group of order five.

17. The answer is in the library.

Set 4.6

1. $(R_{360}, R_{270}, R_{180}, R_{90})$ and (V, H, D_1, D_2).

3. $(1, b^4, b^8), (b, b^5, b^9), (b^2, b^6, b^{10}), (b^3, b^7, b^{11})$.

5. These cosets are just the equivalence classes (mod m). There are m of them: $(0 + km), (1 + km), \ldots, ((m-1) + km)$. $(i + k_1 m) + (j + k_2 m) = ((i+j) + (k_1 + k_2)m) = (j + k_2 m) + (i + k_1 m)$ yields closure and commutativity (since the integers are closed and commutative under addition). Using this notation, the other group postulates can be proven in a similar manner.

7. If $(x, m) = 1$, then $(x + km, m) = 1$, so the problem is meaningful. If $(s + k_1m, m) = 1$ and $(y + k_2m, m) = 1$, then $(xy + k_3m, m) = 1$. $(1 + km, m) = 1$. If $(x + k_1m, m) = 1$, then $((m - x) + k_2m, m) = 1$.

9. Let $a \equiv b \pmod{S}$ mean $a = b \cdot s$ for some s in the subgroup S. Then $a \equiv a$, since $a = a \odot u$ and u is in S. $a \equiv b$ implies $b \equiv a$, since if $a = b \odot s$, then $b = a \odot s^{-1}$ and s^{-1} is in S when s is in S. Transitivity follows from closure in S.

Set 4.8

1. All subgroups of the octic group G are invariant: $S_1 = G$, $S_2 = (I, D_1, D_2, R_1)$, $S_3 = (I, R_1, R_2, R_3)$, $S_4 = (I, H, V, R_1)$, $S_5 = (I, D_1)$, $S_6 = (I, D_2)$, $S_7 = (I, R_1)$, $S_8 = (I, H)$, $S_9 = (I, V)$, $S_{10} = (I)$, where $I = R_{360}$, $R_1 = R_{180}$, $R_2 = R_{270}$, $R_3 = R_{90}$. Then $G/S_1 = (S_1)$, $G/S_2 = (S_2, (H, V, R_2, R_3))$, $G/S_3 = (S_3, (H, V, D_1, D_2))$, $\ldots$, $G/S_9 = (S_9, (R_1, H), (R_2, D_2), (R_3, D_1))$, $G/S_{10} = ((I), (R_1), (R_2), (R_3), (V), (H), (D_1), (D_2))$.

CHAPTER 5

Set 5.3

1. $A \cdot B = \begin{bmatrix} 47 & 274 \\ 29 & -82 \end{bmatrix}$, $B \cdot A = \begin{bmatrix} 2 & 143 \\ 82 & -37 \end{bmatrix}$.

3. $A \cdot B = \begin{bmatrix} 9 & 21 \\ 41 & 99 \end{bmatrix}$, $B \cdot A = \begin{bmatrix} 28 & 26 \\ 85 & 80 \end{bmatrix}$.

5. $A \cdot B = \begin{bmatrix} 7 & 59 \\ 26 & 169 \end{bmatrix}$, $B \cdot A = \begin{bmatrix} 11 & 57 \\ 38 & 165 \end{bmatrix}$.

7. $A \cdot B = \begin{bmatrix} 5 & 390 \\ 23 & 156 \end{bmatrix}$, $B \cdot A = \begin{bmatrix} 171 & 120 \\ 54 & -10 \end{bmatrix}$.

9. $A \cdot B = \begin{bmatrix} 2 & 9 \\ 2 & 9 \end{bmatrix}$, $B \cdot A = \begin{bmatrix} 11 & 0 \\ -4 & 0 \end{bmatrix}$.

11. (a) $\begin{bmatrix} 5 & 1 \\ 1 & 2 \end{bmatrix}, \begin{bmatrix} 2 & 3 \\ 5 & 5 \end{bmatrix}$.   (b) $\begin{bmatrix} 3 & 2 \\ 1 & 2 \end{bmatrix}, \begin{bmatrix} 2 & 3 \\ 2 & 3 \end{bmatrix}$.

13. (a) $\begin{bmatrix} 2 & 0 \\ 6 & 1 \end{bmatrix}, \begin{bmatrix} 0 & 5 \\ 1 & 3 \end{bmatrix}$.   (b) $\begin{bmatrix} 1 & 1 \\ 1 & 3 \end{bmatrix}, \begin{bmatrix} 0 & 2 \\ 1 & 0 \end{bmatrix}$.

15. (a) $\begin{bmatrix} 0 & 3 \\ 5 & 1 \end{bmatrix}, \begin{bmatrix} 4 & 1 \\ 3 & 4 \end{bmatrix}$.   (b) $\begin{bmatrix} 3 & 3 \\ 2 & 1 \end{bmatrix}, \begin{bmatrix} 3 & 1 \\ 2 & 1 \end{bmatrix}$.

17. (a) $\begin{bmatrix} 5 & 5 \\ 2 & 2 \end{bmatrix}, \begin{bmatrix} 3 & 1 \\ 5 & 4 \end{bmatrix}$.   (b) $\begin{bmatrix} 1 & 2 \\ 3 & 0 \end{bmatrix}, \begin{bmatrix} 3 & 0 \\ 2 & 2 \end{bmatrix}$.

19. (a) $\begin{bmatrix} 2 & 2 \\ 2 & 2 \end{bmatrix}, \begin{bmatrix} 3 & 0 \\ 3 & 0 \end{bmatrix}$.   (b) $\begin{bmatrix} 2 & 1 \\ 2 & 1 \end{bmatrix}, \begin{bmatrix} 3 & 0 \\ 0 & 0 \end{bmatrix}$.

21.

	I	A	B	C	D	E	F	G
I	I	A	B	C	D	E	F	G
A	A	D	C	F	E	I	G	B
B	B	G	D	A	F	C	I	E
C	C	B	E	D	G	F	A	I
D	D	E	F	G	I	A	B	C
E	E	I	G	B	A	D	C	F
F	F	C	I	E	B	G	D	A
G	G	F	A	I	C	B	E	D

23. The subgroup (I, B, D, F) is isomorphic to the cyclic four group.

25. $\begin{bmatrix} 4x + y & x \\ 3x & y \end{bmatrix}.$

27. Yes.

Set 5.4

1. $I_n \cdot A = \begin{bmatrix} 1 & 0 & \ldots & 0 \\ 0 & 1 & \ldots & 0 \\ \vdots & \vdots & \vdots & \vdots \\ 0 & 0 & \ldots & 1 \end{bmatrix} \cdot \begin{bmatrix} a_{11} & a_{12} & \ldots & a_{1n} \\ a_{21} & a_{22} & \ldots & a_{2n} \\ \vdots & \vdots & \vdots & \vdots \\ a_{n1} & a_{n2} & \ldots & a_{nn} \end{bmatrix} =$

$$\begin{bmatrix} 1 \cdot a_{11} + 0 \cdot a_{21} + \ldots + 0 \cdot a_{n1} \ldots 1 \cdot a_{1n} + 0 \cdot a_{2n} \ldots + 0 \cdot a_{nn} \\ 0 \cdot a_{11} + 1 \cdot a_{21} + \ldots + 0 \cdot a_{n1} \ldots 0 \cdot a_{1n} + 1 \cdot a_{2n} \ldots + 0 \cdot a_{nn} \\ \vdots \qquad\qquad\qquad\qquad\qquad \vdots \\ 0 \cdot a_{11} + 0 \cdot a_{21} + \ldots + 1 \cdot a_{n1} \ldots 0 \cdot a_{1n} + 0 \cdot a_{2n} \ldots + 1 \cdot a_{nn} \end{bmatrix}$$

The proof that $A \cdot I_n = A$ is similar.

3. (a) $A \cdot G = \begin{bmatrix} 4 & 0 & 2 \\ 1 & 1 & 0 \\ 2 & 6 & 1 \end{bmatrix}$ (b) $A \cdot X = \begin{bmatrix} d & e & f \\ a & b & c \\ g & h & i \end{bmatrix}$

5. (a) $B \cdot G = \begin{bmatrix} 1 & 1 & 0 \\ 4 & 0 & 2 \\ 3 & 7 & 1 \end{bmatrix}$ (b) $B \cdot X = \begin{bmatrix} a & b & c \\ d & e & f \\ a+g & b+h & c+i \end{bmatrix}$

7. (a) $C \cdot G = \begin{bmatrix} 4 & 4 & 0 \\ -4 & 0 & -2 \\ 2 & 6 & 1 \end{bmatrix}$ (b) $C \cdot X = \begin{bmatrix} 4a & 4b & 4c \\ -d & -e & -f \\ g & h & i \end{bmatrix}$

9. (a) $D \cdot G = \begin{bmatrix} 2 & 6 & 1 \\ 1 & 1 & 0 \\ 0 & 0 & 0 \end{bmatrix}$ (b) $D \cdot X = \begin{bmatrix} g & h & i \\ a & b & c \\ 0 & 0 & 0 \end{bmatrix}$

11. (a) $G \cdot A = \begin{bmatrix} 1 & 1 & 0 \\ 0 & 4 & 2 \\ 6 & 2 & 1 \end{bmatrix}$ (b) $X \cdot A = \begin{bmatrix} b & a & c \\ e & d & f \\ h & g & i \end{bmatrix}$

13. (a) $G \cdot B = \begin{bmatrix} 1 & 1 & 0 \\ 6 & 0 & 2 \\ 3 & 6 & 1 \end{bmatrix}$ (b) $X \cdot B = \begin{bmatrix} a+c & b & c \\ d+f & e & f \\ g+i & h & i \end{bmatrix}$

15. (a) $G \cdot C = \begin{bmatrix} 4 & -1 & 0 \\ 16 & 0 & 2 \\ 8 & -6 & 1 \end{bmatrix}$ (b) $X \cdot C = \begin{bmatrix} 4a & -b & e \\ 4d & -e & f \\ 4g & -h & i \end{bmatrix}$

17. (a) $G \cdot D = \begin{bmatrix} 1 & 0 & 1 \\ 0 & 0 & 4 \\ 6 & 0 & 2 \end{bmatrix}$ (b) $X \cdot D = \begin{bmatrix} b & 0 & a \\ e & 0 & d \\ h & 0 & g \end{bmatrix}$

19. (a) $G \cdot E = \begin{bmatrix} 3 & 9 & -3 \\ 10 & 24 & 8 \\ 17 & 36 & -8 \end{bmatrix}$ (b) $E \cdot G = \begin{bmatrix} 19 & -5 & 9 \\ 14 & -10 & 6 \\ 23 & 39 & 10 \end{bmatrix}$

21. Problem 3b demonstrates that left multiplication of *any* matrix X by A yields a matrix which is X with 1st and 2nd rows interchanged.

23. See Problem 7b.

25. See Problem 11b.

27. See Problem 15b.

29. Multiplication by A interchanges 1st and 2nd rows (columns). Multiplication by B adds the 1st row (column) to the 3rd, etc.

31. Multiply out (as in Problem 1).

33. Multiply out (as in Problem 1).

35. $\begin{bmatrix} A \cdot E & A \cdot F \\ B \cdot G & B \cdot C \end{bmatrix} = \begin{bmatrix} 2 & 4 & -2 & 3 & 6 & 0 \\ 1 & 5 & -1 & 2 & 1 & 4 \\ 3 & 2 & 6 & 7 & 2 & 1 \\ 1 & 1 & 0 & 4 & 0 & 0 \\ 4 & 0 & 2 & 0 & -1 & 0 \\ 3 & 7 & 1 & 4 & 0 & 1 \end{bmatrix}$

37. $\begin{bmatrix} A \cdot E & A \cdot G \\ G & C \end{bmatrix} = \begin{bmatrix} 2 & 4 & -2 & 4 & 0 & 2 \\ 1 & 5 & -1 & 1 & 1 & 0 \\ 3 & 2 & 6 & 2 & 6 & 1 \\ 1 & 1 & 0 & 4 & 0 & 0 \\ 4 & 0 & 2 & 0 & -1 & 0 \\ 2 & 6 & 1 & 0 & 0 & 1 \end{bmatrix}$

39. The matrix has no multiplicative inverse if the identity element is I. However it is a one-element group by itself with the operation multiplication.

Set 5.5

3. Let $M = (a_{rc})$ and $K = (c_{rc})$, where $c_{rr} = k$ and $c_{rc} = 0$ $(r \neq c)$. Then $(c_{rc})(a_{rc}) = (\sum_i c_{ri}a_{ic}) = (c_{rr}a_{rc}) = (ka_{rc})$. Similarly, $(a_{rc})(c_{rc}) = (ka_{rc})$.

Set 5.6

1. Five (plus the identity itself). There is one for each permutation.

3. (a) An infinite number of them.
   (b) Fifteen (plus the identity itself).
   (c) Nine (plus the identity itself).

5. (a) An infinite number of them.
   (b) Thirty-six (plus the identity itself).
   (c) Sixty-six (plus the identity itself).

7. $L_1 = E_1, L_2 = E_2, L_3 = E_3,$ $(4 \times 4$ matrices on pages 178 and 179$)$.

9. $$N = \begin{bmatrix} 1 & 0 & 0 \\ 1 & 1 & 0 \\ -1 & 0 & 1 \end{bmatrix}$$

11. Let $Q = E_{k_m}E_{k_{m-1}}\ldots E_{k_2}E_{k_1}$ such that $E_{k_m}\ldots E_{k_2}E_{k_1}M = I$. Then, if M^{-1} exists, $QMM^{-1} = IM^{-1}$ or $Q = QI = M^{-1}$. Now form (M, I) with the rule $A(M, I) = (AM, AI)$. Then $Q(M, I) = (QM, QI) = (I, Q)$, as desired.

13. $\frac{1}{38}\begin{bmatrix} -20 & -1 & 61 \\ 2 & 2 & -8 \\ 18 & -1 & -15 \end{bmatrix}$

15. $\frac{1}{605}\begin{bmatrix} -331 & 76 & 136 & 37 \\ 76 & -1 & -161 & 103 \\ 140 & 30 & -10 & -65 \\ 91 & -41 & 54 & -12 \end{bmatrix}$

17. $\begin{bmatrix} 1 & 0 & 0 & 0 & 0 \\ 0 & \frac{1}{2} & 0 & 0 & 0 \\ 0 & 0 & \frac{1}{3} & 0 & 0 \\ 0 & 0 & 0 & \frac{1}{4} & 0 \\ 0 & 0 & 0 & 0 & \frac{1}{5} \end{bmatrix}$

21. Yes. $A \cong B$ if and only if there exist P and Q as products of elementary matrices such that $B = PAQ$ (Theorems 5.3 and 5.4). Since $A = IAI$, $A \cong A$, for all A. If $A \cong B$, i.e., $B = PAQ$, then since P^{-1} and Q^{-1} exist as products of elementary matrices (Theorem 5.2 and proof), $A = P^{-1}BQ^{-1}$ or $B \cong A$. If $A \cong B$ and $B \cong C$, i.e., $B = PAQ$ and $C = RBS$, then $C = RPAQS$ and RP, QS are products of elementary matrices, so that $A \cong C$. Hence $\cong$ is an R.S.T. relation, Q.E.D.

23. $\begin{bmatrix} 1 & 0 \\ 0 & 0 \end{bmatrix} \cdot \begin{bmatrix} 0 & 0 \\ 1 & 0 \end{bmatrix} = (0)$, but $\begin{bmatrix} 0 & 0 \\ 1 & 0 \end{bmatrix} \cdot \begin{bmatrix} 1 & 0 \\ 0 & 0 \end{bmatrix} = \begin{bmatrix} 0 & 0 \\ 1 & 0 \end{bmatrix}$.

25. There is only one converse, viz., "If T is row-equivalent to the identity matrix, then T is a triangular matrix as described." This is false, since for example $\begin{bmatrix} 1 & 2 \\ 3 & 4 \end{bmatrix} \overset{\text{row}}{=} \begin{bmatrix} 1 & 0 \\ 0 & 1 \end{bmatrix}$.

Set 5.8

3.
$$F^2 = \begin{bmatrix} 0 & 0 & 0 & -1 \\ 0 & 0 & 1 & -3 \\ 0 & -1 & 2 & -3 \\ 1 & -1 & 1 & -1 \end{bmatrix} = F^{-1}.$$

7. $M^s \cdot M^t = \underset{s \text{ terms}}{(M \cdot M \cdot \ldots \cdot M)} \cdot \underset{t \text{ terms}}{(M \cdot M \cdot \ldots \cdot M)} =$

$\underset{t \text{ terms}}{(M \cdot M \cdot \ldots \cdot M)} \cdot \underset{s \text{ terms}}{(M \cdot M \cdot \ldots \cdot M)}$ (by repeated applications of the associativity law) $= M^t \cdot M^s$. A vigorous proof can be developed using mathematical induction on s.

9.
$$T = \begin{bmatrix} 1 & 0 & 0 \\ 1 & 1 & 0 \\ 0 & 0 & 2 \end{bmatrix} \qquad MT = \begin{bmatrix} 3 & 2 & 6 \\ 9 & 5 & 12 \\ 6 & 3 & 6 \end{bmatrix}$$

11. $T = 1/5 \begin{bmatrix} 2 & -3 \\ -1 & 4 \end{bmatrix} \qquad MT = \begin{bmatrix} 1 & 0 \\ 0 & 1 \end{bmatrix}$

13. $T = \begin{bmatrix} 0 & 0 & 0 \\ 1 & 0 & 0 \\ 1 & 0 & 1 \end{bmatrix} \qquad MT = \begin{bmatrix} 1 & 0 & 0 \\ 4 & 0 & 2 \\ 1 & 0 & 1 \end{bmatrix}$

15. Let $\begin{bmatrix} 1 & 2 \\ 0 & 1 \end{bmatrix} = M$. Any matrix of the form $a_0 m^n + a_1 m^{n-1} + \ldots + a_{n-1} m + a_n I$, where the a_i's are integral, satisfies the conditions for x.

Set 5.9

1. $\begin{bmatrix} 0 & 0 & a \\ 0 & 0 & b \\ 0 & 0 & c \end{bmatrix} \cdot \begin{bmatrix} d & e & f \\ g & h & i \\ 0 & 0 & 0 \end{bmatrix}$

3. Infinitely many solutions

$\begin{bmatrix} 0 & ka \\ k/a & 0 \end{bmatrix}^2 = \begin{bmatrix} k & 0 \\ 0 & k \end{bmatrix}$ for real $a \neq 0$

4. $\begin{bmatrix} -k & m \\ k/m & k \end{bmatrix}$ for $m \neq 0$

7. The cancellation property of integral domains is not satisfied.

13. yes $\{I, B, D, F\}$

Set 5.10

3. $[-10, -32, -72]$

5. $\begin{bmatrix} 7 & 11 & 45 \\ 34 & 67 & 225 \\ 15 & 29 & 99 \\ 19 & 19 & 117 \end{bmatrix}$

7. Not meaningful.

9. Not meaningful.

13. Not meaningful.

15. $\begin{bmatrix} -16 & -8 \\ 8 & 4 \end{bmatrix}$

17. Yes

19. The product matrix indicates that David paid altogether \$10.72 for his purchases.

21. $\begin{bmatrix} 7 & 1 \\ 3 & 17 \end{bmatrix}$

25. $\begin{bmatrix} \sqrt{3}/2 & 1/2 \\ -1/2 & \sqrt{3}/2 \end{bmatrix}$

27. $\begin{bmatrix} a & b \\ c & d \end{bmatrix}^{-1} = \dfrac{1}{ad - bc} \begin{bmatrix} d & -b \\ -c & a \end{bmatrix},$

provided that $ad - bc \neq 0$.

29. $A, B, C, E, F, G.$

31. $[\ast \ \ast \ \ast \ \ldots \ \mathcal{E}] \begin{bmatrix} \ast \\ \ast \\ \ast \\ \vdots \\ \ast \end{bmatrix} = [\ast]$, but $\begin{bmatrix} \ast \\ \ast \\ \ast \\ \vdots \\ \ast \end{bmatrix} [\ast \ \ast \ \ast \ \ldots \ \ast] = \begin{bmatrix} \ast & \ast & \ast & \ldots & \ast \\ \ast & \ast & \ast & \ldots & \ast \\ \ast & \ast & \ast & \ldots & \ast \\ \vdots & \vdots & \vdots & \vdots & \vdots \\ \ast & \ast & \ast & \ldots & \ast \end{bmatrix}$

33. See Problem 15.

35. R must be of the form

$$\begin{bmatrix} a & b \\ c & d \\ e & f \\ g & h \end{bmatrix}, \text{ where } a, \ldots h, \text{ satisfy} \qquad \begin{aligned} 3a + 4c + 7e + 2g &= 1 \\ 3b + 4d + 7f + 2h &= 0 \\ a + 9c + 3e &= 0 \\ b + 9d + 3f &= 1. \end{aligned}$$

These can be solved for four of the variables in terms of the other four (which are arbitrary). With $c, d, e, f,$ chosen arbitrarily as zero, we have

$$R = \begin{bmatrix} 0 & 1 \\ 0 & 0 \\ 0 & 0 \\ 1/2 & -3/2 \end{bmatrix}, \text{ and } R \cdot A = \begin{bmatrix} 1 & 9 & 3 & 0 \\ 0 & 0 & 0 & 0 \\ 0 & 0 & 0 & 0 \\ 0 & -23/2 & -1 & 1 \end{bmatrix}.$$

37. $L = \begin{bmatrix} 5/13 & -1/13 & 0 \\ -2/13 & 3/13 & 0 \end{bmatrix}$, for example (See Problem 35).

Set 5.12

1. $A' = (6 + 5\sqrt{3}, 15 - 2\sqrt{3})$, $B' = (1 + 3\sqrt{3}, -3 + \sqrt{3})$,
   $C' = (11/2, 7/2\sqrt{3})$, $D' = (-15/2 + \sqrt{3}, -1 - 15/2\sqrt{3})$,
   $E' = (73/2, 57/2\sqrt{3})$, $F' = (6 - 2\sqrt{3}, 2 + 6\sqrt{3})$.

3. For $a = 3$, $A' = (1, 4)$, $B' = (3, 13)$, $C' = (2, -1)$.
   For $a = -7$, $A' = (1, -6)$, $B' = (3, -17)$, $C' = (2, -21)$.
   For $a = 1/2$, $A' = (1, 3/2)$, $B' = (3, 11/2)$, $C' = (2, -6)$.

5. New Loci, (a) Circle: $x^2 + y^2 = 49/4$.
              (b) Circle: $x^2 + (y - 5/2)^2 = 25/4$
              (c) Circle: $(x - 3/2)^2 + y^2 = 9/4$
              (d) Line: $y = 3/2$
              (e) Line: $x = 1$.

7. New loci, (a) Rhomboid: $(0, 0)$, $(1, 0)$, $(4, 1)$, $(3, 1)$.
             (b) Rhomboid: $(0, 0)$, $(5, 0)$, $(8, 1)$, $(3, 1)$.
             (c) Same as old locus: $y = 3$.
             (d) Straight line: $x = 3y + 7$.
             (e) Ellipse: $x^2 - 6xy + 10y^2 = 1$.

11. (a) $P(x, y)$ is renamed: $P'(\sqrt{2}/2(x + y + 2), \sqrt{2}/2(-x + y + 8))$.
    (b) $P(x, y)$ is renamed:
         $P'(1/2x + \sqrt{3}/2y - 3/2 + 5/2\sqrt{3}, -\sqrt{3}/2x + 1/2y + 5/2 + 3/2\sqrt{3})$.
    The order in which the transformations are performed is essential.

19. The matrix of a stretching transformation is a scalar and so commutes with all other matrices (such as those which represent rotations or shears).

CHAPTER 6

Set 6.1

1. $(x, y, z) = (1/2, -3, 5)$.

3. $(A, B, Z) = (0, 0, 0)$.

5. $(x, y, q) = (7/3, 14/3, 0) + (-4/3, 1/3, 1)q$.

7. $(J, P, A) = (-11, 12, 0) + (4, -4, 1)A$.

9. No solution.

11. Yes,
$(J, P, A, E, T) = 1/36(32, 4, 107, 16, 0) + 1/36(12, -12, 3, -12, 36)T$.

13. $(P, N, D) = (108, 36, 0) + (5/4, -9/4, 1)D$, with $D = 12$, i.e.,
$(P, N, D) = (123, 9, 12)$.

17. They are equivalent systems.

19. The common solution to all three systems, M, $E_a \cdot M$, $E_b \cdot M$, is
$(x, y, z, w) = (-1, 1, -2, -5)$. See pp. 228–230 of text.

21. Each solution to the system is a solution also to the "enlarged" system,
and conversely, since *any* solution satisfies an identity.

CHAPTER 7

Set 7.1

1. $A = \begin{bmatrix} 1 & 0 & 0 \\ 0 & 1 & 0 \\ 0 & 4 & 1 \end{bmatrix} \cdot \begin{bmatrix} 1 & 0 & 0 \\ 0 & -22 & 0 \\ 0 & 0 & 1 \end{bmatrix} \cdot \begin{bmatrix} 1 & 0 & 0 \\ 0 & 1 & 6 \\ 0 & 0 & 1 \end{bmatrix} \cdot \begin{bmatrix} 1 & 0 & 0 \\ 0 & 0 & 1 \\ 0 & 1 & 0 \end{bmatrix}$

$\cdot \begin{bmatrix} 1 & 0 & 0 \\ 2 & 1 & 0 \\ 0 & 0 & 1 \end{bmatrix} \cdot \begin{bmatrix} 1 & 0 & 0 \\ 0 & 1 & 0 \\ 9 & 0 & 1 \end{bmatrix},$

for example

3. $C = \begin{bmatrix} 1 & 2 & 0 & 0 \\ 0 & 1 & 0 & 0 \\ 0 & 0 & 1 & 0 \\ 0 & 0 & 0 & 1 \end{bmatrix} \cdot \begin{bmatrix} 1 & 0 & -5 & 0 \\ 0 & 1 & 0 & 0 \\ 0 & 0 & 1 & 0 \\ 0 & 0 & 0 & 1 \end{bmatrix} \cdot \begin{bmatrix} 1 & 0 & 0 & 0 \\ 0 & 1 & 4 & 0 \\ 0 & 0 & 1 & 0 \\ 0 & 0 & 0 & 1 \end{bmatrix} \cdot \begin{bmatrix} 1 & 0 & 0 & 0 \\ 0 & 1 & 0 & 0 \\ 0 & 0 & 1 & 7 \\ 0 & 0 & 0 & 1 \end{bmatrix}$

$\cdot \begin{bmatrix} 1 & 0 & 0 & 22 \\ 0 & 1 & 0 & 0 \\ 0 & 0 & 1 & 0 \\ 0 & 0 & 0 & 1 \end{bmatrix} \cdot \begin{bmatrix} 1 & 0 & 0 & 0 \\ 0 & 1 & 0 & -19 \\ 0 & 0 & 1 & 0 \\ 0 & 0 & 0 & 1 \end{bmatrix}$ is one way.

5. $n!$ terms, one for each permutation of $(1, 2, \ldots, n)$.

7. If B is expressed as the product $\begin{bmatrix} 1 & 0 & 0 \\ 1 & 1 & 0 \\ 0 & 0 & 1 \end{bmatrix} \cdot \begin{bmatrix} 1 & 0 & 0 \\ 0 & 1 & 1 \\ 0 & 0 & 1 \end{bmatrix} \cdot \begin{bmatrix} 1 & 1 & 0 \\ 0 & 1 & 0 \\ 0 & 0 & 1 \end{bmatrix}$

$\cdot \begin{bmatrix} 1 & 0 & 1/2 \\ 0 & 1 & 0 \\ 0 & 0 & 1 \end{bmatrix} \cdot \begin{bmatrix} 1 & 0 & 0 \\ 0 & 1 & 0 \\ 0 & 2 & 1 \end{bmatrix} \cdot \begin{bmatrix} 1 & 0 & 0 \\ 0 & 1 & 0 \\ 0 & 0 & 6 \end{bmatrix}$, then the product of the

determinants is $1 \cdot 1 \cdot 1 \cdot 1 \cdot 1 \cdot 6 = 6$ which is the determinant of B.

9. With $D = \begin{bmatrix} 1 & 0 \\ 0 & 1/19 \end{bmatrix} \cdot \begin{bmatrix} 1 & 0 \\ -3 & 1 \end{bmatrix} \cdot \begin{bmatrix} 1 & 0 \\ 0 & 197 \end{bmatrix} \cdot \begin{bmatrix} 1 & -4 \\ 0 & 1 \end{bmatrix} \cdot \begin{bmatrix} -19 & 0 \\ 0 & 1 \end{bmatrix}$

$= 1/19 \cdot 1 \cdot 197 \cdot 1 \cdot (-19) = -197$

11. By Theorem 7.3, $|E \cdot A| = |E| \cdot |A|$. Let A be a matrix with two
identical rows and let E interchange these two rows. Then $|E \cdot A| =
|A|$. But by Theorem 7.2, $|E| = -1$, so that $-1 \cdot |A| = |A|$ or
$|A| = 0$. Now let one row of A be a constant c times another and let
E multiply that row by $1/c$. Then $|A'| = |E \cdot A|$, where A' has two
identical rows. By the above, $|A'| = 0$, so that $|E \cdot A| = |E| \cdot |A| =
1/c \cdot |A| = 0$, or $|A| = 0$.

15. By Theorem 7.6, $1 = |I| = |A \cdot A^{-1}| = |A| \cdot |A^{-1}|$.

17. $132 = 6 \cdot 22$.

19. $0 = 0 \cdot 6$.

21. $1 = 1 \cdot 1$

23. $0 = 0 \cdot 22$

25. $-7645373 = 38809 \cdot (-197)$.

Set 7.2

1. 81.

3. $x^2 - x - 37$.

5. 0.

7. $-x^3 + 2x^2 - 23$.

9. $-xy(x - y) - yz(y - z) - zx(z - x)$.

11. $x = 1, 3, -3$.

Set 7.3

1. The 1st and 3rd rows are proportional.

3. $y^2 - x^2 = (y + x)(y - x)$.

5. (1) Divide 1st row by 2.
   (2) Add 3rd row to 1st.
   (3) Add 2nd row to 3rd.

7. $\ldots = \begin{vmatrix} 7 & -6 & 0 \\ -2 & 3 & 1 \\ -10 & 8 & 0 \end{vmatrix} = \begin{vmatrix} 7 & -6 \\ -10 & 8 \end{vmatrix} = \ldots = 4.$

9. $\ldots = \begin{vmatrix} 0 & -1 & 0 & 0 \\ -1 & -2 & -12 & 4 \\ 2 & -1 & 10 & 1 \\ 3 & 0 & 10 & 5 \end{vmatrix} = \ldots = \begin{vmatrix} -1 & -12 & 8 \\ 2 & 10 & -7 \\ 1 & 0 & 0 \end{vmatrix} = \ldots = 4$

11. $\begin{vmatrix} 6 & 5 & 7 & 4 \\ 0 & -2 & 1 & -1 \\ 3 & 3 & 6 & 2 \\ 3 & 1 & 7 & 1 \end{vmatrix} = \begin{vmatrix} 6 & 5 & 7 & 4 \\ 0 & -2 & 1 & -1 \\ 3 & 1 & 7 & 1 \\ 3 & 1 & 7 & 1 \end{vmatrix} = 0.$

13. $\begin{vmatrix} x & y & 1 \\ 2 & -3 & 1 \\ -4 & 5 & 1 \end{vmatrix} = 0.$

15. (a) 19 sq. units.
    (b) 0 sq. units (a degenerate triangle).

17. $38x^2 + 38y^2 + 140x - 38y + 858 = 0.$

CHAPTER 8

Set 8.1

1. Closure: $(a+bi) + (c+di) = (a+c) + (b+d)i$, $(a+bi) \cdot (c+di) = (ac - bd) + (ad + bc)i$.
   Associativity: (straightforward, assuming the reals are associative.)
   Identity: $z = 0 + 0i$, $u = 1 + 0i$.
   Inverse: $(s + iy)^* = -z - iy$, $(x + iy)^{-1} = (x - iy)/(x^2 + y^2)$, for $(x, y) \neq (0, 0)$.
   Commutativity: (straightforward).
   Distributivity: (straightforward).

3. The field postulates are the same as the integral domain postulates, with the exception of the cancellation law. But existence of inverses implies the latter: If $a \cdot b = a \cdot c$ and $a \neq z$, then there exists a^{-1}, so that (by uniqueness, etc.) $b = u \cdot b = (a^{-1} \cdot a) \cdot b = a^{-1} \cdot (a \cdot b) = a^{-1} \cdot (a \cdot c) = (a^{-1} \cdot a) \cdot c = u \cdot c = c$.

5. $1, 3a, 5, 10$.

7. Define division as $x/y = y^{-1} \cdot x (y \neq z)$. Then division by $y \neq z$ is always possible by Field Postulate 4. It is unique, since multiplication is uniquely defined and, by Theorem 4.3 (page 138), so are inverses.

9. Since the elements of S are *a fortiori* elements of F, Postulates 2 and 5 and the Distributive Laws hold in S. Postulates 1 and 4 are given. Postulate 3 follows from 4 and 1, *provided that S is non-void*. For then there exists an element a in S, hence a^* is in S, so that by closure, $z = a + a^*$ is in S. Likewise, if $b \neq z$ is in S, then so are b^{-1} and u.

11. Closure: $a, b \in S_1 \cap S_2 \Rightarrow a, b \in S_1$ and $a, b \in S_2 \Rightarrow a \circ b \in S_1$ and $a \circ b \in S_2 \Rightarrow a \circ b \in S_1 \cap S_2$, where $\circ$ is either $\cdot$ or $+$.
    Associativity: (automatic).
    Identity: S_1 and S_2 must each contain the u and z of F, hence $S_1 \cap S_2$ contains u and z.
    Inverse: $a \in S_1 \cap S_2 \Rightarrow a \in S_1$ and $a \in S_2 \Rightarrow a^*$, $a^{-1} \in S_1$, a^*, $a^{-1} \in S_2 \Rightarrow a^*, a^{-1} \in S_1 \cap S_2$, if $a \neq z$.
    Commutativity: (automatic).
    Distributivity: (automatic).

13. No, since matrix multiplication is not commutative for this set of matrices. This set also does not have an additive identity, and closure under addition fails to hold.

15. (a) The mod 12 system is not even an integral domain (cancellation law fails), much less a field.
   (b) By Theorem 2.4 (page 75) and Problem 13 (page 141), the mod 13 system forms a field. The elements are the sets $(0) = (x \in J \mid x \equiv 0 \pmod{13})$, $(1) = (x \in J \mid x \equiv 1 \pmod{13})$, $\ldots$, $(12) = (x \in J \mid x \equiv 12 \pmod{13})$, where '$(x \in J \mid x \equiv k \pmod{13})$' is read 'all members x of the set J of integers such that $x \equiv k \pmod{13}$'.

17. (a) $a \cdot (b + c) = (b + c) \cdot a$, $a \cdot b = b \cdot a$, $a \cdot c = c \cdot a$, by commutativity. By the first distributive law and uniqueness, $(b + c) \cdot a = a \cdot (b + c) = a \cdot b + a \cdot c = b \cdot a + c \cdot a$.
   (b) The commutativity asserted in Postulates 3 and 4 could be deleted.

Set 8.2

1. (a) Commutative ring with unity.
   (b) Ring with unity.
   (c) Field.
   (d) Not a ring: additive closure violated and additive identity excluded by requiring $f(0) \neq 0$.
   (e) Commutative ring.

3. Closure: $(a_0 + a_1 x + \ldots + a_m x^m) + (b_0 + b_1 x + \ldots + b_n x^n) = c_0 + c_1 x + \ldots + c_k x^k$, where $c_i = a_i + b_i$ is in the original ring (which is closed under addition) and $k = \max(m, n)$. Assume $a_i = 0$ for $i > m$, $b_i = 0$ for $i > n$. Similarly, $(a_0 + a_1 x + \ldots + a_m x^m) \cdot (b_0 + b_1 x + \ldots + b_n x^n) = d_0 + d_1 x + \ldots + d_k x^k$, where $d_i = \sum a_p b_q$, summing over all p, q such that $p + q = i$, and $k = m \cdot n$.
   Associativity: Follows in a similar way from the associativity of the original ring.
   Additive Identity: Exists, since z exists in the original ring and $a_0 = z$ defines a constant polynomial.
   Additive Inverse: If $a_0 + a_1 x + \ldots + a_n x^n$ is a polynomial, then so is $a_0{}^* + a_1{}^* x + \ldots + a_n{}^* x$, and $(a_0 + \ldots + a_n x^n) + (a_0{}^* + \ldots + a_n{}^* x^n) = (a_0 + a_0{}^*) + (a_1 + a_1{}^*)x + \ldots + (a_n + a_n{}^*)x^n = z$, since for each

a_i in the original ring, there is an additive inverse $a_i{}^*$ in the original ring. Commutativity: Follows like associativity from the (additive) commutativity of the original ring. If the original ring is commutative (in multiplication), then so are the polynomials, for then $\sum a_p b_q = \sum b_p a_q$. Distributive Laws: Follow from those of the original ring.

5. Yes

7. (1) $a + a^* = z$ (Post. 4)
    (2) $(a + a^*) + (a^*)^* = z + (a^*)^*$ (Well-defined property of $+$)
    (3) $a + (a^* + (a^*)^*) = (a^*)^*$ (Post. 2, 3)
    (4) $a + z = (a^*)^*$ (Post. 4)
    (5) $a = (a^*)^*$ (Post. 3)

9. No. 0 is an additive identity, but 1 does not have an inverse.

11. (a), (b), and (c) define commutative rings (without unity, unless $n = 1$). (d) defines a commutative ring without unity.

Set 8.3

1. The set $(x: x = 5k$ for $k \in J)$ is a subset of the set J of integers. It is closed under addition from within $(x + y = 5m + 5n = 5k$, for $k = m + n)$ and under multiplication by elements of J $(xr = (5n)r = 5(nr))$. It is also itself a ring (by Problem 11, Set 8.2).

3. M is a ring and the specified set of multiples is an ideal.

5. No.

7. (a) Yes: the set is just the multiples of five.
    (b) No.

9. (a) Yes.
    (b) No.

11. Since M is closed under multiplication by elements of R, $7 \cdot 12 = 84$ is in M and $5 \cdot 17 = 85$ is in M. But M is closed under subtraction from within so that $85 - 84 = 1$ is in M. Again by multiplicative closure, if $n \in R$, then $1 \cdot n = n \in M$; hence $M = R$, Q.E.D.

Set 8.4

1. (a) First prove that $z \cdot x = z$. (Start with $x \cdot x + z = x \cdot x = (x + z) \cdot x = x \cdot x + z \cdot x$. Then $(x \cdot x)^* + (x \cdot x + z) - (x \cdot x)^* + x \cdot x + z \cdot x$, and finally $z + z = z + z \cdot x$, or $z \cdot x = z$.) Now if M is non-void, it contains some element x. Then $z \cdot x \in M$. But $z \cdot x = z$, so $z \in M$. Hence $a \equiv a \bmod M$, for all $a \in R$, since $a = a + z$.
    (b) If $a = b + m$, then $b = a + n$, with $n = m^*$.
    (c) If $a = b + m$ and $b = c + n$, then $a = c + k$, with $k = n + m$.
    (d) If $a = b + m$ and $c = d + n$, then $a + c = b + d + (m + n)$, and $a \cdot c = b \cdot d + (m \cdot d + n \cdot b + m \cdot n)$.

3. The cosets of $R \bmod M$ are the residue classes mod 5.

5. (a) $M \subset R$, since multiples of 18 are multiples of 6. $x,\ y \in M \Rightarrow x + y \in M$, since $18m + 18n = 18(m + n)$. $x \in M,\ r \in R \Rightarrow x \cdot r = r \cdot x \in M$, since $(18n)k = 18(nk)$.
    (b) $(\ldots -18, 0, 18, 36,\ \ldots)$, $(\ldots -12, 6, 24, 42,\ \ldots)$, $(\ldots -6, 12, 30, 48,\ \ldots)$.
    (c) three
    (d) The mod 3 system.

7. (a) The residue classes 2, 4, 7, 14 (mod 28).
    (b) The residue classes 2, 4, 6, 8, 12 (mod 24).
    (c) None.

9. If $(x - 1)(x - 2)g(x) \in M$, $(x - 1)(x - 2)h(x) \in M$, then $(x - 1)(x - 2)g(x) + (x - 1)(x - 2)h(x) = (x - 1)(x - 2)f(x) \in M$, with $f(x) = g(x) + h(x)$. If $(x - 1)(x - 2)g(x) \in M$ and $p(x) \in R$, then $((x - 1)(x - 2)g(x))p(x) = p(x)((x - 1)(x - 2)g(x)) = (x - 1)(x - 2)q(x) \in M$, with $q(x) = g(x)p(x)$.

Set 8.5

1. (a) $3x + 5$
    (b) $6x + 19$
    (c) $x - 3$
    (d) $7x + 28$

    (e) $-6x + \dfrac{1}{30}$

    (f) $17x - 4$

3. $x^2 = (x^2 + 1) \cdot 1 - 1$ and $-1 = (x^2 + 1) \cdot 0 - 1$.

5. $21i + 62$, $9i + 10$, $24i + 24$, $6i + 1141/30$, $119i + 119$.

7. The isomorphism between $R[x]$ mod $(x^2 + 1)$ and the complex numbers is illustrated by the comparison.

9. (a)

	1	-1	x	$-x$
1	1	-1	x	$-x$
-1	-1	1	$-x$	x
x	x	$-x$	-1	1
$-x$	$-x$	x	1	-1

(b) The Cyclic Group of Order Four.

Student Notes

Student Notes

Student Notes

Student Notes

Student Notes

Student Notes

Student Notes

Index

A_{12}, alternating group on four symbols, 156
Additive identity, 10, 30
Additive inverse, 10, 30
Adjoint matrix, 261, 291
Admittance matrix, 210
Aeronautical engineering, 202, 293, 303
Algebra
  Boolean, 87
  of mappings, 81
Algebraic approach to Boolean algebra, 118
Algorithm, Euclid's, 39
Alternating group (A_{12}) on four symbols, 156
Amicable numbers, 48
And (Boolean), 93, 109
Answers and hints, 311
Archimedes' Division Axiom: $A = B \cdot q + r$, with $0 \leq r < B$, 38, 284
Area of a triangle, determinant form, 260
Arithmetic of classes, 67
Artin, E., 148
Associative Postulate: $a \oplus (b \oplus c) = (a \oplus b) \oplus c$, 10, 30, 96, 119, 124, 125
Associativity
  of matrices, 175
  of substitutions, 124
Atomic structure, 202
Automorphism, 148

Baker, C. L., 33, 35
Ball, W. R., 114
Beckenbach, E. F., 211, 220
Bell, E. T., 162

Binary
  Boolean algebra, 95, 141
  Boolean arithmetic, 89
  system, 49
Birkhoff, G., 82
Boole, George, 87, 117
Boolean
  algebra, 87, 278
    an algebraic approach, 118
    binary, 95, 141
    point set interpretation, 115
  arithmetic, binary, 89
  functions, 90
  polynomials, 93, 107
Bourbaki, 82
Bruck, R., 161
Brun, Viggo, 34
Butrand's Postulate, 317

Cancellation postulate: $c \neq z, c \cdot a = c \cdot b$ imply $a = b$, 10, 30, 272
Cartesian product of two sets, 80
"Casting out nines," 73
Cauchy, 300
  sum, 218
Cayley, A., 267
Cayley's Theorem: Every finite group G of order n is isomorphic to some regular substitution (permutation) group of order n on n symbols, 145
Characteristic equation: $|I \cdot x - A| = 0$, 290
  polynomial, 290
  roots, 292
  vector, 293

Chinese Remainder Theorem, 83
Circuit
  design, 112
  electrical, 107
Classes
  arithmetic of, 67
  equivalence, 66, 78, 149
  residue, 78
Classification of integers, 32
Closure: $a, b \in D$ implies $a \odot b \in D$, 10, 29, 119
Cofactor of an element of a matrix, 250
Collar, A., 202
Common divisor, 37
Commutative Postulate: $a \oplus b = b \oplus a$, 10, 30, 95, 119
  group, 275
  ring, 276
  ring with unity, 276
Complement of an element $(\tilde{\mathbf{B}})$, 119
Complementary relationships in Boolean Algebra, 101
Complete residue system, 79
Complex numbers, 1, 68, 283
Composite numbers, 32
Compression transformation, 212
Computing science, 307
Conditions, necessary and sufficient, 22
Congruence: $a \equiv b \pmod{m}$ implies $\exists\, k$ such that $a = b + k \cdot m$, 2, 6, 69, 149, 150
  group, 150
  linear, 74
Contrapositive, 20, 314
Convergence of an infinite series of matrices, 300
Converse, 20
Coset, 149, 150, 281
Counterexamples, 23
Counting number, 1
Courant, R., 1, 49, 110, 117
Cramer, G., 267
Cramer's Rule, 265, 267
Cross ratio, 134
Crystal structure, 202
Cyclic group, 142, 150, 155
CYPAK, 102

De Morgan's Theorem:
  $(x + y + z + \ldots)' = x' \cdot y' \cdot z' \cdot \ldots$, 103
Derivatives of matrices, 303
Determinant, 243, 250
  of elementary row transformation, 245
Dickson, L. E., 45
Disjoint cosets modulo a subgroup, 152
Distributive Postulate: $\forall a,\ b,\ c \in S$, $a \cdot (b + c) = a \cdot b + a \cdot c$, 10, 30, 96, 119
Divergent series, 34
  of matrices, 305
Divides with remainder zero, 6, 65
Divisibility, tests for by 3 and 9, 73
Division by zero, 36
Divisor
  common, 37
    greatest, 37
  positive, 45
  proper, 45
Domain, integral, 9, 29, 124, 271
Dual, 88
Duality, principle of, 87
Duncan, W., 202

e^A (where A is a square matrix), 303
Eigenvalue, 292
Eigenvector, 293
Eigenwerte, 292
Electrical
  circuit, 90, 107
  engineering, 202, 293
  networks, 107, 209
Elementary row operations for matrices, 178
Elementary row transformations, 245
Equals relation, 5, 9, 61
Equation
  characteristic, 289
  determinantal form, 260
  systems of linear..., 223
Equivalence
  classes, 66, 78, 149
  of matrices, 180
  relation, 5, 61, 80, 124

Equivalent, 22
   statements, 22
Euclid, 33
Euclid's Algorithm, 39
Euclidean
   geometry, 88
   plane transformation, 216
Euler
   circles, 87
   Fermat-Euler Theorem, 4, 83
   Φ function, 83
Exclusive "or," 93

Factor
   analysis, 168
   group, 159
Fermat, 44
   -Euler Theorem, 4, 83
Field: A set of elements having two
   well-defined operations such that each
   forms a commutative group (except
   that the additive identity need have
   no multiplicative inverse), and which
   also satisfy the distributive postulates,
   272
     finite, 285
Finite
   field, 285
   group, 139
Fitch, F. B., 29
Forsythe, G. E., 241
Four-terminal network, 209
Frazer, R., 202
Free groups, 157
Function, 81
   Boolean, 90
   Euler, Φ, 83
   indicator, 83
   recursive, 54

Galois Theory, 161
Gardner, Martin, 47, 58
Gauss, 84
General solution of linear equations, 234
Generator of a group, 142
Geometry, 124
Gillies, D. B., 46, 58

Goldbach's conjectures, 34
Greatest common divisor: $(A, B) = g$
   implies (1) g divides both A and B
   and (2) every common divisor of both
   A and B also divides g, 37
     left, of matrices, 291
Greenspan, D., 184
Group: A set of elements having one
   well-defined operation, and an equivalence relation, which satisfy the
   closure, associative, identity and inverse postulates, 124
     cyclic, 142
     finite, 139
     free, 157
     generator of, 142
     octic, 134, 147
     permutation, 142
     postulates, 124
     substitution, 126, 142
Gruenberger, F. J., 33, 35

Hamilton-Cayley Theorem, 218, 291
Harmonic series, 34
Heaslet, M., 47, 50
Hempel, C. G., 29
Herstein, I. N., 82
Hill, S., 29, 58
Hodgman, C. D., 33
Hohn, Franz, 110, 117
Homogeneous system, general solution,
   234
Howard, R. A., 211
Hypothesis, 19

Ideal M (of ring R): $1 \neq M \subset R$ such
   that $m, n \in M$ implies $(m - n) \in M$
   and also $m \in M$, $r \in R$ implies
   $m \cdot r \in M$, 279
     principal, 280
     of the set of polynomial functions,
     298
Identity: $\exists z \in S$ such that $\forall b \in S$,
   $b \oplus z = z \oplus b = b$, 10, 30, 61, 124,
   271
     matrix, 173
Image set, 81

Inclusive "or," 93
Inconsistent system, 238
Indefinite admittance matrix, 210
Independent postulates, 65
Independent study, 58, 82, 121, 163, 221, 241, 269, 287, 308
Indicator function, 83
Indices modulo p, 84
Indirect proof, 24
Induction, mathematical, 25
Infeld, Leopold, 162
Infinite
matrices, 183
series with matrix elements, 300
Information science, 307
Ingham, A., 33
Integer, 1
classification of, 32
Integral domain: A set of elements having two well-defined operations $+$ and $\cdot$ which satisfy the postulates: Closure $(+, \cdot)$, Commutative $(+, \cdot)$, Associative $(+, \cdot)$, Identity $[+(\text{zero}), \cdot(\text{one})]$, Additive inverse (negative), Distributive, 9, 29, 124, 271
Integrals of matrices, 303
Intersection, 115
Into, mapping, 81
Invariant subgroup, 159, 160
Inverse: $\forall b \in S$, $\exists b^* \in S$ such that $b \oplus b^* = b^* \oplus b = z = $ identity under $\oplus$, 10, 20, 30, 124, 273–75
left, of a non-square matrix, 197
mapping, 81
matrix, 179, 182, 247
right, of a non-square matrix, 197
of a theorem, 20
Isomorphism: A mapping (function) between two sets which preserves the operations, 91, 92, 143, 148, 274, 285

Jacobian, 267
Jacobson, N., 82
Jeraldine problem, 112
Johnstone, H., 29
Jordan-Hölder Theorem, 158, 160
Jordan Product, 194

Kraitchik, M., 114
Kronecker, L., 2, 178
Kronecker's delta, 178
Kuhn, H., 220

Lagrange's Theorem: The order of (number of elements in) a subgroup S of a finite group G is a divisor of the order of G, 149, 154
Lanczos, C., 241
Lander, L. J., 44, 45
Latent roots, 292
Law of conservation
of kinetic energy, 268
of momentum, 268
Lazarsfeld, P. F., 220
Le Corbeiller, P., 211
Left
congruence, 150
divisor of a matrix, 291
inverse of a non-square matrix, 197
Lehmer, D. N., 33, 35
Levens, A. S., 269
Lie, Sophus, 193
Algebra, 193
Lieber, H. G., 162
Lieber, L. R., 162
Linear
algebra, 124
congruences, 74
equations, systems of, 223
Logical syntax, 88
Loop, 135, 162
Lorentz Transformation, 220

MacDuffee, C. C., 38, 161, 220
MacLane, S., 29, 82
McKinsey, J. C. C., 220
Mapping, 81, 211, 215
into, 81
inverse, 81
one-to-one, 81
onto, 81
product of two, 82
Markov Chain, 208

Mathematical
  induction, 25
  system, 123
Matrix polynomial, 289
Matrices
  associativity, 175
  derivatives, 303
  divergent series, 305
  elementary row operations, 178
  square roots, 190
Matrix, 166
  adjoint, 261
  admittance, 210
  cofactor of an element, 250
  eigenvalue, 292
  eigenvector, 293
  equivalence, 180
  identity, 173
  indefinite admittance, 210
  infinite, 183
    series, 300
  integral, 303
  inverse, 179, 182, 247
  left divisor, 291
  left inverse, 197
  maximum profit, 207
  minimum function, 298
  minor of an element, 250
  nonsquare, 194
  Pauli, 168
  product, 167
  right inverse, 197
  singular, 264
  symmetric, 215, 293
  theory, 289
  transfer, 209
  transformation, 238
    of a vector space, 211
  transpose, 254, 255
  triangular, 246
  trigonometric functions, 302
Maximal invariant subgroup, 160
Mersenne primes, 35, 46, 58
Metamathematics, 88
Minimum function of a matrix, 298
Minor of an element in a matrix, 250
Modulo 6 system, 6, 69

Modulo 7 system, 2, 69
Moler, C. B., 241
Monotone, 83
Multiple factor analysis, 202
Multiplicative identity, 10, 30
Murnaghan, F., 220

n-tuple, 211
Necessary conditions, 22
    and sufficient conditions, 22
Negation, 101
Negative relationship in Boolean algebra, 101
Network
  electrical, 107, 209
  four-terminal, 209
Nim, 49
Niven, I., 1
"No function," 93
Noncommutative systems, 165
Nonsquare matrix, 194
Numbers
  amicable, 48
  binary, 49
  complex, 1, 68, 283
  composite, 32
  counting, 1
  natural whole, 1
  perfect, 45, 58
  prime, 23, 32
  rational, 1, 68
  real, 1, 68
  systems, 48

Octic group, 134, 147
One-sided matrix inverses, 197
One-to-one mapping, 81
Onto mapping, 81
Opposite, 20
Or (Boolean), 93, 109
Order
  of an element, 139
  of a group, 139
  relationship, 11
Ordered pairs, 68

Pairs, ordered, 68
Parallel switches, 91, 92
Parameters, 234
Parkin, T. R., 44, 45
Particular solutions of linear equations, 234
Pauli matrices, 168
Peano postulates, 25
Perfect numbers, 45, 58
Period (of a group element b): the least positive integer N such that $b^N =$ the group identity, 139
Permutations, 127
  groups, 142
  regular permutation group of order n on n letters, 145
Petersen, G., 269
Pipping, N., 34
Plimpton Babylonian tablets, 39
Point set interpretation of Boolean algebra, 115
Points, 81
Polynomial, 276
  characteristic, 290
  function, 276
  modulo $(x^2 + 1)$, 283
Population analysis, 207
Positive divisor, 45
Postulates, 18
  field, 272
  independent, 65
  integral domain, 271
  ring, 275
Prime: An integer other than zero or a unit ($P \neq 0, 1, -1$) having no factors other than units (± 1) and $\pm N$, 23, 32
  Mersenne, 35, 46, 58
  pairs, 34
  positive integer, 23
  super-, 52
  triplets, 37
  twin, 34
Principal ideals, 280
Principle of duality, 87
Probability, 302

Product
  matrix, 167
  of two mappings, 82
Proof, 19
  indirect, 24
  structure, 189
Proper
  divisor, 45
  subgroup, 139
  value, 292

Quadratic
  form, 215
  reciprocity law, 84
  residues, 84
Quantum mechanics, 202
Quasigroup, 162
Quaternion, 68
  as Pauli matrices, 169
Quotient
  groups, 158, 159
  ring, 282

Ratio, cross, 134
Rational numbers, 1, 68
Real numbers, 1, 68
Recursive function, 54
Reductio ad absurdum, 24
Reflection, 215
Reflexive postulate: $\forall a$, a E a, 5, 62
Regular permutation group, 145
Relation
  equals, 5, 9, 61
  equivalence, 5, 61, 80, 124
Relay relationship in Boolean algebra, 101
Residue
  class rings, 281, 285
  classes, 78
  quadratic, 84
Right
  congruence, 150
  inverse of a nonsquare matrix, 197

Ring: A set of elements having two well-defined operations $+$ and $\cdot$ which forms a commutative group under $+$ and is closed and associative under $\cdot$ and for which the distributive postulates hold, 275
    postulates, 275
    quotient, 282
    residue class, 281, 285
    with unity, 185, 276
Robbins, H., 1, 49, 110, 117
Roots, characteristic, 292
Row operations of matrices, 178

Scalar, 196
Semigroup, 162
Series
    divergent, 34
    harmonic, 34
    switches, 91
Set of postulates, 124
Shanks, D., 58
Shearing transformation, 212
Sieve of Eratosthenes, 36
Singular matrix, 264
Solution
    general, 234
    of a linear congruence, 75
    by matrix methods, 224
    particular, 234
Square
    matrices, 171
    roots of matrices, 190
Stabler, E. R., 117
Statistics, 302
Stretching transformation, 212
Stoll, R. R., 120
Study, independent, 58, 82, 121, 163, 221, 241, 269, 287, 308
Subgroup, 139
    invariant, 159
    proper, 139
Substitution, 126
    groups, 126, 144

Sufficient conditions, 22
    and necessary conditions, 22
Superprime, 52
    leader, 53
Suppes, P., 29, 58
Switches, 90
    in parallel, 91
    in series, 91
Symmetric matrix, 215, 293
Symmetric postulate: $a \mathrel{E} b$ implies $b \mathrel{E} a$, 5, 62
Systems
    homogeneous, 236
    inconsistent, 238
    isomorphic, 91
    of linear equations, 223
    mathematical, 123
    noncommutative, 165

Taussky, O. (Todd), 297
Tensor, 267
Tests for divisibility by 3 and 9, 73
Theorems, 19
Thurstone, L., 202
Todd, J., 297
Todd, O. Taussky, 297
Topology, 124, 217
Totient, 83
Transfer matrix, 209
Transformations, 211
    compression, 212
    elementary row, 245
    Euclidean plane, 216
    Lorentz, 220
    shearing, 212
    stretching, 212
Transitive postulate: $a \mathrel{E} b$, $b \mathrel{E} c$ imply $a \mathrel{E} c$, 5, 62
Translation, 136, 216
    equivalence, 136
Transpose of a matrix, 254, 255
Triangular matrix, 246
Trigonometric functions of a matrix, 302
Triplets, prime, 37
Tucker, A., 220
Twin primes, 34

Unbounded monotone increasing function, 83
Undefined terms, 18
Union, 115
Uniqueness, 70, 138
Unit, 32
Unity, 10, 30, 32
Universal element $(\widetilde{\mathbf{B}})$, 119
Uspensky, J., 47, 50

Value
  characteristic, 292
  eigen-, 292
  latent, 292
  proper, 292
Vandiver, H. S., 84

Vector, 194
  characteristic (eigen-), 293
Venn diagrams, 87, 117
Vinogradov, I., 34

Well defined, 66, 123
  operation, 124
Weyl, H., 29
Whittaker, J. V., 149
Whole numbers, 1
Wigner, E. P., 125
Wilder, R. L., 29, 58
Wronskian, 267

Zero, 10, 30, 32
  element $(\widetilde{\mathbf{B}})$, 119

Index of Symbols

$\subset$ (is a subset of), 2

$a \equiv b \pmod 7$, 2, 62

$a \equiv b \pmod 6$, 6, 62

$|$ (divides with zero remainder), 6, 32, 65

$A \Rightarrow B$ (implies), 20

$\boxminus$ (is equivalent to), 62

$\not\boxminus$ (is not equivalent to), 62

$\equiv$ (congruent), 62

$=$ (equal), 63

$\leq$, 63

$\cong$, 63

$\sim$, 65 .

$\neq$, 65

$\cdots$, 66

$a \equiv b$
  mod m, 69
  groups, 150
  rings, 282

$\otimes$ (product of two mappings), 82

Φ (Euler Phi function), 83

$\left(\dfrac{b}{p}\right)$ (Legendre symbol), 84

$\cap$, $\cup$ (set intersection and union), 115

$\odot$ (group operation), 123

x^0, 139, 157

x^{-m}, 139

G/H (quotient group), 159

E_i (elementary matrix), 178

$A \cong B$ (matrix equivalence), 180

ξ_n (an n-element vector), 215

$|M|$ (determinant of matrix M), 244

$\in$ (is an element of), 271

$R \bmod m$ or R/M (quotient ring), 282

$R[x]$ (ring of real polynomials), 283

$R[x]/(x^2+1)$ (real polynomials modulo $(x^2 + 1)$, isomorphic to complex numbers), 283

$R[x]/M$, 284